Humic and Fulvic Acids

ACS SYMPOSIUM SERIES **651**

Humic and Fulvic Acids

Isolation, Structure, and Environmental Role

Jeffrey S. Gaffney, EDITOR
Argonne National Laboratory

Nancy A. Marley, EDITOR
Argonne National Laboratory

Sue B. Clark, EDITOR
Washington State University

Developed from a symposium sponsored
by the Division of Industrial and Engineering Chemistry, Inc.

American Chemical Society, Washington, DC

Library of Congress Cataloging-in-Publication Data

Humic and fulvic acids: isolation, structure, and environmental role/
Jeffrey S. Gaffney, editor, Nancy A. Marley, editor, Sue B. Clark, editor.

   p.   cm.—(ACS symposium series, ISSN 0097–6156; 651)

   "Developed from a symposium sponsored by the Division of Industrial
and Engineering Chemistry at the 210th National Meeting of the
American Chemical Society, Chicago, Illinois, August 20–24, 1995."

   Includes bibliographical references and indexes.

   ISBN 0–8412–3468–X

   1. Humic acid—Congresses. 2. Fulvic acids—Congresses.

   I. Gaffney, Jeffrey S., 1949–   . II. Marley, Nancy A., 1948–
III. Clark, Sue B., 1961–   . IV. Series.

QD341.A2H852   1996
631.4′17—dc20   96–38703
                                                           CIP

This book is printed on acid-free, recycled paper.

Foreword

THE ACS SYMPOSIUM SERIES was first published in 1974 to provide a mechanism for publishing symposia quickly in book form. The purpose of this series is to publish comprehensive books developed from symposia, which are usually "snapshots in time" of the current research being done on a topic, plus some review material on the topic. For this reason, it is necessary that the papers be published as quickly as possible.

Before a symposium-based book is put under contract, the proposed table of contents is reviewed for appropriateness to the topic and for comprehensiveness of the collection. Some papers are excluded at this point, and others are added to round out the scope of the volume. In addition, a draft of each paper is peer-reviewed prior to final acceptance or rejection. This anonymous review process is supervised by the organizer(s) of the symposium, who become the editor(s) of the book. The authors then revise their papers according to the recommendations of both the reviewers and the editors, prepare camera-ready copy, and submit the final papers to the editors, who check that all necessary revisions have been made.

As a rule, only original research papers and original review papers are included in the volumes. Verbatim reproductions of previously published papers are not accepted.

ACS BOOKS DEPARTMENT

Contents

Preface

HUMIC AND FULVIC ACIDS, along with other organic colloidal materials, are fascinating substances that can have profound environmental consequences. Their abilities to complex radionuclides and toxic metals have been recognized for some time by researchers interested in the migration and mobilization of nuclear and industrial waste at contaminated sites. The micellar properties of humic and fulvic acids also give them the ability to play important roles in the solubilization and transport of hydrophobic pollutants.

Most of the studies of these naturally occurring organic colloids have been reported in symposia that focused on the contaminants. At the 210th ACS National Meeting held in August 1995 in Chicago, Illinois, a symposium was held on recent studies of humic–fulvic acids and organic colloidal materials in the environment. This symposium brought together researchers from the United States, France, Germany, and England who have been using a wide variety of techniques to characterize the structures and determine the chemical properties and physical roles that humic and fulvic acids play in the environment, particularly in surface and groundwaters. This book is the result of that symposium, which attempted to bring direct attention to humic–fulvic acids and organic colloidal substances.

Numerous advances in analytical methods and separation techniques have enabled researchers interested in humic and fulvic acids to begin to chemically characterize these complex molecular mixtures. These methods have also allowed the interactions of humic and fulvic acids with metals and with organic pollutants in the environment to be explored. We believe that the reader will be intrigued by the wide diversity of the research results presented here and the potential environmental significance of these naturally occurring organic colloidal compounds. Clearly, the environmental chemistry and physics of humic and fulvic acids must be better understood if we are to develop safe and sound waste-storage and contaminated-site-cleanup strategies.

We thank the authors for their contributions to this volume, which represents some of the recent advances in obtaining that understanding. Special thanks are expressed to Dale Perry of Lawrence Berkeley National Laboratory for suggesting the humic–fulvic acid symposium and encouraging the publication of this ACS Symposium Series Book, and to Mary M. Cunningham for her help in completing this volume.

JEFFREY S. GAFFNEY
NANCY A. MARLEY
Environmental Research Division
Argonne National Laboratory
Building 203, 9700 Cass Avenue
Argonne, IL 60439

SUE B. CLARK
Department of Chemistry
Washington State University
P.O. Box 644630
Pullman, WA 99164–4630

August 7, 1996

OVERVIEW

Chapter 1

Humic and Fulvic Acids and Organic Colloidal Materials in the Environment

Jeffrey S. Gaffney[1], Nancy A. Marley[1], and Sue B. Clark[2,3]

[1]Environmental Research Division, Argonne National Laboratory,
Building 203, 9700 Cass Avenue, Argonne, IL 60439–4831
[2]Savannah River Ecology Laboratory, University of Georgia,
P.O. Drawer E, Aiken, SC 29802

Humic substances are ubiquitous in the environment, occurring in all soils, waters, and sediments of the ecosphere. Humic substances arise from the decomposition of plant and animal tissues yet are more stable than their precursors. Their size, molecular weight, elemental composition, structure, and the number and position of functional groups vary, depending on the origin and age of the material. Humic and fulvic substances have been studied extensively for more than 200 years; however, much remains unknown regarding their structure and properties.

Humic substances are those organic compounds found in the environment that cannot be classified as any other chemical class of compounds (e.g., polysaccharides, proteins, etc.). They are traditionally defined according to their solubilities. Fulvic acids are those organic materials that are soluble in water at all pH values. Humic acids are those materials that are insoluble at acidic pH values (pH < 2) but are soluble at higher pH values. Humin is the fraction of natural organic materials that is insoluble in water at all pH values. These definitions reflect the traditional methods for separating the different fractions from the original mixture.

The humic content of soils varies from 0 to almost 10%. In surface waters, the humic content, expressed as dissolved organic carbon (DOC), varies from 0.1 to 50 ppm in dark-water swamps. In ocean waters, the DOC varies from 0.5 to 1.2 ppm at the surface, and the DOC in samples from deep groundwaters varies from 0.1 to 10 ppm (1). In addition, about 10% of the DOC in surface waters is found in suspended matter, either as organic or organically coated inorganic particulates.

Structure and Composition

Humic materials have a wide range of molecular weights and sizes, ranging from a few hundred to as much as several hundred thousand atomic mass units. In general, fulvic acids are of lower molecular weight than humic acids, and soil-derived materials are larger than aquatic materials (1,2). Humic materials vary in composition depending on their source, location, and method of extraction; however, their similarities are more

[3]Current address: Department of Chemistry, Washington State University, P.O. Box 644630, Pullman, WA 99164–4630

0097–6156/96/0651–0002$15.00/0
© 1996 American Chemical Society

pronounced than their differences. The range of the elemental composition of humic materials is relatively narrow, being approximately 40-60% carbon, 30-50% oxygen, 4-5% hydrogen, 1-4% nitrogen, 1-2% sulfur, and 0-0.3% phosphorus (*3*). Humic acids contain more hydrogen, carbon, nitrogen, and sulfur and less oxygen than fulvic acids. Studies on humins have shown that they are similar to humic acids except that they are strongly bound to metals and clays, rendering them insoluble (*4*).

Substantial evidence exists that humic materials consist of a skeleton of alkyl/aromatic units cross-linked mainly by oxygen and nitrogen groups with the major functional groups being carboxylic acid, phenolic and alcoholic hydroxyls, ketone, and quinone groups (*5,6*). The structures of fulvic acids are somewhat more aliphatic and less aromatic than humic acids; and fulvic acids are richer in carboxylic acid, phenolic, and ketonic groups (*7*). This is responsible for their higher solubility in water at all pH values. Humic acids, being more highly aromatic, become insoluble when the carboxylate groups are protonated at low pH values. This structure allows the humic materials to function as surfactants, with the ability to bind both hydrophobic and hydrophilic materials. This function in combination with their colloidal properties, makes humic and fulvic materials effective agents in transporting both organic and inorganic contaminants in the environment.

Colloidal Characteristics

The colloidal state represents a phase intermediate between true solutions, where species are of ionic or molecular dimensions, and suspended particulates, where species are sufficiently large to settle under the influence of gravity. The colloidal range is considered to extend from 0.001 to 1.0 μm or 10 to 10,000 Ångstroms (Figure 1). Chemical and physical reactions are generally enhanced in colloidal systems due to the large surface areas of colloidal particles. At the same time, mobility through surface waters or groundwaters is also enhanced, approaching that for true solutions. The ranges of molecular sizes for the majority of humic and fulvic acids place them in the colloidal range when in aqueous solution. Humic colloidal materials are thought to consist of coiled, long-chain, or three-dimensional cross-linked macromolecules with electrical charges variously distributed on the particle. The presence of charged sites, arising from ionized acidic groups, results in mutual repulsion and causes maximum expansion of the molecule (*2*).

The factors most important in controlling the molecular conformation of humic materials are concentration of the humic, pH, and ionic strength of the system (*8*). At high sample concentrations (>3.5 g/L), low pH (<3.5) and high electrolyte concentrations (>0.05 M), the humic materials are rigid uncharged colloidal particles. At low sample concentrations, high pH, and low electrolyte concentrations, humics and fulvics exist as flexible linear polyelectrolytes. In fresh waters, where both humic and ionic strength would be expected to be low, and the pH is greater than 3.0, humic materials should exist as linear polyelectrolytes.

Free Radicals in Humic Materials

Humic materials have a relatively high content of radicals, presumably of the semiquinone type, which are more prominent in humic acids than in fulvic acids. These radicals can exist as permanent components or as transient species, generated by pH

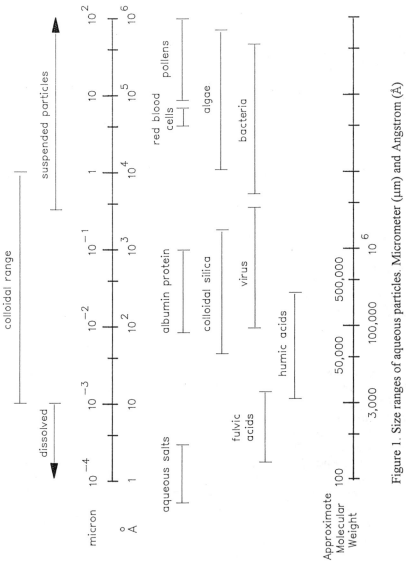

Figure 1. Size ranges of aqueous particles. Micrometer (μm) and Angstrom (Å) units are shown using a logarithmic scale, while the nominal molecular weights are approximate and are not scaled.

changes, chemical reduction, or solar irradiation, that have lifetimes of minutes to hours (*9*). These free radicals most likely play important roles in polymerization or oxidation-reduction reactions. Humic materials can reduce metals with estimated reduction potentials of 0.5-0.7 eV (*10*). This can have a major effect on the migration of reducible cations.

The absorption spectrum of an aquatic humic acid is shown in Figure 2. The absorbance values generally decrease with increasing wavelength with essentially no absorption above about 550 nm. Most of the solar energy absorbed by the humic materials is between 300 and 500 nm. Absorption of light in this region can initiate a number of photochemical processes. This light absorption can produce peroxy radicals and hydroxyl radical as well as hydrated electrons, hydrogen peroxide, singlet oxygen, and superoxide (*11*). These species can also promote redox reactions. In addition, humic acids can photosensitize nonpolar organics such as herbicides and possibly accelerate their decomposition and detoxification.

Separation and Purification

Humic and fulvic acids are traditionally extracted from soils and sediment samples as the sodium salts by using sodium hydroxide solution. The material that remains contains the insoluble humin fraction (Figure 3). The alkaline supernatant is acidified to pH 2 with HCl. The humic acid precipitates and the fulvic acid remains in solution with other small molecules such as simple sugars and amino acids. These molecules can be separated by passing the solution through a hydrophobic resin, such as the methacrylate cross-linked polymer, XAD-8. The fulvic acids will sorb to the resin while the more hydrophilic molecules pass through the column. The fulvic acid can be removed with dilute base.

Aqueous samples are treated similarly beginning with the acidification step. The entire sample is then put through the hydrophobic resin, and the fulvic acids are eluted at pH 7. The humic acids are removed with 0.1 M NaOH (*2*). After extraction, purification of the samples can be accomplished by freeze-drying and dialysis. The use of strong acids and bases has been criticized for several reasons. They can promote degradation, decarboxylation, oxidation, and condensation reactions. Strong acids and bases can also dissolve siliceous materials and lyse cells, resulting in contamination of the sample. Other extractants have been proposed, such as sodium pyrophosphate or sodium fluoride; however, the classical procedure offers the most complete dissolution of humic material from solid samples and is still most often used (*12*).

High-volume ultrafiltration techniques, such as hollow fiber ultrafilters, have been proposed for the separation of humic materials from aqueous samples (*13-15*). This technique avoids exposure to strong acids and bases and maintains the materials in aqueous concentrates closest to their natural states. The hollow fiber filters are available in nominal molecular weight cutoff ratings of 3,000, 10,000, 30,000, and 100,000 atomic mass units. In addition, flat membrane filters, which are used in stirred vortex cells, and spiral-wound cartridge filters are available at ratings of 500 and 1,000 nominal molecular weights. By using a series of filters, aqueous humic materials can be separated into size fractions, with the fulvic acids in the smaller size ranges and the humic acids in the larger.

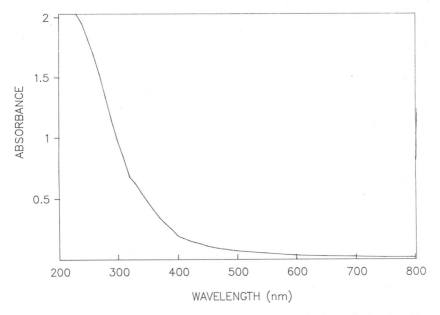

Figure 2. Ultraviolet-visible absorption spectrum of a typical aquatic humic acid from a sample of bog water.

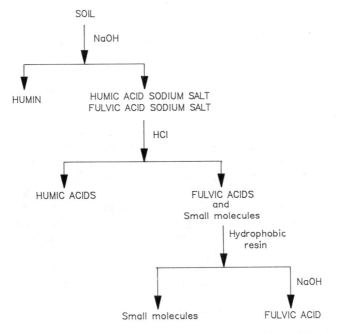

Figure 3. Methods for extraction of humic acids, fulvic acids, and humin from soil.

Methods of Characterization

Nearly every method available to the analytical chemist has been used in an attempt to unravel the complex properties and behavior of humic substances. Some of the more widely used methods are listed in Table I.

Table I. Methods for Analysis and Characterization of Humic Materials

Molecular Weight Determination
> Viscosity
> Vapor pressure osmometry
> Ultracentrifugation
> Gel filtration
> Laser Light Scattering
> Field desorption mass spectrometry

Functional Group Analysis
> Fourier transform infrared spectroscopy
> ^{13}C nuclear meagnetic resonance
> Electron Spin Resonance
> Pyrolysis-gas chromatography
> Pyrolysis-mass spectrometry
> Pyrolysis-Fourier transform infrared spectroscopy
> pH titration

Binding Studies
> Cation exchange
> Fluorescence
> Photoacoustic spectroscopy
> Dialysis
> Potentiometric titration

They include both chemical and physical, degradative and nondegradative methods (*16*). Chemical methods yield information on elemental composition and functional groups. The degradative methods use oxidative, reductive, thermal, or other degradation techniques to break down the complex humic molecules into simpler units. After identification, these simple compounds are related back to the structure of the starting material. nondegradative methods yield data regarding the structure and behavior of the whole molecule.

The techniques that have been used to determine the approximate molecular weights of humic materials are often those used to study biological macromolecules. Such techniques are often prone to artifacts arising from adsorption, precipitation, and degradation. Many of the methods suffer from one common problem, the lack of appropriate standards. Only colligative properties are free of this problem; however, they yield number-averaged molecular weight, with no indication of the polydispersity of the humic material.

A variety of spectroscopic methods have been employed to characterize the functional groups within the humic molecule. Traditionally, the method of choice for studying functional groups of organic molecules would be infrared spectroscopy. An infrared spectrum of an aquatic humic acid is shown in Figure 4. The vibrational bands giving rise to these features are listed in Table II (*15*). The infrared spectrum yields very broad bands and severe band overlap results. This overlap makes band assignments difficult and quantitation nearly impossible. Unfortunately, the broad bands are typical of infrared spectra of humic materials in the solid phase. Recent advances in internal reflectance techniques have made it possible to study humic materials in the aqueous phase (*17,18*). Preliminary results indicate that this technique will yield better resolution than those previously obtained and may provide information concerning metal binding to carboxylates and conformational dynamics.

Table II.Infrared Absorption Bands of Humic Acids

Frequency (cm-1)	Assignment
3400	H-bonded OH
3230(sh)	Aromatic C-H stretch
2970(sh)	Aliphatic C-H stretch
1740(sh)	COOH stretch
1585	COO^{-1} asymmetric stretch
1415	COO^{-1} symmetric stretch
1100-1045	C-C or C-OH stretch

(*15*)

Table III. Chemical Shifts of 13C-NMR Bands of Humic Acid

Shift(ppm)	Assignment
0-40	Aliphatic carbon
40-100	Ethers
100-110	Acetals
110-150	Aromatic carbon
170-190	Carboxylic acids
200-230	Carbonyl carbon

(*15,19*)

Another technique which is widely used for functional group analysis of humic materials is carbon-13 nuclear magnetic resonance specroscopy (^{13}C-NMR). The ^{13}C-NMR solid-state spectrum of an aquatic humic acid is shown in Figure 5. The band assignments for the types of carbon that can be detected by NMR are listed in Table III. Again, bands are broadened due to the presence of free radicals in the structure. More information can be obtained with ^{13}C-NMR regarding the carbon skeleton of the humic

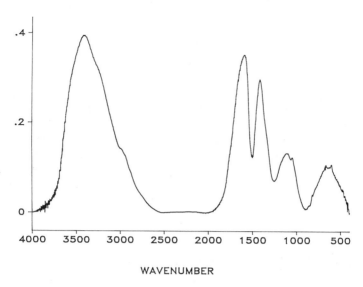

WAVENUMBER

Figure 4. Infrared absorption spectrum of a typical aquatic humic acid from a sample of bog water. The spectrum was obtained on a dry sample in a Kbr pellet.

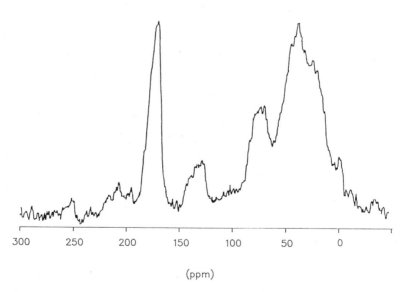

(ppm)

Figure 5. Carbon-13 nuclear magnetic resonance spectrum of a typical aquatic humic acid obtained from a sample of bog water.

materials than can be obtained from other spectroscopic techniques. From the integrated values of aromatic carbons (110-150 ppm) and aliphatic carbons (0-40 ppm), the percent aromaticity and aliphaticity of the humic material can be calculated (*19*).

The most commonly used degradative methods employ pyrolysis in conjunction with one of the other techniques for identifying functional groups such as pyrolysis-gas chromatography, pyrolysis-mass spectrometry, or pyrolysis-Fourier transform infrared spectroscopy. This combination allows for easier identification of the simpler units created from thermal degradation of the humic material. Some major pyrolytic products from the degradation of humic acid, as determined by gas chromatography, are shown in Table IV (*15*). Studies of the pyrolytic products of humic acids as detected by soft ionization mass spectrometry have resulted in the identification of a number of degradation products which are summarized in Table V (*19*). A major problem with this technique is the degradation of carboxyl units to CO_2, which is difficult to detect and results in an underestimation of carboxylate content.

Table IV. Pyrolysis-gas Chromatography Products From an Aquatic Humic Acid. Sample was taken from Volo Bog, Illinois.

Retention Time (min)	Compound
5.30	Benzene
5.96	Acetonitrile
6.10	Thiophene
6.33	Toluene
7.62	Saturated C_{11}
8.27	Unsaturated C_{11}
9.69	Saturated C_{12}
10.44	Unsaturated C_{12}
11.90	Anisole
11.98	Saturated C_{13}
12.78	Unsaturated C_{13}
13.82	Acetic acid
14.31	Saturated C_{14}
15.12	Unsaturated C_{14}
15.64	Pyrrole
15.96	Propionic acid
16.58	Saturated C_{15}
17.38	Unsaturated C_{15}
20.34	Acetamide
25.18	Phenol and o-cresol
26.67	m-Cresol and p-cresol

(adapted from *15*)

Chemical methods of degradation include 1) reduction using metals, hydrogen, or metal hydrides; 2) oxidation using copper oxide, permanganate, or chlorine; and 3) alkaline hydrolysis *(20)*. Prediction of the types of bonds that are attracted by a particular chemical degradation reaction can indicate how the subunits are linked together in the parent humic molecule; however, results of chemical degradation techniques are qualitative because of the low (1-25%) and variable product yields often obtained.

Table V. Pyrolysis-Mass Spectrometry Products (Soft Ionization) From a Soil Humic Acid

Compound Type		MassMarkers(m/z)
Polysaccharides		72-144
Aldohexose		120, 144
Phenols		94, 110, 120
Methoxylated phenols		108
Methylphenols		152-212
Monolignins		152-212
Dilignins		246-378
n-Fatty Acids:	n=11	186
	n=12	200
	n=13	214
	n=14	228
	n=15	242
	n=16	256
	n=17	270
	n=18	284
n-Alkanes:	n=14	198
	n=15	212
	n=16	226
	n=17	240
	n=19	268
	n=20	282
	n=35	492
	n=49	646

(data summarized 19)

As yet, no single analytical method, degradative or instrumental, can provide data for absolute characterization of the structure of humic materials. Therefore, a combination of several techniques, with comparison and confirmation of results from each method, must be used to unravel the complex issues of humic composition and properties.

Metal and Radionuclide Transport and Binding

It has been recognized for some time that humic substances influence the geochemical behavior of metals and radionuclides. For this reason, the transport of these species in the form of humic or fulvic colloids has been emphasized in areas such as metal bioavailability (21) and the safety assessments of nuclear waste disposal facilities (22). Enrichment factors of trace heavy metals in humate sediments were found to be 10^4 to 1 over the supernatant waters (23). In addition, the concentrations of heavy metals in surface waters correlates with the concentration of DOC in the form of humic colloidal materials (24). Other field-scale studies have demonstrated high mobility for DOC(25) and for metals associated with the mobile organic matter (18).

Although there are a variety of functional groups in the humic and fulvic structures, the carboxylate groups are primarily responsible for binding metals and radionuclides under most natural conditions (1). Experimental techniques have been employed to directly probe metals complexed by the carboxylate functional groups; for example, luminescence techniques have been used to relate changes in lanthanide metal hydration to metal binding sites exhibited by a fulvic acid (26). Time-resolved fluorescence techniques are also useful and can provide discrimination between 1) the fluorescence from the DOC and that of fluorescing metals such as trivalent actinides or the uranyl ion (UO_2^{+2}) and 2) changes in the metal fluorescence due to complexation with the DOC (27). These approaches provide experimental data that are useful in developing models of metal binding to humic substances.

Modeling metal binding to DOC is necessarily complex due to the polyelectrolytic nature of these ligand materials. As described previously (21), modeling approaches can be empirical or derived from concepts using thermodynamic descriptions of solution equilibria. One proposed empirical approach treats metal binding as a statistical phenomenon that is adequately described by a Gaussian distribution function (28). The advantage of this type of empirical modeling is that descriptions of discrete metal binding sites in these difficult-to-define mixtures are not required; however, conceptual models that extend beyond purely statistical descriptions and attempt to provide insight into the nature of the metal-ligand interaction have also been proposed. One such approach assumes binding as an electrostatic condensation of cations along the polyelectrolyte structure (29), while the second modeling approach describes binding to discrete sites within the polymeric structure of the organic molecule (30).

The relationship between types of metal interactions and their thermodynamic and kinetic stabilities with conceptual models is not yet clear. Reported stability constants for interactions between some metals and dissolved organic matter (DOM) suggest that the presence of humic substances can dominate the metal's solution speciation in aquatic systems (1). In general, calculation of stability contants for humate complexes requires the use of multiple parameters to adequately account for increased complexation due to the polyelectrolytic nature of humic substances. Studies on the rates of dissociation of metal-humate complexes have also suggested that multiple interactions are possible (31-34). Although thermodynamically stable interactions are expected at equilibrium, complex lability is initially high and decreases with time. Slow conformational changes and diffusion of cations into the tertiary structure of the humic substances are believed to lead to increased metal binding by discrete internal sites exhibiting decreased lability (35); however, experimental techniques that can provide the necessary direct molecular-level evidence are not yet available.

Despite the absence of molecular-level evidence, such interpretations are consistent with conceptual models that envision discrete sites for binding, as well as with the majority of the experimental evidence. A common point of all descriptions of metal binding to DOC involves increased binding with a higher degree of ionization of the humic material. Although changes in the complexation capacity as a function of pH, ionic strength, or other experimental parameters have been suggested to be an artifact of the models used in the theoretical description (*36*), increasing the ionization of the organic polyelectrolytes is generally agreed to cause greater intermolecular repulsion, which results in conformational changes and disaggregation or uncoiling of the polymer chain. The binding of metal cations neutralizes these repulsive forces and promotes contraction of the polymer. At extensive metal loading, the humic molecule exists as a compact unit whose exterior is hydrophobic. This results in coagulation and precipitation and occurs at various ratios of bound cations to ionized carboxylates, depending on the cation charge (*21,37*). At low metal-to-carboxylate ratios, the metal-humic complexes are soluble in the aqueous systems.

Humic and fulvic materials are extremely important in the mobilization and concentration of toxic metals and radionuclides in the environment. Due to size changes induced by conformational rearrangement and aggregation/dissociation arising from intermolecular hydrogen bonding, humic and fulvic acids can form soluble complexes that can migrate long distances or precipitate, carrying bound cations with them. This depends on the metallic ion, the cation charge, the degree of ionization of the organic molecule, the ionic strength of the media, and the metal loading (1). Therefore, any model predictions of migration must include binding kinetics, as well as thermodynamics, to be successful.

Hydrophobic Transport

The presence of humic materials can also promote the solubilization of nonpolar hydrophobic compounds. This acts to decrease the sorption of these materials (e.g., DDT) to the soils or sediments or to decrease the volatility rate of the more volatile organics (e.g.. polychlorinated biphenyls). The mechanism of this enhancement is not well understood. One theory is that the nonpolar groups within the humic molecule form a micellar or double-layer structure that traps the nonpolar organic in a microscopic hydrophobic environment similar to the behavior of surface active micelles (*38*). Others have shown evidence for the formation of ionic and hydrogen bonds between humic materials and hydrophobic compounds (*39*). As with the binding of humic materials to metal cations, both mechanisms may be important, depending on the type of organic involved. The extent of this binding depends on ionization of the humic material, ionic strength, and counterion concentrations and is most pronounced for the least-soluble organic compounds (*40*). This is consistent with a mechanism that is sensitive to the conformational changes within the humic molecule.

The ability of humic substances to bind hydrophobic organics can affect not only their mobility, by decreasing the sorption to sediments, but also the rate of chemical degradation, photolysis, volatilization, and biological uptake of these organics. This interaction can serve to lengthen the lifetimes and transport distances of these contaminants in the environment.

Future Implications

The presence of humic and fulvic acids in surface waters and groundwaters will have a significant influence on the transport and fate of metals, radionuclides, and organic contaminants in the environment. These natural organic acids can either transport or immobilize contaminants, depending on the environmental conditions. Humic and fulvic substances can also retard or enhance the photochemical decomposition of pesticides or toxic organics. Therefore, to be sucessful any remediation strategies must consider the effects of humic materials. If properly understood, this behavior can be used to manipulate pollutant solubilization and facilitate containment or cleanup of contaminated sites.

In addition to complexing metals in the aqueous phase, humic materials can also remove metals and radionuclides contained within the mineral matrix (*18*). The factors that control this behavior are not well understood; however, it has direct implications on waste storage and containment strategies. The binding of organic and inorganic contaminants to humic substances is known to alter their bioavailability (*41*). Organic contaminants that are associated with humic substances are essentially unavailable for uptake by biota. In most cases studied, toxic metals associated with humic materials also have reduced uptake. With the new focus on bioremediation of polluted are as, the effect of the association of pollutants with humic materials on their phytotoxic properties must be considered, particularly for bound metals and radionuclides.

Acknowledgement

The authors (J.G. and N.M.) wish to acknowledge the support of the U.S. Department of Energy, Office of Energy Research, Office of Health and Environmental Research, under contract W-31-109-ENG-38. This work was performed at Argonne National Laboratory and the Savannah River Ecology Laboratory.

Literature Cited

1. Choppin, G. R.; Allard, B.In *Handbook on the Physics and Chemistry of theActinides*; Freeman, A. J.; Keller, C. Eds. ; Elsevier Science Publishers: Amsterdam, the Netherlands, B.V. **1985**, pp 407-429.
2. Stevenson, F. J. Humus Chemistry: *Genesis, Composition, Reactions*; Wiley: New York, NY, **1982**; pp 285-248.
3. *Aquatic Humic Substances: Influence on Fate and Treatment of Pollutants*; MacCarthy, P.; Suffet, I. H., Eds.; Advances in Chemistry Series 219; American Chemical Society: Washington, DC, **1989**; pp xvii-xxx
4. Schnitzer, M.; Kahn, S. U. *Humic Substances in the Environment*; Marcel Dekker, Inc.: New York, NY, **1972**
5. Livens, F. R. *Environ. Pollut.* **1991**, *70*, 183-208.
6. Schulten, H.-R.; Plage, B. *Naturwissenschaften* **1991**, *78*, 311-312.
7. Shulten, H.-R.; Schnitzer, M. *Naturwissenschaften* **1995**, *82*, 487-498.
8. Gosh, K.; Schnitzer, M. *Soil Sci.* **1980**, *129*, 266-276.

9. Senesi, N.; Schnitzer, M. In *Environmental Biogeochemistry and Geomicrobiology*; Krumbein, W. E. Ed.; Ann Arbor Science: Ann Arbor, MI, **1978**; pp 467-481.
10. Skogerboe, R. K.; Wilson, S. A. *Anal. Chem.* **1981**, *53*, 228-231.
11. Cooper, W. J.; Zika, R. G.; Pastasne, R. G.; Fischer, A. M. In *Aquatic Humic Substances: Influence on Fate and Treatment of Pollutants*; MacCarthy, P.; Suffet, I. H Eds.; Advances in Chemistry Series 219; American Chemical Society: Washington, DC, **1989**; pp 333-362.
12. Kononova, M. M. *Soil Organic Matter*; 2nd Ed. Pergamon Press: Oxford, England, **1966**; pp 544.
13. Marley, N. A.; Gaffney, J. S.; Orlandini, K. A. *Hydrol. Proc* **1991**, *5*, 291299.
14. Dearlove, J. P. L.; Longworth, G.; Ivonovich, M.; Kim, J. I.; Delakowitz, B.; Zeh, P. *Radiochim. Acta* **1991**, *52/53*, 83-89.
15. Marley, N. A.; Gaffney, J. S.; Orlandini, K.A.; Picel, K. C.; Choppin, G. R. *Sci. Total Environ.* **1992**, *113*, 159-177.
16. Boggs, S.; Livermore, D. G.; Seitz, M. G. *Macromol. Chem. Phys,* **1985**, *C25*, 599-657.
17. Marley, N. A.; Gaffney, J. S.; Cunningham, M. M. *Spectroscopy* **1992**, *7*, 4453.
18. Marley, N. A.; Gaffney, J. S.; Orlandini, K. A.; Cunningham, M. M. *Envi ron. Sci. Technol.* **1993**, *27*, 2458-2461.
19. Schnitzer, M. *Soil Sci.* **1991**, *151*, 41-58.
20. Sonnenberg, L. B.; Johnson, J. D.; Christman, R. F. In *Aquatic Humic Sub stances: Influence on Fate and Treatment of Pollutants*; MacCarthy, P.; Suffet, I. H., Eds.; Advances in Chemistry Series 219; American Chemical Society: Washington, DC, **1989**; pp 3-23.
21. Buffle, J. In *Metal Ions in Biological Systems*, Sigel H., ed., Marcel Dekker: New York, NY **1984**, pp 165-221.
22. Moulin, V.; Ouzounian, G. *Appl. Geochem.* **1992**, *1*, 179-186.
23. Choppin, G. R. *Radiochim. Acta* **1992**, *58/59*, 113-120.
24. Nelson, D. M.; Penrose, W. R.; Karttunen, J. O.; Mehlhaff, P. *Environ. Sci. Technol.* **1985**, *19*, 127-131.
25. McCarthy, J. F.; Williams, T. M.; Liang, L.; Jardine, P. M.; Jolley, L. W.; Taylor, D. L.; Palumbo, A. V.; Cooper, L. W. *Environ. Sci. Technol.* **1985**, *19*, 127-131.
26. Dobbs, J. D.; Suseyto, W.; Knight, F. E.; Castles, M. A.; Carreira, L. A.; Azarraga, L. V. *Anal. Chem.* **1989**, *61*, 483-488.
27. Moulin V.; Tits, J.; Moulin, C.; Decambox, P.; Mauehien, P.; Ruty, O. D. *Radiochim. Acta* **1992**, *58/59*, 121-128.
28. Purdue, E. M.; Lytle, C. R. *Environ. Sci. Technol.* **1983**, *17*, 654-660.
29. Manning, G. S. *J. Phys. Chem.* **1981**, *8*, 870-877.
30. Marinsky, J. A.; Anspach, W. M. *J. Phys. Chem.* **1975**, *79*, 433-439.
31. Hering, J. G.; Morel, F. M. M. *Environ. Sci. Technol.* **1990**, *24*, 242-252.
32. Rate, A. W.; McLaren, R. G.; Swift, R. S. *Environ. Sci. Technol.* **1992**, *26*, 2477-2483.
33. Rate, A. W.; McLaren, R. G.; Swift, R. S. *Environ. Sci. Technol.* **1993**, *27*, 1408-1414.

34. Choppin, G. R.; Clark, S. B. *Marine Chem.* **1991**, *36*, 27-38.
35. Rao, L.; Choppin, G. R.; Clark, S. B. *Radiochim. Acta* **1994**, *66/67* 141-147.
36. Purdue, E. M. In *Aquatic Humic Substances: Influence on Fate and Treatment of Pollutants*, Suffet, I.H., ed. American Chemical Society: Washington, DC, **1989**; pp 281-295.
37. Carlsen, L.; Lassen, P.; Moulin, V. *Waste Management* **1992**, *12*, 1-6.
38. Wershaw, R.I. *J. Contam. Hydrol.* **1986**, *1* 29-45.
39. Senesi, N.S.; Testini, C.; Miano, T.M. *Org. Geochem.* **1987**, *11*, 25-30.
40. Carter, C.V.; Suffet, I. H. *Environ. Sci. Technol.* **1982**, *16*, 735-740.
41. McCarthy, J.F.; Jimenez, B.D. *Environ. Toxicol. Chem.* **1985**, *4*, 511-521.

Molecular Properties and Sampling

Chapter 2

Micellar Nature of Humic Colloids

T. F. Guetzloff[1] and James A. Rice[2]

Department of Chemistry and Biochemistry, South Dakota
State University, Brookings, SD 57007–0896

Humic and fulvic acids are ill-defined and heterogeneous mixtures of naturally-occurring organic molecules that possess surface active properties. The molecules that comprise this mixture are also known to form aggregates of colloidal dimensions. Humic and fulvic acids are shown to be able to solubilize hydrophobic organic compounds (HOC) in a manner that is consistent with known, micelle-forming surfactants, but not at organic carbon concentrations that are environmentally relevant. In addition, it is found that some HOCs are not solubilized to the same extent as other HOCs. Some implications of the micellar nature of humic materials are briefly discussed.

Micelles are colloidal particles formed by the concentration-dependent aggregation of surfactant molecules (1). In an aqueous environment micelles form when the hydrophobic portions of the surfactant molecules begin to associate at a surfactant concentration that is referred to as the "critical micelle concentration", or CMC, as a result of hydrophobic effects. In water, a micelle has a hydrophobic core and a charged surface that is the result of the orientation of ionizable or hydrophilic functional groups out into the bulk solution. At concentrations prior to the CMC the surfactant molecules migrate to the solution-air interface which disturbs the structure of the water molecules and results in a decrease in the solution's surface tension (2). At concentrations greater than the CMC, increasing

[1]Current address: Chemistry Department, Mount Marty College, Yankton, SD 57078
[2]Corresponding author

amounts of surfactant result in the formation of additional micelles but the surface tension remains constant. The surface tension of a surfactant will typically undergo an abrupt transition to a constant value at the CMC.

Because aqueous micelles have a hydrophobic core they can, in effect, act as a second, nonaqueous phase in a system and greatly enhance the apparent water solubility of relatively insoluble hydrophobic organic compounds (HOC). Because this solubility enhancement is only observed at, or after, the onset of micelle formation, it is a criterion for identifying the formation of a micelle (3). The coincidence of the onset of a constant surface tension and the abrupt solubilization of a HOC is a definitive test for micelle formation.

It has been recognized for some time that the presence of even small amounts of humic or fulvic acid in an aqueous solution can significantly enhance the apparent water solubility of a hydrophobic organic compound (eg., 4). This observation has been extrapolated so that the ability of humic and/or fulvic acid to effect this solubilization in aqueous solutions is often attributed to the presence of micelles (4-12). Guetzloff and Rice (13) first demonstrated that humic acid will form a micelle in alkaline, aqueous solutions but not at concentrations that are likely to be encountered in natural environments. Though the solution conditions in the experiments performed to date limit the extrapolation of these results to natural systems, there are important aspects of humic substances chemistry that can be studied by understanding their process of micelle formation. For example, the hydrophobic core of an aqueous micelle provides a nonaqueous environment into which hydrophobic organic contaminants can partition. While the alkaline conditions that are necessary for micelle formation to occur in the experiments reported here make this an unlikely mechanism in a natural environment, understanding the solubilization of HOC by humic and fulvic acid may provide insight into HOC interactions with humic materials in the solid state. In an aqueous system, a micelle forms as a result of hydrophobic interactions between lyophilic portions of surfactant molecules which produces a colloidal aggregate whose surface is studded with hydrophilic functional groups. This implies a certain juxtaposition of hydrophilic and hydrophobic regions of the molecules that comprise a humic acid or fulvic acid. In order to form a micelle the molecules must be amphiphilic, that is, one portion of the molecule must be more hydrophobic while the other must be more hydrophilic. This gives indirect information on the arrangement of functional groups that might be present.

Materials and Methods

Materials. Humic acid (HA) and fulvic acid (FA) were isolated from a soil known as the Poinsett silt-loam using a traditional alkaline extraction procedure. Solutions of HA or FA were made by dissolving enough material in aqueous NaOH (pH=10.6) to give a range of concentrations upto ~12.3 gm HA or FA per liter.

Surface Tension Measurements. The surface tension of each HA or FA solution was measured using a du Nouy ring tensiometer as previously described (13).

HOC Solubilization. Carbon-14 labeled DDT (2,2-bis(4-chlorophenyl)-1,1,1-trichloroethane) or pyrene were used as HOC probes to study the micelle formation phenomena. The probe dissolved in toluene was added to a culture tube, and the solvent evaporated under nitrogen. Ten milliliters of each HA or FA solution were added to the tube which was then capped with a teflon-lined lid. The tube was transferred to a water bath (25.0 °C±0.5 °C) where the solutions were equilibrated for 24 hours with periodic shaking. After equilibration the tubes where centrifuged to minimize the amount of suspended or particulate probe molecule. An aliquot of the supernatant was then withdrawn and transferred to a scintillation vial. Scintillation fluid was added to the vial and the β-emission counted at a 0.5% counting error. Additional methodological details can be found elsewhere (13).

Small-angle X-ray Scattering. Small-angle x-ray scattering (SAXS) measurements of the PSL humic acid were performed on the 10-meter SAXS camera at Oak Ridge National Laboratory (14). Fractal dimensions were calculated from log-log plots of the scattering intensity as function of the scattering vector (I(q)) versus the scattering vector (q). The procedure used to characterize humic acid, and determine its fractal dimension, has been described by Rice and Lin (15-16). The radius of gyration (which is defined as the root mean square of the distance from the electrons in a particle to the center of charge) of the HA sample was obtained from the slope of of a plot of ln I(q) versus q^2 (17).

Results and Discussion

HOC Solubilization. The effect of increasing HA and FA concentrations on the solubility of DDT and pyrene are shown in Figures 1 and 2. The effect of increasing HA or FA concentration on the surface tension of the resulting solution is also shown. The surface tension of each solution decreases with increasing HA or FA concentration until at concentrations of 7 gm HA/L or 6.8 gm FA/L the surface tension becomes constant. The surface tension of the the FA solution is higher than that of the HA which indicates that FA is a weaker surfactant than humic acid under these conditions. Combined with the solubility data of the HOC probes discussed later, these values can be taken as CMC values for the formation of micelles by HA and FA. The surface tension of the FA solution (Figure 2) shows a pronounced drop just prior to becoming constant. This type of behavior is typical of mixed surfactant solutions (18). Given the heterogeneous nature of fulvic acid, the presence of this drop in the surface tension would be expected.

The solution concentration of DDT is found to increase at the CMC for both HA and FA, and it increases dramatically in HA solutions with concentrations above the CMC. The coincidence of the abrupt change in slope of the surface tension plots, and the abrupt increase in the DDT concentration is direct evidence for the formation of micelles in aqueous alkaline solutions of humic materials (13). It should be reemphasized,

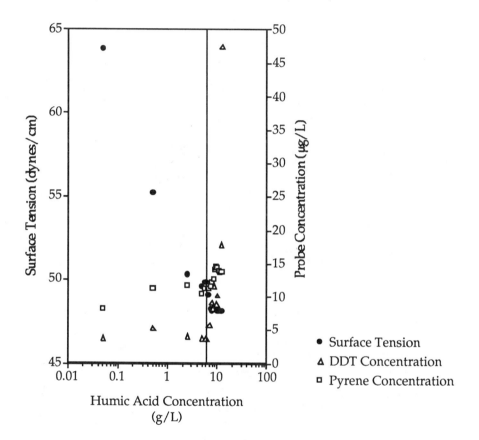

Figure 1. Effect of increasing humic acid concentration on the surface tension and apparent solubility of DDT and pyrene. The solid line indicates the position of the HA CMC. Standard deviations are less than the height of the symbols.

however, that the concentrations at which the CMC was observed in these experiments (7 gm HA/L and 6.8 gm FA/L) make it unlikely that micelle formation would take place in a natural system.

The solubilization behavior of pyrene is different than that exhibited by DDT. The apparent water solubility of pyrene does increase at the CMC in the HA solutions, but not as abruptly as in the case of DDT. Apparently pyrene is not as readily taken up by HA micelles as is DDT. Careful examination for Figure 1 also shows that there is a gradual solubility enhancement of pyrene even before the CMC. Figure 2 shows a solubility enhancement for pyrene in the presence of even low concentrations of FA, but there is no abrupt concentration increase that can be attributed

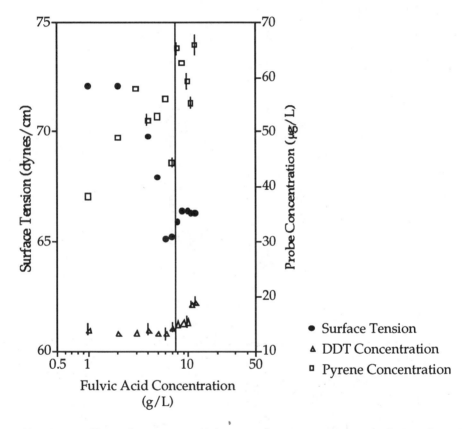

Figure 2. Effect of increasing fulvic acid concentration on the surface tension and apparent solubility of DDT and pyrene. The solid line indicates the position of the FA CMC. When present errors bars represent ± 1 standard deviation, all others are less than the height of the symbol.

to the formation of a micelle. Thus, while FA is able to enhance the water solubility of pyrene it does not do so through micelle formation. Gauthier *et al.* (19-20) concluded that polyaromatic hydrocarbon binding to HA or FA was a result of van der Waals interactions between the aromatic rings of polyaromatic hydrocarbons and the aromatic components of the humic material; higher HA or FA concentrations should result in higher solution pyrene concentrations. While this mechanism can be used to explain the solubility behavior of pyrene in HA and FA solutions prior to the CMC (the HA and FA concentrations used in their studies were on the order of milligrams of humic material/L), it does not seem to explain the concentration increase at the onset of HA micelle formation which is attributed to the solubilization of pyrene in the hydrophobic core of the micelle.

The difference in the ability of HA and FA to solubilize the probes could be related to the size of the molecules which comprise each humic material. Humic acid is generally believed to consist of larger molecules than does fulvic acid isolated from the same environment (21-22). The micelles formed during FA aggregation could be smaller and less able to accommodate DDT and pyrene in their interior than larger HA micelles. Estimates of the molecular dimensions of DDT and pyrene using HYPERCHEM gave lengths of 10.3 Å and 9.1 Å, respectively. Using SAXS measurements Thurman *et al.* (23) estimated the size of FA aggregates and found that every sample in the suite that they characterized had a radius of gyration of less than 10Å. This gives an indication that the probe molecules being investigated are on the order of the size of the FA micelles which may interfere with the probe's solubilization. In contrast, the radius of gyration of HAs from a variety of environments have been reported in the range of 8 Å to 100 Å (23-24).

Table I gives the radius of gyration of the HA solutions at several concentrations and the corresponding fractal dimension. The values

Table I. Radius of gyration and the fractal dimension of HA at various solution concentrations.

HA Concentration	Radius of Gyration*	Fractal Dimension*
2.0 gm/L	5.5 Å	nd
6.0 gm/L	4.8 Å	2.8
8.3 gm/L	4.8 Å	2.9
11.0 gm/L	5.2 Å	2.8

nd - could not be determined because scattering was very weak.
*The uncertainty associated with each measurement is ± 0.1.

reported here are smaller than those previously reported in the literature (23-24), but given the heterogeneity of humic materials and the effect of solution parameters on their size, this is not surprising. As the solution concentration increases the particle size remains essentially constant. The fractal dimension also remains constant as the concentration increases. Neither of these measurements, however, show an abrupt change as the concentrations go from below the CMC (2.0 and 6.0 gm HA/L) to above the CMC (8.3 and 11.0 gm/L) which would be expected if all of the components were involved in the micellization process. The lack of this abrupt transition suggests that only a portion of the molecules that comprise HA are responsible for the HOC solubilization observed in Figures 1 and 2.

Summary

Both HA and FA have been shown to form micelles, but not at concentrations that are environmentally relevant. The inability of FA to solubilize pyrene at concentrations above its CMC, and the lower solubility enhance of DDT in FA micelles compared to HA suggests that the smaller size of the molecules which comprise FA, and consequently the micelles that form from it, affects the solubilization phenomena. The SAXS analysis of HA does not show an abrupt change in size or fractal dimension as the solution concentration increases beyond the CMC which suggests that only a portion of the molecules which comprise HA are involved in the micellization phenomena.

Acknowledgement

This work was supported by the US Department of Agriculture, National Research Initiative Competitive Grants program through agreement no. 91-37102-6864.

References

1. Tanford, C. *The Hydrophobic Effect: Formation of Micelles and Biological Membranes*; Wiley: New York, 1980.
2. Popiel, W. *Introduction to Colloid Science*; Exposition Press: New York, 1974.
3. Mukerjee, P.; Mysels, K. *Critical Micelle Concentrations of Aqueous Surfactant Systems*; Natl. Bureau Standards Data Ser., 36, National Bureau of Standards: Washington, DC, 1971, pp. 1-21.
4. Wershaw, R.; Burcar, P.; Goldberg, M. *Environ. Sci. Technol.* **1969**, *3*, 271-273.
5. Piret, E.; White, R.; H.; Walther, H.; Madden, A. *Sci. Proc. R. Dublin Soc.* **1960**, *A1*, 69-79.
6. Visser, S. *Nature* **1964**, *204*, 581.
7. Boehm, P. D.; Quinn, J. G. *Geochim. Cosmochim. Acta* **1973**, *37*, 2459-2477.
8. Tschapek, M.; Wasowski, C. *Geochim. Cosmochim. Acta* **1976**, *40*, 1343-1345.
9. Chen Y.; Schnitzer, M. *Geoderma* **1978**, *26*, 87-104.
10. Rochus, W.; Sipos, S. *Agrochim.* **1978**, *22*, 446-454.
11. Hayano, H.; Shinozuka, N.; Hyakutake, M. *Yukagaku* **1982**, *31*, 357-362.

12. Hayase, K.; Tsubota, H. *Geochim. Cosmochim. Acta* **1983**, *47*, 947-952.
13. Guetzloff, T. F.; Rice, J. A. *Sci. Total Environ.* **1994**, *152*, 31-35.
14. Wignall, G. D.; Lin, J. S.; Spooner, S. *J. Appl. Crystallog.* **1990**, *23*, 241-246.
15. Rice, J. A.; Lin, J. S. *Environ. Sci. Technol.* **1993**, *27*, 413-414.
16. Rice, J. A.; Lin, J. S. IN *Humic Substances in the Global Environment and Implications on Human Health*; Senesi, N.; Miano. T. M., Eds., Elsevier B. V.: Amsterdam, 1994, pp. 115-120.
17. Chen, S.; Lin, T. IN *Methods of Experimental Physics*, v. 23B, Academic Press: New York, 1987, pp. 489-543.
18. Schott, H. *J. Phys. Chem.* **1966**, *70*, 2966-2973.
19. Gauthier, T. D.; Shane, E. C.; Guerin, W. F.; Seitz, W. R.; Grant, C. L. *Environ. Sci. Technol.* **1986**, *20*, 1162-1166.
20. Gauthier, T. D.; Seitz, W. R.; Grant, C. L. *Environ. Sci. Technol.* **1987**, *21*, 243-248.
21. Stevenson, F. J. *Humus Chemistry*; Wiley: New York, 1982, pp. 285-308.
22. Wershaw, R. L.; Aiken, G. R. IN *Humic Substances in Soil, Sediment, and Water: Geochemistry, Isolation, and Characterization*; Aiken, G. R.; McKnight, D. M.; Wershaw, R. L.; MacCarthy, P., Eds., J. Wiley: New York, 1985, pp. 477-492.
23. Thurman, E. M.; Wershaw, R. L.; Malcolm, R. L.; Pinckney, D. J. *Org. Geochem.* **1982**, *4*, 27-35.
24. Wershaw, R. L.; Burcar, P. J.; Sutula, C. L.; Wiginton, B. J. *Science* **1967**, *157*, 1429.

Chapter 3

The Use of Hollow-Fiber Ultrafilters for the Isolation of Natural Humic and Fulvic Acids

Jeffrey S. Gaffney, Nancy A. Marley, and Kent A. Orlandini

Environmental Research Division, Argonne National Laboratory, Building 203, 9700 Cass Avenue, Argonne, IL 60439

Hollow-fiber ultrafiltration can be used to isolate natural humic and fulvic acids from surface and groundwaters for further chemical and physical characterization. Ultrafilters are particularly useful in the approximate sizing of colloidal humic and fulvic acids with effective diameters below 0.45 μm. By first using hollow-fiber filters and then stirred-cell flat ultrafiltration techniques, these naturally occurring organics can be separated into size fractions down to approximately 500 molecular weight. Sufficient material can be obtained by using these size-sampling methods to apply a number of chemical and physical characterization techniques. Examples are presented of size-specific data obtained for fulvic and humic acids with a variety of spectroscopic techniques (e.g., infrared, ultraviolet-visible, ^{13}C nuclear magnetic resonance, and mass spectroscopy) and pyrolysis gas chromatography. Inorganic trace element analysis and radiochemical characterization of the materials bound to the humic and fulvic size fractions can yield information on the geochemical importance of these natural organics in the migration of low-level wastes.

Humic and fulvic acids compose a significant fraction of the dissolved organic carbon in surface and groundwaters (1). These naturally occuring materials are macromolecular colloids that are typically in the sub-0.45 μm fraction of these waters and can act as transporting agents for both organic and inorganic contaminants (including radionuclides) in a wide variety of hydrologic situations. Aqueous humic and fulvic acids are of particular concern because of their potential to mobilize of actinides and other radionuclides associated with stored wastes. This potential is largely due to the ability of humic and fulvic acids to strongly complex radionuclides and metals (2-7) and to their small size, in the macromolecular to colloidal range, which allows them to migrate through the subsurface media (8-10). Indeed, recent studies have shown that americium, plutonium, uranium, radium, and thorium can all be mobilized in groundwaters when colloidal organic matter is present or when humic and fulvic acids are injected into an aquifer (11,12)

0097–6156/96/0651–0026$15.00/0
© 1996 American Chemical Society

Humic and fulvic acids are particularly interesting because they contain both polar and nonpolar substituents in their molecular structures and thus can transport both water-soluble and -insoluble species (*13,14*). Indeed, colloidal organic matter, such as the soluble humic and fulvic acids found in ground and surface waters, has been refered to as a "third" phase in the traditional two-phase system of solids and water used in hydrologic chemical modeling (*13*). Because humic and fulvic acids in the colloidal size ranges are key actors in chemical waste migration and other hydrologic processes, sampling methods that can yield sufficient humic and fulvic materials for chemical and physical characterization as a function of size are necessary to improve our understanding of their role in aqueous geochemistry.

Classical methods for isolating humic and fulvic acids from soil and water samples have been based on the chemical separation of the materials by precipitation at a pH of approximately 2 (*1,15*). Many studies attempted to remove metals from humic and fulvic acids by treatment with HCl and HF (*16*). These methods suffer because the strong chemical treatments can alter the chemical structure of the humic and fulvic acids in the water samples through acid catalyzed reactions (decarboxylation, deamination, etc.), as well as by causing physical changes in the shape(and therefore the effective diameter) of the colloidal organics. Other methods to minimize these problems by using XAD-resins or sodium hydroxide processing suffer from the potential for cation exchange reactions and other means of artifact generation. These are important concerns, particularly if one is interested in isolating the humic and fulvic materials that are responsible for the binding and mobilization of inorganic cations (including radionuclides and toxic metals) in their natural geochemical state.

A method for sampling and size-fractioning of humic and fulvic acids and other organic and inorganic colloidal materials from ground or surface waters should be reproducible with minimal artifacts (both chemical and physical). The method should be amenable to processing large volumes of water to obtain sufficient material for a wide variety of analytical procedures (infrared, ^{13}C nuclear magnetic resonance, pyrolysis gas chromatography, mass spectrometry, etc.). The method should also be relatively simple and easy to use for convenience in the laboratory and particularly in the field. Hollow-fiber ultrafiltration methods can satisfy many of these requirements.

Hollow-Fiber Ultrafiltration Techniques

The sizes of concern in characterizing colloidal material range from less than 1 to 450 nm (0.45 μm) (*13*). Hollow-fiber ultrafilters have been developed that are capable of size-fractionating samples in these size ranges. These filters differ from conventional flat ultrafilters in that they do not suffer from significant polarization when processing large volumes of materials. Figure 1 shows a schematic diagram of the features and advantages of the hollow-fiber ultrafiltration approach. In the case of the flat ultrafilter, processing of large amounts of water to obtain sufficient material for analysis leads to polarization of the filter. That is, the trapped molecules tend to pile up and act as a filter themselves. Thus, physical sizing artifacts can interfere with attempts to isolate material in the colloidal size ranges, because smaller materials can be trapped in the pores of the layered, larger molecules. In the case of the hollow-fiber filters, the polarization effects are minimized because the flow of water to be processed is parallel

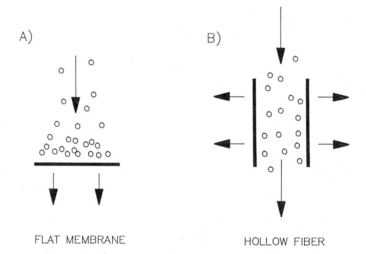

Figure 1. Diagram of filtration mechanisms for A) flat-membrane filters and B) hollow-fiber filters, showing the concentration polarization on the flat filters.

to the hollow-fiber filter; the water is pumped from the inside of the filter to the outside during filtration.

Table I summarizes typical size and flow characteristics for hollow-fiber ultrafilters (Amicon, polysulfone). Flow rates for hollow-fiber filters with size cutoffs above 300 molecular weight are greater than 0.2 L min^{-1}. To minimize pile up or polarization effects, the samples are first passed through a 35 μm screen to remove large particulate matter and then prefiltered with a 0.45 μm filter to remove large particles. Samples are size-fractionated by first using the larger filters then sequentially smaller ones. If required, the sample can be filtered to a 500 molecular weight size cutoff by using stirred flat ultrafilters or spiral wound cartridge filters. The filters are typically 20-60 cm long making them a reasonable size for laboratory and field studies. The filters can be used in a cascaded series for isolation of colloidal materials and for automated field analysis (*17*).

Table I. Molecular weight cutoff of hollow-fiber ultra filters, as related to pore size and solution output

Nominal Molecular Weight (a.m.u.)	Microns	Nanometers	Ångstroms	Output (L min^{-1})
1,000	0.001	1.0	<10	0.05
3,000	0.001	1.0	10	0.2
10,000	0.002	2.0	20	0.3
30,000	0.0025	2.5	25	0.8
100,000	0.005	5.0	50	1.0
>1,000,000	0.10	100.0	10000	1.0

Size-Specific Chemical Characterization

In the following sections, examples of size-specific information obtained with hollow-fiber ultrafiltration are presented. Surface waters (50-60 gallons) were taken from Volo Bog (Illinois), Saganashkee Slough (Illinois), and from Lake Bradford (Florida) for the analyses presented.

Volo Bog is a small, glaciated, sedge peat bog in northern Illinois. The bog has no surface inlet or outlet, and the water is at pH 4-5 with low nutrient concentrations. Saganashkee Slough drains into the Calumet Sag Channel and into the Des Plaines River in northern Illinois. It is at pH 7 and has a higher nutrient content than the bog. The Lake Bradford samples, collected near Tallahassee, Florida had a pH of 5-6. Humic and fulvic acids from this lake have been used in radionuclide binding studies (*18*). The lake is fed by a spring and by surface runoff.

Figure 2 is a color photograph of size-fractionated concentrates obtained by filtration of a water sample taken from Volo Bog. Differences in color and concentrations of the size fractions are quite apparent, as is typically the case for surface waters with high organic contents. In Figure 3, the dissolved organic carbon (DOC) distribution for the same sample is shown. The DOC measurements were made by injecting 1-10 mL of the concentrates into a Sybron PHOTOchem Organic Carbon Analyzer (Model E3500). The colors in Figure 2 correlate with the relative amounts of DOC (Figure 3). Some information about the bioavailability of the humic material can be obtained by analyzing samples after storage. Figure 3 includes DOC data for the same sample after three weeks of storage. The figure shows that the sample distribution has not changed significantly, although some degradation of the 0.1 μm size range has apparently occurred, with a corresponding increase in the fraction below 3000 molecular weight. Analysis of size fractions provides this type of information on the stability of the organic material, where as analysis of unsized samples would not.

Sufficient sample can be obtained by hollow-fiber processing for a wide variety of analytical measurements as a function of size in addition to simple DOC characterization. For example, Figure 4 shows data obtained for DOC, silica, iron, magnesium, and manganese in the various size fractions of the Volo Bog sample. The inorganic analyses were performed by inductively coupled plasma spectroscopy (Instruments SA, Model JY 86 spectrometer) with an HF-compatible torch. The organic fraction less than 3000 molecular weight contained appreciable silica; this could be small colloidal material, dissolved silica, or organically bound silica (19). Upon evaporative concentration of this fraction for other analyses, the material was found to etch the Pyrex beaker in which the low-temperature evaporation was performed (see Figure 5).* This etching was not observed for the larger fractions, indicating the potential chemical activity of fulvic acids with glass materials and quartz sands. Indeed, research examining the injection of humic- and fulvic-containing waters into a sandy aquifer showed that silica and bound actinides could be leached from the sands by these materials (10). This is another example of information that can be obtained on the reactivity of specific types of organic molecules by size fractionation of samples.

Actinide Analysis

Coupling hollow-fiber separations to radiochemical analysis provides a great deal of information about the active humic and fulvic binding agents that are responsible for groundwater transport of actinides and other radionuclides (10,11). Sufficient size-fractionated material can be obtained to measure the background and low-level radionuclides (10, 11) and to determine isotopic ratios for key species such as thorium (20). In addition, radiochemical tracers can be added to samples to evaluate exchange rates and processes occurring within the colloidal size fraction (17). Figure 6 presents the plutonium, americium, thorium and uranium distributions observed in the size fractions of Volo Bog samples. These analyses used low-background alpha-counting techniques. The actinide levels were all at natural background levels (uncontaminated except for atmospheric fallout). These results show the strong correlation typically observed between the smaller size fractions and the actinides. The distribution pattern of the Volo Bog sample is very similar to that of groundwater samples, probably because the bog is a stagnant, rain-fed system with little material deposited from runoff. The data also show that the fulvic acid and small humic materials are apparently binding the actinides in the water. Hollow-fiber ultrafiltration allows this type of analysis to be

NOTE: Please see color illustration, page 32.

Figure 2. Size-fractionated concentrates from water samples taken at Volo bog. Size ranges are (from left to right) 0.45-0.1 μm, 0.1 μm-100K, 100K-30K, 30K-3K, and 3K-500 molecular weight.

Figure 5. Pyrex beaker (right) used to concentrate aqueous size fractions of Volo Bog samples, showing etching observed by the smaller organic fraction (< 3000 molecular weight). The clear beaker (left) is a new one unexposed to the organics.

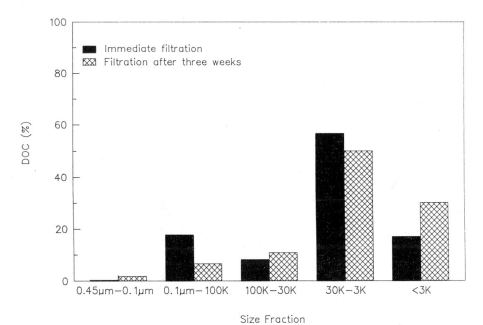

Figure 3. Dissolved organic carbon (DOC) size distribution of water samples taken at Volo Bog filtered immediately after sampling and after storage for three weeks before filtration.

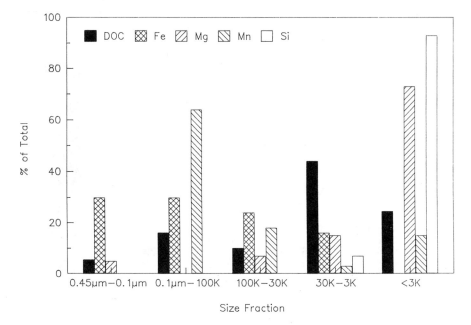

Figure 4. DOC, silica, and metals distributions as a function of colloidal size for samples taken at Volo Bog.

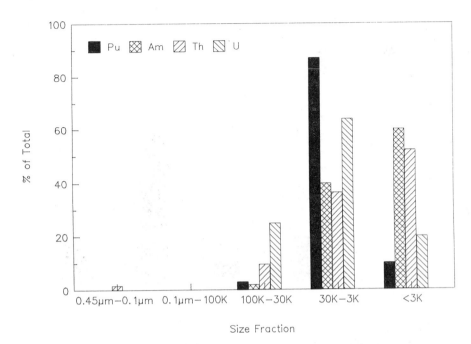

Figure 6. Percent contributions of actinides in the various size fractions for water from Volo Bog.

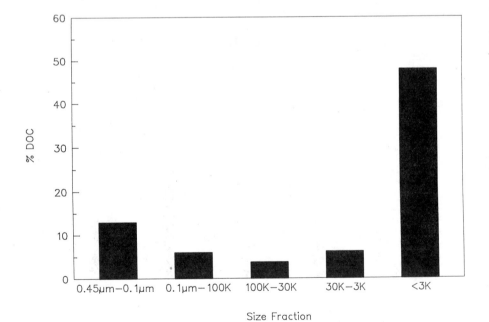

Figure 7. DOC distribution in size fractions of a water sample taken from Saganashkee Slough.

performed with minimal artifacts, because the samples are not treated with strong acids for size separation, and changes in the ionic equilibria are thus unlikely.

The observation that the smaller organic fraction is associated with actinides could have significant applications in the mitigation and cleanup of low-level radioactive contamination in ground and surface waters, in that hollow-fiber ultrafiltration methods could be used to physically concentrate the smaller fractions and remove the actinides (*10,21*).

Pyrolysis Gas Chromatography, NMR, and Infrared Analysis

Size fractionation of colloidal humic and fulvic acids can be coupled to more sophisticated analytical procedures to give a better insight into the chemical and physical properties of these important materials. For example, pyrolysis gas chromatography can provide useful information about the functional groups that make up these naturally occurring macromolecules. Figure 7 shows a DOC profile for a sample of water collected from Saganashkee Slough in Illinois. The data clearly indicate that all of the size fractions contain organics, with the fractions at 0.45-0.1 μm fraction and less than 3000 molecular weight containing the bulk. However, the DOC results do not provide any further insight on whether the organics are similar or dissimilar in chemical composition.

Figure 8 shows the pyrolysis gas chromatographs obtained by evaporating a portion of the concentrated humic and fulvic acids from the 0.45 to 0.1 μm fraction and the fraction of 3000-1000 molecular weight, which contained the bulk of the carbon compounds in this slough surface water. The chromatograms in Figure 8 were obtained from 20 μg of material by using a heating rate of 20 C ms^{-1} and a maximum pyrolysis temperature of 700 C. A Carbowax 20 m coated fused-silica capillary column (50 m, 0.2 mm i.d.) and a standard gas chromatograph equipped with a flame ionization detector (Hewlett Packard Model HP5880A) were used to obtain the pyrolysis gas chromatographic data. The figure clearly shows significant differences in the chromatograms for the two size fractions. Closer examination reveals that the levels of acetonitrile, acetamide, and indole are significantly larger in the 0.45 to 0.1μm fraction than in the smaller colloidal material, indicating that the larger materials have more nitrogen-containing functional groups and are similar to soil-derived humic material. The chromatogram for the 3000-1000 molecular weight fraction is consistent with a composition for this material that is predominantly fulvic acids.

Analysis of a fraction with ^{13}C-NMR allows further characterization of the types of carbon functional groups that the fraction contains. Figure 9 shows the ^{13}C-NMR spectrum obtained for the 1000-3000 molecular weight size fraction of the sample from Saganashkee Slough. This spectrum was obtained on approximately 50 mg of sample using a 200 MHz spectrometer (Bruker) operated at 50.27 MHz for carbon, with magic-angle spinning techniques (*22*). The spectral signal is measured as a shift from a standard carbon resonance in parts per million. This spectrum clearly shows that this size fraction contains appreciable carboxylate groups and has the larger aliphatic-to-aromatic content typical of aqueous fulvic acids.

Another powerful tool that can be applied successfully to size-fractionated samples from hollow-fiber ultrafiltration is Fourier transform infrared spectroscopy. After fractionation, samples can be evaporated, dried, and analyzed by using the

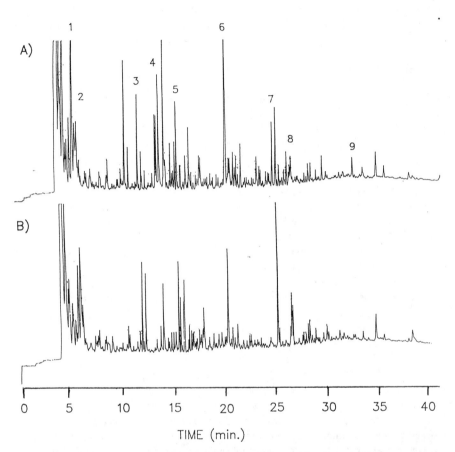

Figure 8. Pyrolysis gas chromatographs of size fractions A) 0.45-0.1 micrometers and B) 3K-1K molecular weight of samples obtained from Saganashkee Slough. Marker compounds and retention times are 1) acetonitrile, 5.97 min; 2) thiophene, 6.10 min, and toluene, 6.33 min; 3) anisole, 11.90 min; 4) acetic acid, 13.82 min; 5) pyrolle, 15.64 min; 6) acetamide, 20.34 min; 7) phenol and o-cresol, 25.18 min; 8) m-cresol and p-cresol, 26.67 min; and 9) indole, 32.53 min.

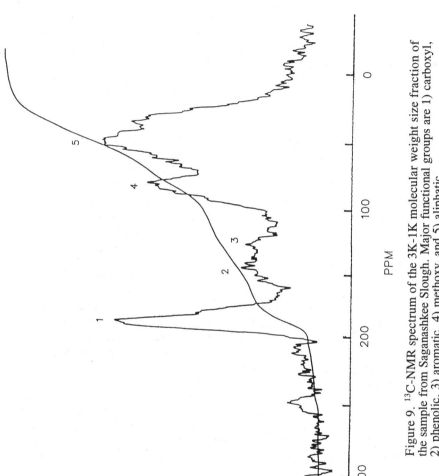

Figure 9. ^{13}C-NMR spectrum of the 3K-1K molecular weight size fraction of the sample from Saganashkee Slough. Major functional groups are 1) carboxyl, 2) phenolic, 3) aromatic, 4) methoxy, and 5) aliphatic.

traditional KBr pellet technique or diffuse-reflectance spectroscopy (23). Another technique, cylindrical internal reflectance (CIR) can be performed directly on the aqueous concentrated fractions gives structural information on the dissolved sample(10,24). Figure 10 shows CIR spectra for two size fractions of humic and fulvic acids isolated from Lake Bradford. The spectra were obtained by using ZnSe rods and a CIRCLE cell (SpectraTech) interfaced to a Fourier transform infrared spectrometer (Mattson Polaris) equipped with a cooled Hg/Cd/Te detector. Spectra were obtained at a resolution of 2 cm^{-1} and at pH 5. The results show the that the two size fractions are similar in that they both contain significant carboxylate functional groups. However, the spectra reveal that the smaller fraction (3000-1000 molecular weight) has significantly more carboxylate-bound metals than the larger fraction (100K-30K molecular weight). This observation is consistent with the inorganic analysis of these samples.

Conclusions

Hollow-fiber ultrafiltration can be a very useful tool in obtaining size-fractionated samples for chemical and physical analyses of humic and fulvic acids in surface and groundwaters. The technique is very reproducible; because it is strictly a physical separation, it minimizes the potential artifacts associated with the more classical chemical separation techniques. The separation method assumes a spherical structure for the cutoffs and therefore is an empirical approach. Nevertheless, hollow fibers can be used much as 0.45 μm filters are used to separate particulate versus dissolved material with a great deal of success.

The hollow-fiber approach is readily adapted to laboratory and field sampling. Large volumes of water (10-100 L) can be processed with minimal polarization, particularly if the samples are processed in a sequential fashion (from the large sizes to the small). Because such large samples can be readily processed, sufficient material can be obtained in the various size fractions for analysis with a wide variety of analytical methods, including routine DOC, metals characterization, and spectroscopic and chromatographic characterization of chemical functional groups. Indeed, adequate amounts (milligrams to grams) can be obtained for practically any available sophisticated analytical techniques. Besides the ^{13}C-NMR, CIR, and pyrolysis gas chromatography size-specific analyses presented here, techniques such as XAFS, XANES, and mass spectrometry can further characterize the binding of radionuclides and metals to humic and fulvic acids, as well as the sorption of nonpolar organics to these important natural organic colloids. One of the main advantages of hollow-fiber ultrafiltration is that the humic and fulvic acids and other colloidal materials are not chemically altered during size separation and concentration. This is, of course, very important if we are to improve our understanding of how these natural organics behave geochemically.

Our observations to date indicate that the humic and fulvic acids can be reasonably separated by size, with humics being in the fractions greater than 10,000 molecular weight and the fulvics in the smaller fractions. Coupling of size fractionation to various analytical tools has shown that the small fulvic acid fractions in surface and groundwaters are the most active metal and radionuclide binding agents. Such information could be exploited in remediation of low-level contaminated wastes by using larger hollow-fiber ultrafiltration systems to isolate and concentrate the small size

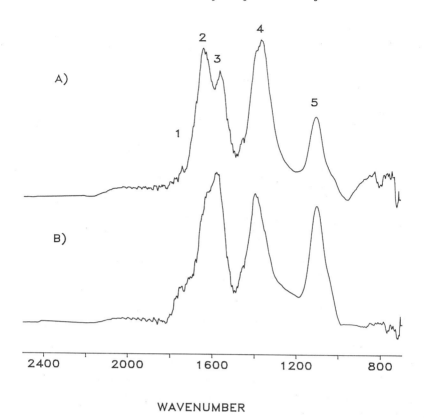

WAVENUMBER

Figure 10. Cylindrical internal reflectance (CIR) infrared spectra of humic and fulvic acid size fractions A) 3K-1K molecular weight and B) 100-30K molecular weight isolated from water samples from Lake Bradford, Florida. The major bands are assigned as 1) C=O asymmetrical stretch, 2) metal-bonded carboxyl asymmetrical stretch, 3) free carboxyl asymmetrical stretch, 4) carboxyl sym metrical stretch, and 5) C-C stretch.

fractions containing the bulk of the radionuclides or metals of concern and concentrating these fractions for disposal. Size-specific information of this sort is vital for an understanding of the migration of contaminants in natural waters. Hollow-fiber ultrafiltration methods can clearly help us understand transport processes and develop mitigation strategies when applied together with modern chemical and physical analyses.

Acknowledgments

We wish to thank Professor G.R. Choppin and Mr. Lin Feng Rao of Florida State University for obtaining the NMR data. This work was performed at Argonne National Laboratory and was supported by the United States Department of Energy, Office of Health and Environmental Research, under contract W-31-109-ENG-38.

Literature Cited

1. Frimmel, F.H.; Christman, R.F. *Humic Substances and their Role in the Environment*; Wiley-Interscience, New York, NY, 1988.
2. Choppin, G.R.; Clark, S.B. *Mar. Chem.* **1991**, *36*, 27-38.
3. Kim, J.I.; Sekine, T. *Radiochim. Acta* **1991**, *55*, 187-192.
4. Moulin, V.; Ouzounian, G. *Appl. Geochem.* **1992**, *Supplement 1*, 179-186.
5. Kim, J.I.; Rhee, D.S.; Buckau, G. *Radiochim. Acta* **1991**, *52/53*, 49-55.
6. Miekeley, N.; Kuchler, I.L. *Inorg. Chim. Acta* **1987**, *140*, 315-319.
7. Maes, A.; de Bradandere, J.; Cremers, A. *Radiochim. Acta* **1991**, *52/53*, 41-47.
8. Kim, J.I. *Radiochim. Acta* **1991**, *52/53*, 71-81.
9. Dearlove, J.P.L.; Longworth, G.; Ivanovich, M.; Kim, J.I.; Delakowitz, B.;Zeh, P. *Radiochim. Acta* **1991**, *53/53*, 83-89.
10. Marley, N.A.; Gaffney, J.S.; Orlandini, K.A.; Cunningham, M.M. *Environ.Sci. Tech.* **1993**, *27*, 2456-2461.
11. Penrose, W.R.; Polzer, W.L.; Essington, E.H.; Nelson, D.; Orlandini, K.A. *Environ. Sci. Tech*. **1990** *24*, 228-234.
12. Marley, N.A.; Gaffney, J.S.; Orlandini, K.A.; Cunningham, M.M. *Environ.Sci. Technol.* **1993**, *27*, 2456-2461.
13. McCarthy, J.F.; Zachara, J.M. *Environ. Sci. Technol.* **1989**, *23*, 496-502.
14. LaFrance, P.; Banton, O.; Campbell, P.G.C.; Villeneuve, J.P. *Wat. Sci. Tech* **1990**, *22*, 15-22.
15. Thurman, E.M.; Malcolm, R.L. *Environ. Sci. Technol.* **1981**, *15*, 463-466.
16. Lobartini, J.C.; Tan, K.H.; Asmussen, L.E.; Leonard, R.A.; Himmelsbach, D.; Gingle, A.R. *Commun. Soil Sci. Plant Anal.* **1989**, *20*, 1453-1477.
17. Marley, N.A.; Gaffney, J.S.; Orlandini, K.A.; Dugue, C.P. *Hydrological Processes* **1991**, *5*, 291-299.
18. Kim, J.I.; Buckau, G.; Bryant, E.; Klenze, R. *Radiochim. Acta* **1989**, *48*,135-143.
19. Marley, N.A.; Bennett, P.; Janecky, D.R.; Gaffney, J.S. *Org. Geochem.* **1989**, *14*, 525-528.
20. Gaffney, J.S.; Marley, N.A.; Orlandini, K.A. *Environ. Sci. Technol.* **1992**, *26*, 1248-1250.
21. Gaffney, J.S.; Marley, N.A.; Orlandini, K.A. In *Manipulation of Groundwater Colloids for Environmental Restoration*; Lewis Publishers; Boca Raton, FL,1993, pp. 225-228.
22. Hatcher, P.G.; Schnitzer, M.; Dennis, L.W.; Maciel, G.E. *Soil Sci. Soc. Am. J.* **1981**, *45*, 1089-1094.
23. Marley, N.A.; Gaffney, J.S.; Orlandini, K.A.; Picel, K.C.; Choppin, G.R. *Sci. Total Environ.* **1992**, *113*, 159-177.
24. Marley, N.A.; Gaffney, J.S.; Cunningham, M.M. *Spectroscopy* **1992**, *7*, 44-53.

CHEMICAL CHARACTERIZATION
AND STRUCTURAL DETERMINATION

Chapter 4

A New Approach to the Structural Analysis of Humic Substances in Water and Soils

Humic Acid Oligomers

Hans-Rolf Schulten

Chemical and Biological Laboratories, Institut Fresenius,
Im Maisel 14, 65232 Taunusstein, Germany

For humic acid (HA) oligomers in water and soils, a novel three-dimensional structural concept is developed which is based on comprehensive investigations combining geochemical, wet-chemical, biochemical, spectroscopic, agricultural and ecological data with analytical pyrolysis. Direct temperature-programmed pyrolysis in the ion-source of the mass spectrometer combined with soft ionization in very high electric fields (Py-FIMS) and Curie-point pyrolysis-gas chromatography/mass spectrometry (Py-GC/MS) are the principal analytical methods for the proposed humic acid monomer and oligomers (n=1-15) with molecular masses in the range of 5,500 to 84,600 g mol^{-1}. Emphasis is put on molecular modeling and geometry optimization of humic complexes using modern PC software (HyperChem) in order to determine low energy conformations, space requirements, voids, as well as inter- and intramolecular hydrogen bonds. The dynamic process of developing an optimal conformation of geomacromolecules can be observed and controlled at nanochemistry level. Strategic ecological aspects for the prediction and molecular-chemical explanation for binding and trapping of biological and anthropogenic substances and inorganics such as heavy metals are put forward.

Novel structural molecular-chemical models for humic substances such as monomer humic acid (HA) were proposed (1-5) in an integrated approach using a wide variety of analytical methods. The HA model is important because humic substances constitute 70-80% of soil organic matter (SOM) and all chemical reactions of latter (large surface, voids, high adsorption capacity, good metal complexer, good medium for microorganisms, can store nutrients and especially water) can be explained on the basis of HAs, which are the principal humic compounds in the environment (6) From the analytical aspect, main emphasis was put on pyrolysis-mass spectrometry and pyrolysis

0097–6156/96/0651–0042$15.00/0

gas chromatography/mass spectrometry (7). In addition, modern techniques of molecular modeling (8) were employed to develop three-dimensional structures of humic acids (9) and organo-mineral soil complexes (*10, 11*). In particular trapping and bonding of biological (peptides, carbohydrates) and anthropogenic substances such as pesticides (*9, 12*) and plastizisers (13) in the voids of the geometrically optimized structures were investigated.

Analytical pyrolysis studies so far have been published mostly as two-dimensional (2D) chemical structures and reaction schemes. However, recent progress in powerful, relatively low cost software and personal computers now allows three-dimensional (3D) displays and computer-assisted design (CAD) of structures and model reactions (*8*). For molecular modeling, geometry optimizations and semi-empirical calculations of complex macromolecules which are often the target of thermal degradation studies, virtually a new dimension is opened up. This is demonstrated in the following for humic acid complexes with molecular weights between about 5,500 and 84,500 g mol^{-1}. The aims of this study were :
- to illustrate the possibilities for display of humic acid macromolecules as 3D structures;
- to utilize semi-empirical calculations for geometry optimization and thus energy minimization; and
- to demonstrate the capacity of computational chemistry for simulation of humic acid interactions and formation of large humic complexes.

Analytical Pyrolysis

The basis for the structural concept of humic acids were geochemical, wet-chemical, reduction-oxidation reactions, ^{13}C-NMR spectroscopic data (*5*) and results of analytical pyrolysis (*4*) using the following two, complementary method described below.

Pyrolysis-Field Ionization Mass Spectrometry (Py-FIMS). For temperature-resolved Py-FIMS, about 100 µg of humic substances such as humic acid (HA), fulvic acid (FA), humin or 5 mg of whole soil samples, respectively, were thermally degraded in the ion source of a MAT 731 (Finnigan, 28127 Bremen, Germany) modified high performance (AMD Intectra GmbH, 27243 Harpstedt, Germany) mass spectrometer. The samples were weighed before and after Py-FIMS (error $\pm$ 0.01 mg) to determine the pyrolysis residue and the produced volatile matter. The heatable/coolable direct introduction system with electronic temperature-programming, adjusted at the +8 kV potential of the ion source and the field ionization emitter, was used. The slotted cathode plate serving as counter electrode was on -6 kV potential. Thus, at 2 mm distance between the emitter tips and the cathode, in total a potential difference of 14 kV is applied resulting in an extremely high electric field strength which is the essential basis for the described soft ionization method. All samples were heated in high vacuum (1.3 10^{-4} Pa) from 323 K to 973 K at a heating rate of approximately 0.5 K s^{-1}. About 60 magnetic scans were recorded for the mass range m/z 16 to m/z 1,000. In general, at least three replicates were performed for each sample. The total ion intensities (TII) of the single spectra were normalized to 1 mg sample weight, averaged for replicate runs, and plotted versus the pyrolysis temperature, resulting in Py-FIMS thermograms. For the selection of biomarkers and quantitative evaluations, in particular of humic substances. whole soils and soil particle-size fractions, recently detailed descriptions of the method have been given (7).

Curie-point Pyrolysis-Gas Chromatography/Mass Spectrometry (Py-GC/MS). The humic substances and soils were pyrolyzed in a type 0316 Curie-point pyrolyzer (Fischer, 53340 Meckenheim, Germany). The samples were not pretreated except drying and milling. The final pyrolysis temperatures employed were 573 K, 773 K and 973 K, respectively. The total heating time was varied between 3 and 9.9 s.

Following split injection (split ratio 1:3; flow rate 1 mL 20 s^{-1}) the pyrolysis products were separated on a gas chromatograph (Varian 3700, 64289 Darmstadt, Germany), equipped with a 30 m capillary column (DB5), coated with 0.25 μm film thickness and an inner diameter of 0.32 mm). The starting temperature for the gas chromatographic temperature program was 313 K, and the end temperature was 523 K, with a heating rate of 10 K min^{-1}. The gas chromatograph was connected to a thermoionic nitrogen-specific detector (TSD) and a double-focusing Finnigan MAT 212 mass spectrometer. Conditions for mass spectrometric detection in the electron ionization mode were +3 kV accelerating voltage, 70 eV electron energy, 2.2 kV multiplier voltage, 1.1 s/ mass decade scan speed and a recorded mass range between m/z 50 and m/z 500. A detailed description of the principle, potential and limitations of Py-GC/MS of humic fractions and soils has been given (10). Furthermore, studies of organic nitrogen-containing compounds in soils using the TSD detector recently were reported (14) which showed the identification of a wide range of aliphatic and, in particular, heterocyclic nitrogen compounds which should contribute to a better understanding of the so-called unknown organic nitrogen in soil organic matter and soils.

Structural Modeling and Geometry Optimization

The humic acid 2D structure (1) drawn by hand was converted to the three-dimensional (3D) structural model. For all described 2D and 3D work, model construction, chemical interaction studies and semi-empirical calculations the HyperChem software (release 4) for Microsoft Windows 95 (8) was used. In the present text some main software commands are indicated in brackets (in italics). The original program output in Angstrom and kcal was given in nm and kJ, respectively. The employed IBM-compatible personal computer consisted of a tower 486DX2/66, VLB 34 in combination with 32 MB memory, 17" color monitor with AVGA VLB/1MB graphic card, 815 MB disk, and peripheral hardware (e.g., Epson Stylus color printer) plus utility programs.

Association of Humic Acid Molecules

Formation and Structure of Humic Acid Oligomers. The basis in each case was the 3D model of the monomer HA molecule ($C_{308}H_{335}N_5O_{90}$, 738 atoms) previously reported (*9, 10*). In order to provide the necessary links for modeling of HA oligomers and humic particles, the HA structure was completed by adding seven CH_2 groups at the $\rightarrow$ signs in the initial 2D HA model (I, *3*). This allows bonding flexibility and structural variability which are essential features in our concept of humic structures. The elemental composition of this HA is $C_{315}H_{349}N_5O_{90}$ (759 atoms) and an elemental analysis of 67.02% C, 6.23% H, 25.51% O, and 1.24% N was calculated. The corresponding molecular mass is 5,645.192 g mol^{-1}. Preliminary investigations of the nanochemistry of a HA pentamer (*15*) indicated clearly the space requirements and relevant reaction pathways of this HA model.

Oligomeric structures were constructed by modeling and geometry optimization (in vacuo, united atoms) with HyperChem (*8*). Three main processes that lead to humic macromolecules were considered: First, elimination of small molecules such as hydrogen or water in bimolecular reactions and formation of covalent C-C or C-O-C bonds. Second, bonding of HA subunits by formation of intermolecular hydrogen bonds between the initially free starting molecules. Third, aggregation of humic substructures such as monomers dimers, trimers, etc. by weak intermolecular forces, e.g. van der Waals, steric trapping.

Humic Acid Trimer. According to the first reaction pathway, a trimer HA structure was produced by connecting three HA molecules (2,277 atoms) by two water eliminations and subsequent formation of covalent bonds. Figure 1 shows the color plot of the resulting HA trimer (2,271 atoms). The mol. mass of 16,899.41 g mol^{-1}, elemental composition of $C_{945}H_{1043}N_{15}O_{268}$ and elemental analysis of 67.16% C, 6.22% H, 1.24% N, and 25.37% O were calculated using HyperChem (*Molinfo*). Geometry optimization gave the total energy of the HA oligomer as 8,014.2650 kJ $(0.1\ nm)^{-1}\ mol^{-1}$ at a gradient (derivative of the energy with respect to all Cartesian coordinates) of 0.4185 kJ $(0.1\ nm)^{-1}\ mol^{-1}$. The spatial dimensions of this HA oligomer conformation are x = 8.68 nm (width); y = 6.38 nm (height); and z = 5.37 nm (thickness).

According to the default parameters of the used HyperChem software, a hydrogen bond is formed if the hydrogen-donor distance is less than 0.32 nm and the angle made by covalent bonds to the donor and acceptor atoms is less than 120 degrees (*8*). Under these conditions in total six *intramolecular* hydrogen bonds are observed in the displayed HA trimer. Side chains are often immobilized by intramolecular hydrogen bonds as demonstrated in a selected section (69 atoms) of the HA trimer in Figure 2. The location of this hydrogen bond is indicated in Fig. 2a by the atom symbols and in Fig. 2b by the corresponding atoms numbers. The hydrogen bond holding two side chains together is marked by the dashed line between O(1,157) of a carboxyl moiety and O(847) of the hydroxyl function of a phenol moiety. The bond distance of 0.2662 nm and the bond angle of 72.88 degree with H(1,193) were determined. Semi-empirical calculation gave an energy of the hydrogen bond of 3.8851 kJ $(0.1$ nm$)^{-1}$ mol^{-1} at a gradient of 0.0328 kJ $(0.1$ nm$)^{-1}$ mol^{-1}. The H(1,157) - O(791) hydroxyl bond is linked to a carboxyl function with O(792) and C(784) which leads to the aromatic C(778). On the other side of the hydrogen bond, the H(1,193) - O(847) hydroxyl group is bound in position 6 to the benzene ring C(838) to which in 3-position another hydroxyl with H(1194) and O(849) and in 2-position a carboxyl group with C(848), H(1,195) and the two oxygen atoms O(850) and O(851) are connected. In addition, hydrogen H(1,192) in 4-position and an propyl rest in 5-position are bound to the benzene ring which itself is fixed to the HA oligomer via C(837).

Thus, a distinct and reliable description of the intramolecular hydrogen bond in the HA trimer is given as all bond distances and angles can be determined using HyperChem. Moreover, spatial requirements and energies in the described oligomer section can be examined and reactions can be predicted in water and soils.

Humic Acid Decamer. Figure 3 shows the color plot of a HA decamer (7,578 atoms) which has been constructed using the monomer and trimer HA molecules described above. In the first step, the trimer HA was connected to the HA monomer by an *intermolecular* hydrogen bond to form a HA tetramer (n= 4; 3,030 atoms). In the second step, two of these tetramers (molecules 1 and 3) were positioned in left and right side of the workspace. In between, two monomers (molecules 2 and 4) were placed on the upper and lower side, so that in total four molecules are displayed (see Fig. 3). The distances between the molecules 1 - 4 were chosen to be larger than 1 nm which means that no intermolecular hydrogen bond between the molecules was possible under these starting conditions. For the HA decamer, the mol. mass of 56,381.55 g mol^{-1}, elemental composition of $C_{3,150}H_{3,482}N_{50}O_{896}$ and elemental analysis of 67.10% C, 6.22% H, 1.24% N, and 25.43% O were calculated. Geometry optimization using the Polak-Ribiere algorithm (conjugate gradient) allowed the determination of the total energy as 20,057.0533 kJ $(0.1$ nm$)^{-1}$ mol^{-1} after about three days calculation time. The determination condition for the calculations was set at a gradient of < 4.19 kJ $(0.1$ nm$)^{-1}$ mol^{-1} (8). The finally obtained conformation of the humic complex had the spatial dimensions of x = 23.29 nm (width); y = 11.18 nm (height); and z = 5.31 nm (thickness) which can be visualized as a long and fairly wide but very thin ribbon. Within the available space the content of humic material is high and results in association and surface reactions. At this stage of geometry optimization, altogether 31 *intramolecular* and 2 *intermolecular* hydrogen bridges are observed. The voids in the HA complex are in the order of 20 nm in the x-axis and between 2.5 and 7.0 nm in the y-axis. This demonstrates again the characteristic feature of the proposed network (sponge-like) structure of humic substances (*1–5*) and the outstanding capacity for trapping and binding biological (*10, 16*), anthropogenic substances (*9, 12, 13*), and inorganics such as silica, clays and metals (*11, 17, 18*).

In Figure 4a one *intermolecular* hydrogen bond which combines two parts of the molecules that form the HA decamer is shown. The hydrogen bridge between H(76) and O(38) in the selected section (150 atoms) has a length of 0.15 nm and a bond angle of 66.79 degree. The total energy of 1.5992 kJ $(0.1$ nm$)^{-1}$ mol^{-1} at a

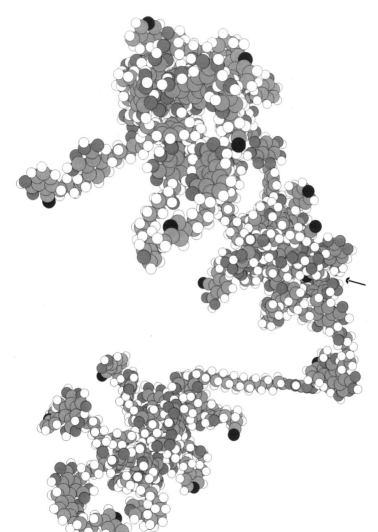

Figure 1 : Color 3D-structure of an open HA trimer (n = 3; mol. mass 16,899.41 g mol^{-1}, 2,271 atoms; convergence limit < 4.19 kJ (0.1 nm)$^{-1}$ mol^{-1}; yes) following geometry optimization. In this plot hydrogen atoms are white; carbon light blue, oxygen red, and nitrogen dark blue.

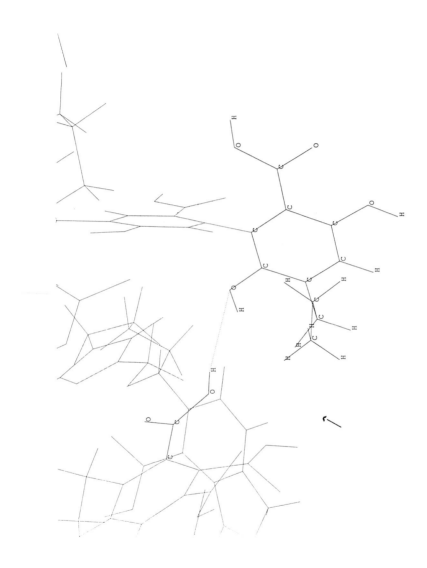

a)

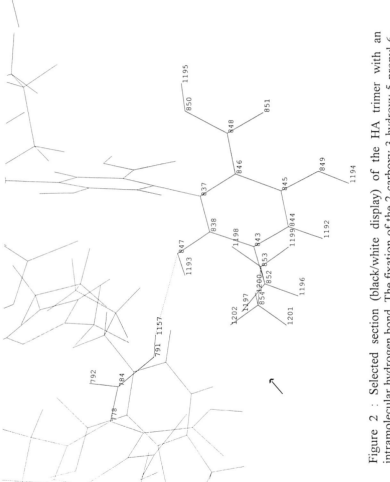

b)

Figure 2 : Selected section (black/white display) of the HA trimer with an intramolecular hydrogen bond. The fixation of the 2-carboxy-3-hydroxy-5-propyl-6-hydroxy benzene side-chain in the trimer HA molecule by a vicinal aryl carboxyl group is illustrated by showing a) the element symbols and b) the atom numbers . The location of this intramolecular hydrogen bond in the trimer HA structure is marked by an arrow (↑) in Fig. 1.

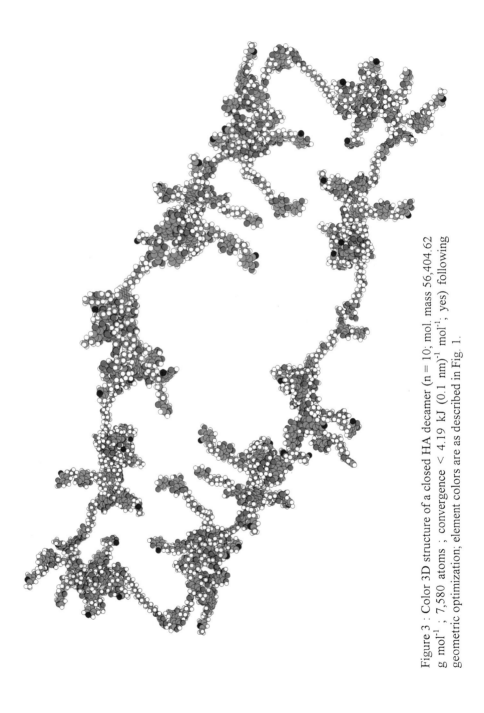

Figure 3 : Color 3D structure of a closed HA decamer (n = 10; mol. mass 56,404.62 g mol^{-1} ; 7,580 atoms ; convergence < 4.19 kJ (0.1 nm)$^{-1}$ mol^{-1}, yes) following geometric optimization; element colors are as described in Fig. 1.

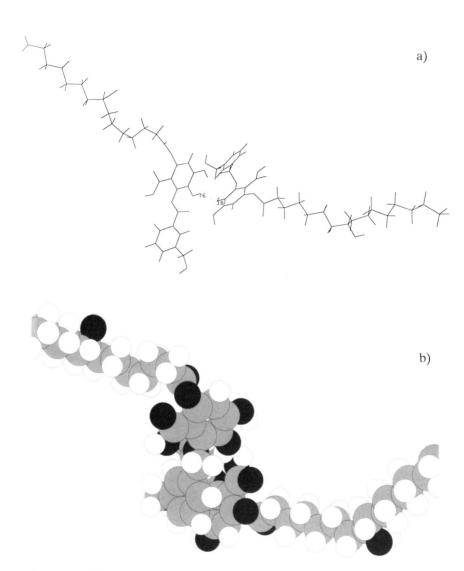

a)

b)

Figure 4 : Selected sections (black/white display) of the HA decamer with a) intermolecular hydrogen bond and b) linking of two side chains by van der Waals forces after extensive molecular mechanical calculations.

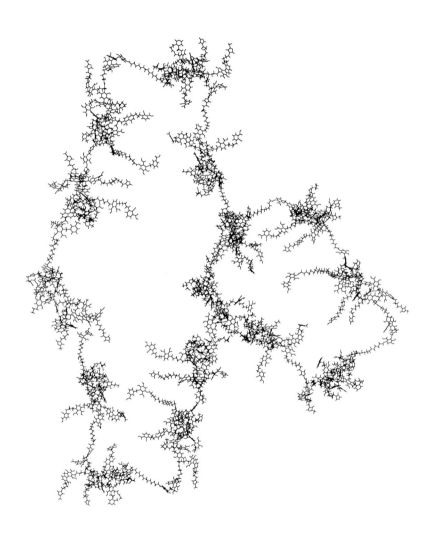

a)

b)

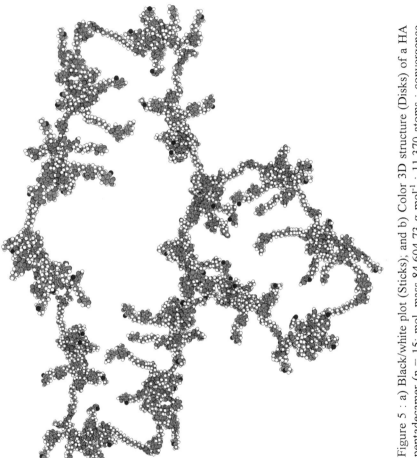

Figure 5 : a) Black/white plot (Sticks); and b) Color 3D structure (Disks) of a HA pentadecamer (n = 15; mol. mass 84,604.73 g mol⁻¹ ; 11,370 atoms ; convergence limit < 4.19 kJ (0.1 nm)⁻¹ mol⁻¹; yes) following geometry optimization. The element colors are as described in Fig. 1

gradient of 0.0360 kJ (0.1 nm)$^{-1}$ mol^{-1} was obtained following geometry optimization and semi-empirical calculations. This display in the sticks mode does not show double bonds and spatial expansion but allows to follow association/dissociation processes during geometry optimization (and thus energy minimization). These bimolecular reactions forming hydrogen bridges between HA molecules require high HA oligomer concentrations and long calculation times at the preset molecular distances.

As illustrated in Fig. 4b, intermolecular linking of HA oligomers can also occur due to trapping by van der Waals forces. Even after more than 2,000 calculation cycles and 4,300 points using the conjugate gradient method < 0.200 kJ (0.1 nm)$^{-1}$ mol^{-1}, the two conditions (< 0.32 nm distance, < 120 degrees angle) for hydrogen bonding were not achieved and dipole/dipole interactions resulted in the displayed intricate bonding situation.

Humic Acid Pentadecamer. In Figure 5a the black/white plot of the initial conformation of a pentadecamer HA (*Sticks, Thick Lines*) is displayed which was constructed from three pentamer HA units (*15*). Clearly the voids and clefts are visible in the C-C skeleton which are formed mainly by isolated aromatic rings linked by aliphatic chains of different lengths. The geometrically optimized model of this HA polymer or humic complex (n = 15; 11,370 atoms) is shown in Fig. 5b (in color). The molecular mass of 84,607.88 g mol^{-1}, elemental composition of $C_{4,728}H_{5,223}N_{75}O_{1,344}$ and elemental analysis of 67.12% C, 6.22% H, 1.24% N, and 25.42% O were determined. With the termination limit of the convergence gradient of < 4.19 kJ (0.1 nm)$^{-1}$ mol^{-1}, the total energy of the obtained pentadecamer conformation was 30,432.4590 kJ (0.1 nm)$^{-1}$ mol^{-1} at an actual gradient of 4.0224 kJ (0.1 nm)$^{-1}$ mol^{-1}. Although the HyperChem software allows a maximum of 32,720 atoms, the described PC equipment is at its limits as weeks of calculation time are required for comprehensive geometry optimizations for molecules of this and larger size and complexity.

Conclusions

1.) Since Berzelius (*19*) a wide range of models for humic substances has been put forward and the difficulties with the modeling approach have been expressed very clearly (*20*). It is interesting that with modern methods of instrumental analytical chemistry such as solid state NMR, Py-GC/MS and Py-MS in combination with efficient data treatment by computational chemistry, previously postulated basic structural features are confirmed and validated. For instance, the concept and importance of hydrogen bonds between carboxyl groups of aromatic rings in FAs and HAs have been emphasized by Schnitzer in the early seventies (*21*) and indeed are verified by novel methods as relevant within and between humic substances.

2.) A new approach in the definition of "molecular structure" is introduced for natural humic substances and soil organic matter by proposing a model HA which represents a characteristic structural pattern for a large number of possible and likely structures. The crucial point is the adaptation, flexibility, and variability to decomposition and synthesis processes in the natural environment.

3.) The enormous interest in humic substances by scientists working in environmental science is reflected by the large number of papers in water and soil research focused on structural properties and quality subjects (e.g. *22-24*). Better knowledge of the molecular-chemical structure of humic substances and organic matter appears to be of fundamental importance for understanding and predicting the composition, dynamics and reactivity of natural (*25, 26*) and anthropogenic organics in water (*13, 27*) and soils.

4.) Computational chemistry offers relatively economical, powerful tools to observe binding of humic substances to biological molecules, mineral surfaces, metal ions, and anthropogenic compounds at nanometer level using exact bond distances, bond angles, torsion angles, non-bonded distances, van der Waals forces, hydrogen bonds and ionic charges. Therefore, the proposed methodology for the environmental and soil sciences should be well suited for fundamental and applied investigations of physical and biological properties, chemical reactivity and molecular dynamics of humic substances.

Acknowledgments. The work described here is financially supported by the Deutsche Forschungsgemeinschaft (projects Schu 416/3; 416/18-3) and the Ministry of Science and Technology, Bonn-Bad Godesberg, Germany.

Literature Cited.

1. Schulten, H.-R.; Plage, B. ; Schnitzer, M. *Naturwissenschaften* **1991**, 78, 311-312.
2. Schulten H.-R. ; Schnitzer, M. *Sci. Total Environ.* **1992**, 117/118, 27-39.
3. Schulten, H.-R. ; Schnitzer, M. : *Naturwissenschaften* **1993**, 80, 29-30.
4. Schulten, H.-R. : In Senesi, N.; Miano, T. M., Eds. ; *Humic Substances in the Global Environment and Implications on Human Health*, **1994**, Elsevier, Amsterdam, The Netherlands, pp. 43-56.
5. Schnitzer, M. : In Senesi, N.; Miano, T. M., Eds.; *Humic Substances in the Global Environment and Implications on Human Health*, **1994**, Elsevier, Amsterdam, The Netherlands, pp. 57-69.
6. Schnitzer, M.; Khan S. U. *Humic Substances in the Environment*, **1972**, Marcel Dekker, New York, pp. 192-193.
7. Schulten, H.-R. *J. Anal. Appl. Pyrolysis* **1993**, 25, 97-122.
8. HyperChem, Hypercube Inc., 419 Phillip Street, Waterloo, Ontario N2L 3X2, Canada.
9. Schulten H.-R. *Fresenius J. Anal. Chem.* **1995**, 351, 62-73.
10. Schulten H.-R. *J. Anal. Applied Pyrolysis* **1995**, 32, 111-126.
11. Leinweber, P.; Schulten, H.-R. *Mitteilgn. Dtsch. Bodenkundl. Gesellsch.* **1994**, 74, 383-386.
12. Schulten H.-R. In *Mass Spectrometry of Soils*; Yamasaki, S.; Boutton, T. W., Eds.; Marcel Dekker, New York, NY, **1996**; pp. 373-436.
13. Leinweber, P. ; Blumenstein, O.; Schulten, H.-R. *Europ. J. Soil Sci.*, **1996**, 47, 71-80.

14. Schulten H.-R.; Sorge, C.; Schnitzer, M. *Biol. Fertil. Soils* **1995**, 20, 174-184.
15. Schnitzer M. ; Schulten H.-R. *Adv. Agron.* **1995**, 55, 167-217.
16. Schulten, H.-R. ; Schnitzer, M. *Naturwissenschaften* **1995**, 82, 487-498.
17. Leinweber, P., Paetsch, C. ; Schulten, H.-R. *Arch. Acker- Pfl. Boden.* **1995**, 39, 271-285.
18. Schulten, H.-R. ; Leinweber, P. *Soil Sci. Soc.* Am. J. **1995**, 59, 1019-1027.
19. Berzelius, J. J. *Lehrbuch der Chemie*, 1839, Arnoldische Buchhandlung, Dresden, Leipzig.
20. Ziechmann, W. In Frimmel, F. H.; Christman, R. F.; Eds.; *Humic Substances and Their Role in the Environment*, 1988, Wiley-Interscience, Chichester, UK, pp. 113-132.
21. Schnitzer, M. Agron. *Abstr., Amer. Soc. Agron.*, **1971**, p.77.
22. Hayes, M. H. B., MacCarthy, P., Malcolm, R. L., Swift, R. S., Eds.; *Humic Substances II; Search of Structure*, **1989**, Wiley, New York.
23. Senesi, N.; Miano, T. M., Eds.; *Humic Substances in the Global Environment and Implications on Human Health*, **1994**, Elsevier, Amsterdam, Netherlands.
24. Boutton, W, Yamasaki, S (Eds.); *Mass Spectrometry of Soils*, **1996**, Marcel Dekker,   New York, pp.1-517
25. Buffle, J.; Leppard, G. G. *Environ. Sci. Technol.* **1995**, 29, 2176-2184.
26. Wlkinson, K. J., Stoll, S., Buffle J. *Fresenius J. Anal. Chem.*, **1995**, 351, 54-61.
27. Schulten, H.-R. J. *Environ. Anal. Chem.*, **1996**, 64, 101-115.

Chapter 5

NMR Techniques (C, N, and H) in Studies of Humic Substances

Jacqueline M. Bortiatynski[1], Patrick G. Hatcher[1,2], and Heike Knicker[2]

[1]Fuel Science Program, [2]Energy and Fuels Research Center, Pennsylvania State University, University Park, PA 16802

Nuclear magnetic resonance (NMR) spectroscopy has been used to characterize soil organic matter and study soil humification processes for nearly two decades, but only recently it has been considered a valuable analytical method in modern soil science. Technological and theoretical advances in the early development of NMR resulted in increased sensitivity and resolution of the NMR spectra of humic materials in soils. In turn, the increased utility of NMR led to the widespread initial use of this technique which is documented in the scientific literature by numerous NMR studies to obtain structural information on soil organic matter. Since those early applications spanning the 1970's and 1980's, a general hiatus has limited new developments in the field of soil science and the technique became routine in its application. Recently, some new applications involving isotope labeling combined with both ^{13}C and ^{15}N NMR has renewed interest in NMR as the premier technique for modern soil science research. This article reviews these recent developments.

Nuclear magnetic resonance, a spectroscopic technique originally suited for the structural analysis of molecules in solution, was not a method employed extensively for the analysis of humic materials in soils until about the late 1970's. Two major instrumental developments led to the expanded application of this structural tool for the analysis of soil components: 1. the application of Fourier transform methods of data acquisition, and 2. the ability to examine solid samples. The former development is responsible for the improved sensitivity which is necessary to analyze complex macromolecular organic materials in soils as well as the relatively insensitive inorganic nuclei. The latter development allowed for the direct examination of solid samples thus eliminating any changes in the physical structure of humic substances due to solubilization. Since the introduction of these two new capabilities, some significant strides were made in the characterization of humic substances and other inorganic substances in soils by NMR.

The first application of NMR for the study of humic substances in soil was made by Neyroud and Schnitzer(*1*) using continuous wave NMR spectroscopy. Later, Gonzalez-Vila et al.(*2*) were the first to apply Fourier transform NMR for the analysis of dissolved humic substances extracted from soil. The development of

0097–6156/96/0651–0057$15.25/0

solid-state NMR, in particular ^{13}C NMR, immediately opened the door for numerous investigators to examine solid samples directly.(*3,4*) From this point on, NMR spectroscopy became one of the most widely used techniques for analysis of all forms of soil organic matter, both the soluble and insoluble humic substances. The ability to examine other nuclei in the solid-state also led to a large amount of research activity directed towards soil inorganic constituents. Most of the research on NMR of soil constituents conducted between this early discovery phase and 1987 has been extensively reviewed in a book by Wilson.(*5*) The purpose of this review paper is to focus on the advances made since in the application of NMR spectroscopy to the study of soil humic substances.

The review is intended to highlight the major advances in understanding humic substances by use of both solution and solid-state NMR. Obviously, the technique has been applied in a routine sense in many more studies than can be described here. What we hope to convey to readers of this review are the most recent innovative applications which have or will make a significant impact our future knowledge of the structure and reactivity of humic substances.

Recent Applications of ^{13}C NMR

^{13}C NMR is a technique well suited for the examination of complex structures such as those present in soil organic matter. Although details of the presence of specific compounds is not always possible with NMR, it does provide a measure of the average distribution of the various types of carbons and this information is sometimes more valuable as average information than as an exhaustive list of individual compounds. Figure 1 shows a typical ^{13}C NMR spectrum for humic acid from a histosol (Minnesota peat). The various peaks in the spectrum can usually be assigned to specific carbon functional groups, but some overlap of resonance assignments can be observed. This is due mostly to the fact that there is a large variety of different environments for carbon functional groups in materials so complex as humic substances. It is only when one type of structure (i.e., polysaccharides, lignin, etc.) is predominant that we observe well defined signals which can be assigned to the carbons of these specific structures. Nonetheless, we can integrate various regions of most spectra and assign specific types of functionalized carbons to those regions in a quantitative manner for the most part. Thus, direct measures of the contents of carboxyl, aromatic and aliphatic carbons can often be made by integration of specific regions, taking into account any possible overlap of the regions. Table I lists various regions of a typical ^{13}C NMR spectrum with associated assignments which can be made. These assignments are usually made on the basis of known chemical shifts of various functionalized carbons in pure standards.

Solution NMR

Although solution-state ^{13}C NMR has revealed much regarding the chemical structural composition of humic substances and dissolved organic matter (*5*), the development of more sophisticated software, pulse sequences, probes in recent years have not been exploited much in the field of soil science. Most of the novel applications have strictly relied on conventional one-dimensional spectroscopy. Two-dimensional NMR methods have become the mainstay of the chemical research field, especially for the structural elucidation of complex soluble biopolymers. The complexity of humic substances along with other inherent physical properties such as limited solubility and the presence of paramagnetics

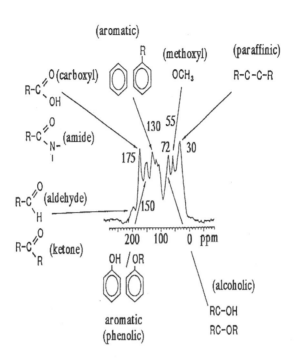

Figure 1. A typical ^{13}C CPMAS NMR spectrum of humic acid from a Minnesota peat showing the various assignments for the peaks.

Table I. Chemical shift assignments for various spectral regions in ^{13}C NMR spectra.

Chemical shift region	Assignment
0-45 ppm	paraffinic structures (C, CH, CH$_2$, CH$_3$)
40-60 ppm	aliphatic C-N
50-58 ppm	methoxyl
60-95 ppm	alkyl-O (carbohydrates, alcohols)
95-110 ppm	acetal and ketal carbon (carbohydrates)
100-140 ppm	aryl-H and aryl-C carbons
140-160 ppm	aryl-O and aryl-N carbons
160-180 ppm	carboxyl and amide carbon
180-220 ppm	aldehyde and ketone carbons

have been, in part, responsible for the slow application of two-dimensional NMR to this area of research. The accessibility of instrumentation has also limited the application of two-dimensional NMR techniques. In addition, the type of information that is gained from a two-dimensional experiment may not always be necessary to obtain specific structural information. Sometimes increased sensitivity can be provided by NMR-specific isotope labeling which also leads to simplification of the NMR spectral data and is site specific enough to extract valuable structural information such as the binding of pollutants to humic substances.(6-9)

One-Dimensional ^{13}C NMR Spectral Techniques. As mentioned above one-dimensional ^{13}C NMR can provide very valuable information concerning the structure of humic substances. Progress beyond the standard one pulse experiments such as broad-band decoupled and inverse-gated ^{13}C NMR spectroscopy to multiple pulse ^{13}C NMR spectroscopy provides a means of determining carbon substitution as well as resolution and sensitivity enhancement. The use of spin-echo experiments which are commonly known as J-resolved multiplicity-sorting (J-SEFT), spin-echo Fourier transform (SEFT), attached proton test (APT), and gated spin-echo (GASPE) have provided resolution enhancement and carbon multiplicity information concerning humic substances.(10,11) When polarization transfer is combined with the spin-echo experiment the resulting pulse sequence known as DEPT (distortionless polarization transfer spectroscopy) can provide up to four times the signal-to-noise of the previously discussed spin-echo experiments. In addition since the spin magnetization from the DEPT experiment is that of the attached protons the delay between scans is reduced and the spectrum can be obtained in a much shorter period of time. The utility of the DEPT experiment has been demonstrated in the study of both soil and aquatic humic acids.(10-14)

Buddrus and coworkers(13) have also shown that quantitative information concerning carbon types can be calculated using the results from DEPT spectra in combination with quaternary-only spectra (QUAT) and comparing them to the complete spectrum obtained using a normal spin echo experiment. The results from the DEPT spectra represent the carbons bearing one, two and three carbons while the QUAT spectra are used to identify carbons without any attached protons. The integrals of the signals originating from the DEPT spectra are summed with those from the QUAT spectra. The summation of these integral should be equal to the integral of the signals found in the spin-echo experiment. Quantitation of DEPT and QUAT spectra provides a means of determining ratios of the functional groups found in humic acids.

The importance of these studies is that a detailed quantitative accounting of ^{13}C NMR spectral regions can be made. They discovered that for aquatic humic substances dominated by aliphatic signals, approximately 78% of the carbons are quaternary carbons, a surprising finding. They concluded that most of the aromatic signals in these aquatic humic substances were derived from lignin and that the alkyl-O region normally attributed to carbohydrate carbons is in fact dominated by lignin-derived carbons. Unfortunately, the NMR spectra shown by Buddrus et al.(13) show practically no resemblance to similar spectra of lignin.(15)

^{13}C-Labeling Studies. In addition to spin-echo experiments such as DEPT, spectral editing and sensitivity enhancements can also be achieved by combining site specific ^{13}C-labeling with standard ^{13}C NMR experiments such as ^{13}C inverse-gated spectroscopy. This approach is probably one of the more novel and exciting new areas of soil science where NMR is expected to have a major impact. In the case of humic substances two types of labeling experiments have been carried out.

The first involves a known functional group derivatization to quantify specific types of carbons found in humic substances, the derivatization being carried out by chemical reactions with ^{13}C-labeled reactants.(16-21) For example, methylation with ^{13}C-labeled diazomethane or methyl iodide has been used to distinguish between and quantify hydroxyl functionalities in humic acids. The second labeling methodology involves the use of ^{13}C-labeled reactants to follow the course of a complex reaction or association such as the interaction of pollutants with humic acids.(6-9,22,23) The first structural evidence for the type of interaction of pollutants with humic substances was provided using ^{13}C and ^{15}N site specific labeling in combination with ^{13}C and ^{15}N NMR, respectively.

The most recent results of experiments which were designed to examine the ^{13}C labeling of phenolic pollutants and their interaction with humic acids will be reviewed briefly. A similar study with ^{15}N labeled aniline was carried out by Thorn et al.(6) The interaction of pollutants with soil organic matter has long been an area of research that relied heavily on data that contained either very limited or no structural information. Therefore, the study of reaction mechanisms concerning these processes was only speculative even in terms of the type of interaction whether it be covalent or noncovalent. Covalent binding interactions were thought to form through enzymatic oxidative coupling of phenolic pollutants to soil organic matter. If indeed this is true, and it has since been shown to be the case, all the carbons in the phenolic ring will reflect the resulting alteration in structure due to the formation of a new bond as a change in the ^{13}C chemical shift.

[13]C-labeled 2,4-dichlorophenol containing 100% [13]C at either the C-2 and C-6 or C-1 positions was enzymatically bound with horseradish peroxidase to three humic acids derived from different depositional environments.(7,9) A similar set of experiments was also carried out with phenol itself which was labeled with 100% [13]C at the C-1 position.(8) Prior to the binding experiment the ^{13}C-labeled phenols were examined by ^{13}C inverse-gated NMR and it was found that the spectra only contained signals from the enriched carbons. Following the enzymatic binding reactions, the isolated humic acids were dissolved in 0.5 M NaOD and the resulting ^{13}C inverse-gated NMR spectra revealed a dispersion of ^{13}C chemical shifts. One such spectrum, C-2,-6 ^{13}C-labeled 2,4-dichlorophenol enzymatically bound to a Minnesota peat humic acid, is shown in Figure 2. The complexity of the spectrum suggested that some of the signals may have resulted from the natural abundance carbons of the humic acid. Upon further examination however, it was clear that signals due to the aliphatic functional groups of the humic acid were not present, thus confirming that the spectrum was the result of the ^{13}C-labeled carbons of the covalently bound 2,4-dichlorophenol. The fact that the signals are broad in this high field (125 MHz) spectrum is indicative of the numerous electronic environments that surround each type of binding interaction. The spectrum was tentatively assigned by regions to specific types of binding interactions (i.e. carbon-carbon, carbon-oxygen, etc.). Carbon-carbon bonds formed between the C-4 and C-6 carbons of the 2,4-dichlorophenol and aromatic and aliphatic carbons of the humic acid were thought to be the preferred bonding interactions based on the intensity of signals that were tentatively assigned to these types of covalent bonds.

[13]C-labeling in combination with ^{13}C NMR has also provided some very valuable insight into the non-covalent bonding association of pollutants with humic acids. The weak non-covalent bonding of pollutants to humic acids serves as a means of transport and subsequent release of pollutants into fresh water and uncontaminated soils. It is critical to understand the mechanism of the binding reaction to predict the movement of such pollutants and to develop mitigation

strategies. Until very recently, the information that has been obtained for the association of phenols with humic acids primarily consisted of concentration measurements obtained indirectly from column effluents. ^{13}C NMR has now been used to directly examine the associative interaction of C-1 ^{13}C-phenol with an Armadale humic acid using spin-relaxation measurements.(8) It was found that the experimentally determined spin-lattice relaxation time, T_1, of the C-1 ^{13}C-labeled carbon of phenol reflects the amount of phenol that is associated with the humic acid. The T_1 values for pure phenol are 35 sec. and progressively decrease with addition of the Armadale humic acid. After 1.3 mg Armadale humic acid is added

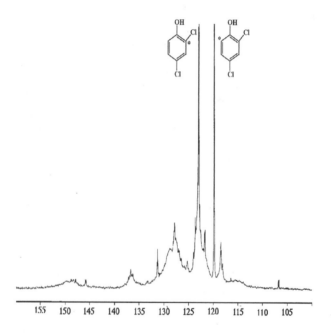

Figure 2. High resolution ^{13}C NMR spectrum of C-2 and C-6 labeled 2,4-dichlorophenol enzymatically bound to a Minnesota Peat humic acid. The signals due to unreacted ^{13}C-labeled 2,4-dichlorophenol are indicated on the spectra and the asterisk identifies the appropriate ^{13}C labeled carbon. The remaining signals in the spectrum result from the covalent binding of ^{13}C-labeled carbons of DCP to functional groups in the humic acid and to other DCP molecules. (7)

to a solution containing 2 mg of C-1 ^{13}C-labeled phenol the T_1 is approximately 1.7 seconds which approaches that of the humic acid carbons. A linear relationship exists between the amount of associated phenol and the concentration of the added humic acid. In addition the relationship between the experimentally obtained T_1 and the concentration of added humic acid was used to obtain a fractal dimension for the humic acid reaction surface. The calculated fractal dimension of 1.2 suggests an open single dimension geometry for the reaction surface.

Solid-State ^{13}C NMR

The great flurry of activity that emerged in the 1980's as the result of the development of solid-state ^{13}C NMR for use in soil science led to some important findings concerning soil organic matter. Probably the most important of these is the discovery that soil humic materials and organic matter in general contain significant quantities of aliphatic macromolecular structures which are not carbohydrates, proteins and low-molecular weight waxes. These components, termed polymethylenic or paraffinic structures, form an important part of soil organic matter from many locales. There seems to be great variation in the relative amounts of such structures from differing soils. Poorly-drained soils and histosols appear to have more paraffinic carbons in their NMR spectra than highly oxidized soils from arid climates, suggesting that conditions favoring the growth of aquatic microorganisms enhances paraffinic structures possibly originating from these microorganisms.

The second important discovery concerning soil humic substances made prior to 1987 was that phenolic carbons are not as dominant structural features of these materials as previously thought. NMR signals for such phenolic structures are generally minor or nonexistent in all except the least humified soils such as histosols. This finding has major implications regarding the origin of humic substances and the mechanisms by which they form from plant precursors such as lignin.

The great majority of the work related to the characterization of humic substances by solid-state ^{13}C NMR was done prior to 1987 and most of this literature is reviewed by Wilson.[5] Since these earlier works, the technique has been applied by numerous individuals to examine the nature of humic substance carbon from various locales and to investigate the reliability of the technique for quantifying the various forms of carbon in soil organic matter. A few limited studies have been conducted to investigate further the major findings discussed above. Perhaps the most important of these involved use of a new technique which was discussed above in solution NMR studies- use of ^{13}C labels to trace soil decomposition processes.

Establishing Quantitative Measurements. Solid-state ^{13}C NMR spectra are usually obtained by use of a special pulse sequence involving a technique called cross polarization and magic angle spinning(CPMAS) to shorten the time interval between pulses, high power decoupling to remove the broadening effect of strong C-H dipolar coupling in solids, and magic angle spinning to primarily remove broadening caused by chemical shift anisotropy. The CP can induce distortions brought about by competitive processes of signal intensification by magnetization transfer between [1]H's and [13]C's and loss of signal due to loss of ^{1}H magnetization caused by spin-lattice relaxation in the rotating frame. The former process builds signal intensity while the latter diminishes it. Both processes are affected by the presence of paramagnetic materials as would be found in soil and mineral-rich humic isolates from soil. If all carbons are affected equally, then either no spectrum is observed, or the relative areas of the peaks will quantitatively represent the types of structures present in a quantitative way. If the different carbons are affected in a differential manner, then the spectra will not quantitatively represent the structures in the humic materials.

Because measurements of spin dynamics for soil organic matter made on whole soils where the signal is very weak due to dilution, are problematic (the carbon contents of less than 10% pose serious detectability problems), few measurements have been made of the quantitative nature of solid-state ^{13}C NMR in soils. One can determine the spin dynamics of humic extracts(24) and establish the fact that spectra of these materials are quantitative, but spectra of whole soils cannot be globally validated as quantitative at this time. Recently Fründ and Lüdemann(25,26) have examined the behavior of soil humic substances in CPMAS experiments at two spectrometer field strengths and in solution. They concluded from studies of the spin dynamics and quantitative solution spectra, that quantitative CPMAS ^{13}C NMR spectra are possible if attention is given to proper experimental conditions. Recently, Kinchesh et al.(27) repeated this work and examined the quantitative application of solid-state ^{13}C NMR for whole soils and found that the degree to which CPMAS is quantitative depends largely on the concentration of paramagnetics, not too surprisingly.

One must even be careful in the preparation of humic extracts for NMR studies. Hatcher and Wilson(28) showed that even a small amount of water in fulvic acids can significantly affect the spectral characteristics. This is due to the fact that the added water induces added molecular mobility to the humic substances, especially fulvic acids, and this added mobility affects the spin dynamics in CPMAS spectra, degrading the ability to obtain quantitative ^{13}C NMR spectra. The effect is most pronounced for carboxyl carbons, presumably because they associate with water more strongly.

Structural Studies of Humic Substances. Solid-state ^{13}C NMR has played a major role in past studies of the chemical structure of humic substances isolated from soil as mentioned above. Recent studies have focused on the examination of humic substances from various environments to establish variabilities which might be indicative of structure and structural evolution.(29-34) Also, combined use of NMR with other analytical methods such as analytical pyrolysis, chemical degradative methods, and solution NMR studies has proved valuable in elucidating structural components. However, the complexity of humic substances has precluded all but inferred structures based on these combined studies.

Humic acids from soils of widely varying type show spectra with features varying from mostly aliphatic to mostly aromatic. Some humic acids are nearly entirely aromatic in structure, showing spectra with signals for only aromatic (130 ppm), carboxyl (175 ppm), and carbonyl (200 ppm) carbons. Using dipolar dephasing to estimate the average degree of aromatic ring substitution, Hatcher et al.(31) calculated some structures for these humic acids from highly oxidized volcanic ash soils and paleosols. The importance of these studies relates to the ability to use the quantitative information generated from the NMR spectra in combination with other chemical characterization methods to construct three dimensional models of these humic acids by computer assisted techniques.(34) Recently, Schulten and Schnitzer have used an identical approach to develop 3-D models of humic acids.(35)

The pressing question regarding the aromatic structures in these oxidized humic acids relates to the mode by which they originate from plant precursors. From studies of humic acids and whole soil in forest ecosystems by the combined use of ^{13}C NMR and CuO chemical degradation/gas chromatography, Kögel-Knabner et al.(36) showed that aromatic structures in surface layers were enriched in lignin-derived phenolic carbons and that increasing decomposition in the soil profile significantly decreased the phenolic carbon content. Inbar et al.(37) also showed a similar transformation in ^{13}C NMR studies of the decomposition in composts, while deMontigny et al.(38) showed the similar loss of phenolic carbon in lignin of woods degraded in a forested ecosystem of northern Vancouver Island. One possible explanation for such changes includes destruction of the aromatic rings by soil fungi.(*39*)   However, the aromatic structures do not diminish in relative concentration with increasing soil depth or increasing degradation, and this implies that the aromatic structures are retained from the original lignin in some altered form or derived from additional structures added to the soil by microorganisms.

The latter explanation is relatively straightforward as soil microorganisms are known to synthesize aromatic compounds. The former explanation is a bit more complex in that it requires transformation of very specific oxygen substituted aromatic structures in lignin to benzene carboxylic acids with little or no directly attached oxygens.   Wang and Huang(*40*) showed that pyrogallol, a trihydric phenol, could be easily transformed abiotically in the presence of manganese oxides, a soil component, to structures showing ^{13}C NMR spectra very similar to oxidized humic acids where the phenolic carbon contents were significantly reduced and carboxyl contents increased.   In this study, increased amounts of aliphatic carbons and the presence of aromatic carbons not substituted by oxygen suggests extreme chemical modification of the aromatic ring with possible ring-opening. Thus, one may conclude that oxidative processes, no doubt partially abiotic, could oxidize lignin-like aromatic ring systems and remove oxygenated substituents while preserving aromatic character and increasing the aliphatic content. DeMontigny et al.(*38*) observed a similar increase in aliphatic structures in degraded wood, suggesting that ring opening plays a role in the modification of the lignin structures.

In less oxidized soils including those from forested regions, the aromatic structures are less dominant than in the oxidized soils. The greater proportion of aliphatic components is manifested in intense signals for paraffinic carbons and alkyl-O carbons as would be found in carbohydrates and lignin side-chains. The aromatic region is usually characterized by a peak for aromatic carbons showing fine structure with a specific shoulder or peak at 150 ppm representing phenolic carbons. Combined with a peak at 55 ppm for methoxyl carbons, this is an indication of the presence of lignin or altered lignin residues. This information is consistent with analytical pyrolysis data which show the presence of lignin residues in humic substances.(*41*)

Recently, Schulten et al.(*42,43*) have proposed a new model for humic substances in soil based partly on ^{13}C NMR data but mostly on pyrolysis studies. This model proposes that the core structure of humic acids is alkyl-aromatic and alkyl-poly-aromatic. The lack of oxygen functionality and the fact that pyrolysis data for highly oxygenated humic acids may be biased in favor of more volatile alkyl-aromatics dictates that the model needs extreme revision or is simply incorrect.   The recent studies of Saiz-Jimenez et al.(*44*) and Hatcher et al.(*34*) appear to refute the original model presented by Schulten et al.(*47*)   The ^{13}C NMR data appears to be in accord with the model presented by Schulten et al.(*47*), however, NMR does not represent carbon structures with specific molecular-level detail.   It only provides average information with a sometimes added benefit of added fine structure.

Perhaps the most sought after question since 1986 has been the search for the origin of the paraffinic structures in humic substances. The model of Schulten et al.(*47*) answers this by proposing that they are part of an alkyl-aromatic core. Unfortunately, there are some serious misgivings about the Schulten model. The ^{13}C NMR and pyrolysis studies conducted by Kögel-Knabner and Hatcher(*45*) and Kögel-Knabner et al.(*46*) suggest that the paraffinic structures in whole forest soils originate from cross-linking of cutin and/or suberin biopolymers during humification.   These could then be incorporated into various humic isolates by degradative processes.   It is clear that future characterization of the origin and formation of paraffinic structures in humic substances will have to rely heavily on a combined spectroscopic and chemical approach using pyrolysis or chemical degradative reactions.

^{13}C Labeling Studies. At natural abundance levels of approximately 1.1% of all carbons, ^{13}C is a relatively dilute nucleus. If one enriches samples with ^{13}C in organic structures by growing biomass on ^{13}C-enriched CO_2 or feeding ^{13}C-labeled biomass precursors to organisms, then a significant gain in signal strength and sensitivity is realized. Probably the most exciting new research in the area of soil science involves the use of ^{13}C enrichment techniques combined with ^{13}C NMR, as described above in solution NMR studies, to examine the structure and reactivity of soil organic matter. Few such studies have been made with solid-state ^{13}C NMR as the detection technique.

Baldock et al.(47,48) incubated soil with ^{13}C-labeled glucose to stimulate uptake of the label by microorganisms. The ^{13}C NMR spectra of the soil incubated with the labeled glucose exhibited signal intensification primarily in the aliphatic region, implying that the microorganisms were using the labeled glucose carbon as a food source and synthesizing new soil biomass whose NMR signals were primarily in the paraffinic region of the spectrum. The enhancement in signal due to the enriched biomass allowed for determination of spin dynamics.(48) In another previous study of the degradation of algal biomass enriched with ^{13}C (the algae were grown on ^{13}C-enriched CO_2), Zelibor et al.(49) showed that the most refractory carbon during degradation was that of paraffinic structures. Thus, in both studies, the changes in the solid-state ^{13}C NMR spectra as a function of increased degradation showed that the macromolecular paraffinic structural components of microbial residues are refractory components likely to be preserved and incorporated in sedimentary organic matter.

Recent Applications of ^{15}N NMR

The chemical structures of nitrogen bearing components in humic substances have long been of great interest, due to their importance as nutrient sources for soil microbial and plant growth and their role in nitrogen cycling. Most of the organic nitrogen enters the soil in the form of decaying plant and microbial material and it is mainly present in the form of amino acids and peptides. Peptides are known to be highly susceptible to microbial biodegradation and chemical hydrolysis. However, some of the organic nitrogen survives early microbial degradation and humification processes and becomes part of the refractory organic matter pool which is highly resistant to microbial and chemical degradation. The chemical structure of this refractory organic nitrogen and the processes involved in its formation are a subject of much controversy. Using standard wet chemical degradative techniques, only 30-60 % of the total nitrogen in soils have been identified, mainly as soluble amino acids and amino sugars.(50) An alternative to these standard degradative techniques is ^{15}N NMR spectroscopy, which allows the non-intrusive characterization of the forms of nitrogen. While, ^{13}C NMR does not yield in any specific information about nitrogen functionality, ^{15}N NMR, on the other hand, does but poses some problems which are not encountered in ^{13}C and ^{1}H NMR. ^{14}N, the most abundant nitrogen isotope in chemical compounds does not yield high resolution spectra because of its large quadrupole moment. ^{15}N NMR is also NMR active. However, it is about 50 times less sensitive than ^{13}C NMR. This is due to the low natural abundance of 0.37 % for the ^{15}N isotope and its low and negative magnetogyric ratio. Therefore, it was assumed for a long time, that natural abundance ^{15}N NMR spectra of humic material could not be obtained within feasible experiment times. Consequently, the first applications of ^{15}N NMR on humic materials were carried out with ^{15}N-enriched composts and soils incubated with ^{15}N-enriched compounds(*51-57*), ^{15}N-enriched melanoidins(*58*) and modelpolymers(*59*).

The first natural-abundance ^{15}N NMR spectrum on humic material was obtained by the solid-state CPMAS technique on a peat having a nitrogen content of 4% by Preston et al.(*53*). After systematic determination of important ^{15}N NMR spectral parameters, it was possible to obtain solid-state ^{15}N NMR spectra of humic material in soil and sediments at natural ^{15}N levels(*60-62*) with an acceptable signal-to-noise ratio. Recently solid-state ^{15}N NMR spectroscopy was extended to the investigation of the nitrogen functionality in coals.(*63*)

Solid-State ^{15}N NMR

In ^{13}C NMR, extensive studies have shown that solid-state ^{13}C CPMAS NMR spectra of most soils can be analyzed quantitatively(*5,25,26*) if one considers the relaxation and cross polarization kinetics of the compounds observed in the spectra. We can extend these results to imply that solid-state ^{15}N CPMAS NMR spectra are also quantitative. Comparable to ^{13}C CPMAS NMR, in a ^{15}N CPMAS NMR experiment, quantification problems may arise from 1) incomplete relaxation of the spins during the experiment, 2) incomplete polarization transfer because of the isolation of protons from nitrogen atoms and 3) spinning side bands. The effects of these parameters on a ^{13}C CPMAS NMR spectrum are extensively discussed by Wilson.(*5*)

The first studies to estimate relaxation behavior and cross polarization kinetics on ^{15}N-enriched plant material composts and humic acids in soil organic matter was carried out by Almendros et al.(*55*), Knicker(*64*) and Knicker and Lüdemann(*65*). While, in undegraded plant material, proton spin-lattice-relaxation-times (T_{1H}) vary from 0.2 to 2 sec. Much shorter T_{1H}'s are observed in degraded compost and humic material. With higher degrees of humification the T_{1H} decreases to values lower than 55 ms. In humic substances, cross polarization of different nitrogen components does not occur at the same rate. Systematic measurements of ^{15}N-enriched plant material, plant composts and humic extracts showed, that for all observable signals (amides, free amino groups, pyrrolic-N), with the exception of the resonances of nitrate and ammonium, quantitative data can be obtained with a contact time of 0.7 to 1 ms.(*61,65*) However, heterocyclic aromatic compounds were not identified in these spectra. Because in heterocyclic aromatic compounds protons are not in direct proximity to nitrogen, these substances may not be observed quantitatively in these solid-state ^{15}N CPMAS spectra. In contrast, with adequate precautions to prevent the Nuclear-Overhauser-Enhancement (NOE) and saturation effects, solution ^{15}N NMR spectra can be integrated over the various chemical shift ranges and interpreted quantitatively. Comparison of the signals from the solution ^{15}N NMR spectra of the soluble fraction of ^{15}N-enriched plant composts with the CPMAS ^{15}N NMR spectra of the same sample as a solid indicates that the two techniques yield similar data within experimental error.(*61*)

Figure 3 compares the solid-state ^{15}N CPMAS NMR spectra of undegraded ^{15}N-enriched fresh wheat material (*Tritium sativum*) and its 631-day-old compost incubated at 25°C and at 60 % of water-holding capacity. Assignments of ^{15}N

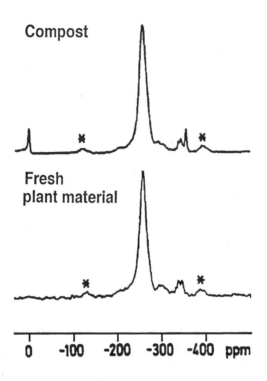

Figure 3. Solid-state ^{15}N NMR spectra of ^{15}N-labeled wheat and its compost incubated for 631 days at 25° C and moisture content equal to 60% of water-holding capacity.(*64*)

signals to specific chemical structures are presented in Table II. The positions of the resonance lines are given relative to the nitromethane scale (CH_3NO_2 = 0 ppm).(*66-68*). Similar spectra are reported from other samples of biogenic material such as the algae, fungi and different plant materials.(*61,62,69-71*) Both spectra, presented here, show their main signal intensity at -257 ppm, in the chemical shift region typically assigned to secondary amides. However, nitrogen in acetylated amino sugars, lactams, unsubstituted pyrroles, indoles and carbazoles may contribute to the intensity of the main peak.(*67,68*) Further, less intense signals are seen in both spectra in the region of primary aliphatic amines (-325 to -350 ppm), NH_2 or NR_2 derivatives (-285 to -325 ppm). The shoulder in the region which can be assigned to indole, imidazole (i.e., histidine) or pyrrole like compounds (-145 to

-220 ppm). The signals between 25 and -25 ppm and -350 and -375 ppm in the ^{15}N NMR of the plant compost reveal the presence of inorganic nitrate and ammonia and are probably the result of ammonification and nitrification processes during composting. The signals denoted by asterisks are spinning sidebands of the amide signal.

Table II. Assignments for the various chemical shifts in ^{15}N NMR spectra.

Chemical shift region	Assignment
25 to -25 ppm	nitrate and nitrite, second order sidebands
-25 to -90 ppm	oximes, imines, phenazine, pyridines
-90 to -145 ppm	first order sidebands of amides, N-7 of purines, nitriles
-145 to -220 ppm	indoles, imidazoles, uric acid, nucleotides, substituted pyrroles
-220 to -285 ppm	amides, lactams, carbamates, unsubstituted pyrroles
-285 to -300 ppm	NH in guanidines
-300 to -325 ppm	primary and secondary amines not bound to carbonyl
-325 to -350 ppm	terminal amino acid groups and free amino sugars
-350 to -375 ppm	ammonium ion
-375 to -430 ppm	first order sideband for amides

The dominance of the signal, assigned to amide functional groups, was also found in the ^{15}N NMR spectra of other microbial degradation experiments of ^{15}N-enriched plant material(55,65), studies of the incorporation of inorganic ^{15}N-enriched compounds in degrading plant material(56,61) and ^{15}N-enriched algal composts.(62) These results strongly indicate, that amide functional groups comprise the main nitrogen form not only in fresh biological precursors of humic material, but also in their microbially degraded residues. The surprisingly high similarity of the solid-state ^{15}N NMR spectra of the degraded and undegraded plant material, further leads to the conclusion that most of the nitrogen in the decomposed sample probably originates from biogenic remains(55,65), most likely proteinaceous materials.

Knicker et al.(61) compared the solid-state ^{15}N NMR spectra of grass composts with their NaOH extracts and their insoluble residues. They found that some minor fractionation has taken place during the extraction procedure. Although the signals were identified at the same chemical shift in all spectra, small changes were observed in the relative intensity distribution. The soluble fraction was enriched in free amino acids, and the signals between -325 and -375 ppm were more pronounced. In the insoluble residues, the high and low field shoulder of the main peptide signal were clearly visible but were almost completely suppressed in the extract. An interesting result of this study is that no nitrogen was detected in the chemical shift region of pyridinic compounds (-40 to -90 ppm). Such compounds were suggested to result from the formation of refractory nitrogen via abiotic polymerization/condensation reactions of lignin degradation products with ammonia produced during microbial degradation of plant residues in soils.(50,72) Signals resulting from nitrile groups (-90 to -145 ppm), recently proposed by Schulten and

Schnitzer(*35,73*) as components of humic material were not identified in any ^{15}N NMR spectra of decomposed biogenic materials and their humic fractions, presented so far.

An explanation for nitrogen resistance in humic substances is given, by the formation of melanoidins from the condensation of carbohydrates and amino groups of amino acids and sugars.(*74*) Benzing-Purdie et al.(*58*) showed in model experiments that these insoluble polymers result in ^{15}N NMR signals in the region of secondary amides (-220 to -285 ppm), pyrrolic-N (-145 to -240 ppm) and pyridinic-N (-40 to -90 ppm). An increase in signal intensity of pyrrolic-N and pyridinic-N with an increasing degree of humification of plant material, would support stabilization of humic nitrogen by this pathway. This result was not observed in the decomposition experiments by Knicker et al.(*61*) and Knicker and Lüdemann(*65*), mentioned above. From these results the authors concluded that the melanoidin pathway cannot be considered as a major pathway for the formation of refractory material in microbially degraded organic material. The feature of the solid state ^{15}N NMR spectra of decomposed plant material and their humic fraction suggests that amide and amine functional groups, most likely of proteinaceous nature, survive microbial degradation. The stabilization may be due to the fast turnover of nitrogen from dead to growing microbial biomass, but it can also be explained by their refractory nature due to chemical or physical protection.

All of the above-mentioned ^{15}N-NMR studies involving ^{15}N-enriched plant materials incubated in the laboratory were initiated to learn more about the stabilization and reaction of nitrogen in humic substances and to obtain more information regarding the nature of the unidentified nitrogen in humic substances. While a major fraction of the humic substances in soil and sediments is several hundred to several thousand years old, laboratory produced material has only been incubated for, at most, 2 years. The solid-state ^{15}N NMR spectra of several native soils and soil fractions at natural ^{15}N abundance(*60-62*), however, show comparable features to the ^{15}N-enriched compost material. They were dominated (80 % of the relative signal intensity) by the amide resonance and show smaller signals for primary aliphatic amines, NH_2 groups and a shoulder in the region of possible indoles and pyrroles. No signals are observed in the area of pyridine and phenazine resonances. Based on the data, the authors conclude that, in these samples and considering the low signal/noise ratio, the broad lineshape of the resonances and the shoulder in the region of pyrroles, heterocyclic aromatic N could not contribute more than 10 % of the total nitrogen signal intensity. No signals were detected in the chemical shift region of nitriles. The results of these studies indicate, that, even after prolonged pedogenesis, nitrogen in humic material is mainly preserved in amide functional groups, most probably in proteinaceous material.(*65*)

This finding was also supported by ^{15}N NMR spectroscopic investigations of a sapropel from Mangrove Lake (Bermuda), deposited under marine conditions 4000 yrs ago.(*62*) ^{15}N NMR spectra of the acid and base extracted residues of an underlying layer, deposited under freshwater conditions revealed that some of this material will even resist chemical degradation.(*75*)

It is likely that such nitrogen survives in soils and sediment for extended periods of time and is the organic nitrogen that has been referred to as unknown nitrogen in past studies of humic substances simply because it could not be released by acid hydrolysis. If this nitrogen is indeed bound in proteinaceous material its resistance during sediment/soil diagenesis is quite astonishing, since proteins are known to decompose quickly in the presence of microbes and during acid hydrolysis. In addition, due to the low mineral and clay content of the examined sapropel layers a protection via the mineral mesopore hypothesis(*76*) cannot completely explain the long-term stability of proteinaceous material in the sediments of Mangrove Lake.

In a recent ^{15}N NMR study, Derenne et al.(*77*) obtained evidence for non hydrolyzable amide structures in refractory fractions of the algae *Scenedesmus quadricauda*. The solid-state ^{15}N NMR spectra showed a major peak around -260 ppm for amides accompanied by a peak around -195 ppm for substituted pyrroles and a shoulder at -235 ppm assigned to unsubstituted pyrroles of the insoluble residue. The authors suggest, that the amides are protected by association with long polymethylenic chains within a macromolecular network. Knicker and Hatcher offer an alternative explanation,- that protection of amide functional groups as part of proteinaceous material are affected by encapsulation within the macromolecular matrix forming sedimentaryhumic material(*75*).

Solution ^{15}N NMR

In solution NMR, the spin lattice relaxation T_{1N} of small molecules can vary from seconds to hours.(*66*) Thus this contributes to even more serious sensitivity problems when one considers the need to wait 4-5 T_{1N} between pulses. In a recent study of the relaxation behavior of nitrogen containing organic material in NaOH-extracts of plant composts with an incubation time up to 541 days, it was found that the extracted compounds behave in a similar manner to biomacromolecules.(*61,66,78*) At a resonance frequency of 30.4 MHz, the T_{1N}'s of humic substances derived from degrading plant material range from 1 to 3 s.(*61*)

NMR spectra of organic material are normally strongly influenced by proton coupling. The splitting caused by such coupling can be eliminated by broadband decoupling. The decoupling phenomenum leads to the Nuclear-Overhauser-Enhancement (NOE) which magnifies the signals dependent on the magnetogyric ratio and the spin-lattice relaxation mechanism.(*68,79*) The ^{15}N nucleus with its a large and negative magnetogyric ratio has a negative NOE, which is equal to $1 + \eta$, where η is the NOE factor. The maximum signal enhancement obtained from proton decoupled nitrogen is approximately -4, provided that spin lattice relaxation occurs completely through a dipolar spin-lattice relaxation mechanism and within the extreme narrowing limits of fast molecular motion. Otherwise a range of NOE factors from 0 to - 4.93 may be observed. While a NOE factor between 0 and -2.0 decreases the signal intensity. A NOE factor of -1.0 eliminates the resonance entirely.(*68*)

A study of the NOE in ^{15}N-enriched plant composts revealed that secondary and primary amides in their organic material produces an enhancement factor of approximately -2.5 (at 30.4 MHz). The band at -306 ppm is assigned to NH_2 groups in amino acids, citrulline or uric acid and nucleotide derivatives and shows enhancement factors of approximately -3.5.(*64*)

Because the application of proton broad band decoupling in solution ^{15}N NMR causes quantification problems, the inverse-gated decoupling technique often is used. With this technique no enhancement is observed, and the spectra can be quantified. However, a problem that arises in NMR experiments with nuclei with low sensitivity such as ^{15}N is acoustic ringing. Ringing occurs because the rf pulse causes brief vibrations in the probe which masquerade as a signal. This signal overlaps the entire spectra which results in a distortion of the baselines. Quantification of these distorted spectra is impossible and in many cases the spectra become useless.

Another attempt to circumvent the problem of nonquantitative signals involves the use of the INEPT pulse sequence which serves as both a sensitivity enhancement and a subspectral editing technique. The theoretical maximum enhancement of ^{15}N signals obtained through this polarization transfer technique is approximately 10.(80) Because only protonated nuclei can be observed, this technique allows us to distinguish between protonated and non protonated nitrogen. A limitation of polarization transfer technique is that ^{15}N nuclei attached to protons which are undergoing rapid exchange will not be polarized.

Using this pulse sequence to estimate the nature of derivatization of Suwannee River fulvic acid with ^{15}N-enriched hydroxylamine to learn more about the carbonyl functionality of fulvic acid, Thorn et al.(16) obtained signals for the primary products as oximes. Additional signals of secondary products arising from Beckmann rearrangements of the initial oxime derivatives were identified as nitriles, secondary amides and lactams. The bands assigned to hydroxamic acid result from a reaction of esters with NH_2OH and are evidence for the presence of esters in the fulvic acid.

In a continuing study(20), these experiments were applied to five fulvic and humic samples. In addition, the DEPT pulse sequence was used. This technique allows for a further discrimination between singly and doubly protonated nuclei. They found that all five samples react in the same manner. Reaction products of hydroxylamine with esters were observed. Other resonances discovered were attributable to the tautomeric forms of the nitrosophenol and monooxime derivatives of quinones. Thus, this study provides indirect evidence for the presence of quinones in humic material and also suggests the possible presence of cyclic ketones and ketones α to quaternary carbons.

In another study, ammonia fixation of ^{15}N-labeled ammonium hydroxide with Suwannee River fulvic acid, IHHS peat and leonardite humic acid were examined by solution ^{15}N NMR with the application of INEPT and DEPT pulse sequences.(23) Similar reaction of ammonia with all three samples is reported. Most of the nitrogen incorporated seems to be in the form of indole and pyrrole followed by pyridine, pyrazine, amide and aminohydroquinone nitrogen. The authors also suggest a possible reaction mechanism to explain the formation of the heterocyclic compounds identified in the spectra. They also claimed that these results need to be substantiated through further work with model compounds and experiments with the reaction conditions, i.e., in which phenols will undergo oxidation to quinones when reacted with ammonia.

No pyridine, phenazine and imine-like structures can be identified by solution ^{15}N NMR in a recent study designed to obtain more information concerning the degradation products of ^{15}N-enriched decomposed plant material incubated up to 541 days.(61,64) The narrow lineshape in solution ^{15}N NMR provides a clear discrimination between pyrrole and amide signals and a quantification of pyrrole and indole derivatives. No increase in signal intensity was observed in the region of pyrrole or aniline derivatives. Only an increase of the amide signal intensity was observed over the course of composting. This trend confirms the observations obtained from the solid-state ^{15}N NMR studies of other composts(55,65) and can most probably be explained by the fact that aromatic rings are not significantly changed during degradation and that proteinaceous substances dominate the pool of nitrogenous compounds throughout the degradative process.

Solution ^{15}N NMR spectra of ^{15}N-enriched fungal material isolated from the composts were measured with broadband decoupling, inverse gated decoupling and DEPT pulse sequences.(61,64) A comparison of these spectra shows a high degree of similarity with the ^{15}N NMR spectra of whole compost discussed above, which leads the authors to the conclusion that during microbial degradation of plant material, nitrogen seems to be predominantly stabilized in biogenic secondary amides.

Grinwalla et al.(*22*) examined the reaction of Suwannee River fulvic acids with chloramine using solution ^{15}N NMR. They report, that while chloramine itself may not form nitrogen containing moieties with Suwannee River fulvic acid, under the conditions examined, the chlorination coproduct, ammonia, most certainly does. They explain that "the assertion that chloramine is a safe replacement to chlorine in the treatment of potable water supplies should be tempered, at least until the toxicity of aminated and chlorinated humic products are examined."

During the last few years solid state and solution ^{15}N NMR have proven to be powerful techniques for the examination of nitrogen chemistry in humic substances. The findings obtained up to now are likely to have a tremendous impact on our understanding of nitrogen functionality, nitrogen fertility and stabilization in humic substances.

Recent Applications of ^{1}H NMR

Although ^{1}H NMR is probably the most widely used NMR technique in chemical applications, it has only been used sparingly in studies of humic substances. Much of the literature in which this technique is used was published prior to 1987 and is well covered in the review by Wilson.(*5*) Use of ^{1}H NMR is mostly limited to solutions, and it is thus necessary to dissolve humic substances in aqueous solutions. Of course, one must reduce the amount of [1]H's from solvent molecules by use of deuterated water, but it is often difficult to reduce these background signals to negligible amounts. Wilson(*5*) described the application of water suppression techniques which have been necessary to obtain ^{1}H NMR spectra of humic isolates from soil. In the case of solid-state ^{1}H NMR of whole soils and humic substances, the development of a new technique called CRAMPS (combined rotation and multiple pulse spectroscopy) has not received the attention it has in the field of coal science, primarily because of the lack of resolution. Thus, recent applications of either solution or solids ^{1}H NMR has been very limited in humic substance science since Wilson's(*5*) review.

Solution ^{1}H NMR Studies

Some of the first NMR spectroscopic applications in humic substance research involved the use of ^{1}H NMR for the examination of soluble humic substances.(*5*) The information obtained from ^{1}H NMR spectra of humic and fulvic acids dissolved in alkaline or neutral aqueous solutions has been quite useful in characterizing the aromatic and aliphatic structures of these complex materials. In as much as humic materials are inherently complex, the spectra show only broad signals with some fine structure which can be related to specific methyl, methylene, methine, and aromatic/olefinic hydrogens. Because ^{13}C NMR spectroscopy contains inherently more structural information for humic substances, the use of ^{1}H NMR has been limited since much of the work conducted prior to 1987. This is unfortunate considering the details observable in ^{1}H NMR spectra of humic and fulvic acids published recently by Thorn et al.(*11*) and Malcolm(*33*).

Solid-State ^{1}H NMR Studies

The strong dipolar coupling of ^{1}H's to each other in solid samples provides a formidable decoupling challenge which has only been recently met by use of a special line narrowing method called CRAMPS. For coal samples, this method is capable of resolving aromatic from aliphatic ^{1}H's, albeit with significant spectral overlap. Add the dispersion caused by the increased variety of spectrally different ^{1}H's, and the spectra cannot be interpreted easily. Frye et al.(81) published the first, and to our knowledge the only, CRAMPS spectra of humic substances. The complexity of the technique and the limited information provided for humic materials and soil organic matter have made it of rather limited use.

Concluding Remarks

Nuclear magnetic resonance spectroscopy has become one of the most powerful analytical tools for the characterization of humic substances. The quantitative representation of different types of structures for ^{13}C, ^{1}H, and ^{15}N atoms within the macromolecular framework of humic substances allows one to obtain average structural information that is not possible with other techniques which are by-and-large destructive and non quantitative. Much of the initial work on applications of new NMR techniques such as the CPMAS techniques or spectral editing occurred prior to the excellent overview by Wilson.(5) Since this time, NMR spectroscopy has not progressed as rapidly as one might have thought, probably because many of the initial discoveries have been made and the next phase of study would involve grinding out detailed spectroscopic information on substances which have poorly resolved spectra at best. The next phase of research would undoubtedly bring other techniques to bear on shedding more light on some of the initial discoveries. In this regard, the work of Kögel-Knabner and associates and Baldock and associates stand out as ones specifically designed to doggedly pursue knowledge concerning the origin of paraffinic structures in soils. Even M. Schnitzer and associates appear to have adopted NMR as a primary investigative technique combined with pyrolysis/field ionization methods to seek an understanding of the structure of humic acids.

Perhaps the most exciting new applications of modern NMR spectroscopy are focused in two areas use of ^{15}N and ^{13}C-enrichments to trace the reactivity of soil carbon with regard to degradation and interaction with pollutants. Using labels by incorporating them into plant materials subjected to degradation or into pollutants which are interacted with humic materials will chart a new course in NMR spectroscopy in humic substance research of the future. In the case of ^{15}N NMR studies using labels, a new understanding of the reactivity of soil nitrogen may have been uncovered, because the long thought of scapegoat for lack of knowledge regarding forms of soil nitrogen-heterocyclic nitrogen, may be nonexistent. One can now explain recalcitrant nitrogen simply as recalcitrant protein nitrogen on the basis of the ^{15}N NMR work of Knicker et al.(64) The labeling studies of Baldock et al.(47) have directly demonstrated to us that paraffinic carbons in soil probably originate from microbial biomass.

Finally, the use of ^{13}C and ^{15}N-labeled pollutants and corresponding NMR spectroscopy are emerging as techniques showing great potential in the examination of bonding and non bonding associations humic materials have with them. The ability to examine pollutant humic interactions in soil will have a major impact on our assessment of the environmental survivability and transport of such pollutants.

It is now through the application of this technique of labeling and NMR that we will eventually learn about the types of pollutant/humic bonding associations.

Literature Cited

1. Neyroud, J.A.; Schnitzer, M. *Can. J. Chem.* 1972, 52, 4123-4132.
2. Gonzalez-Vila, F.J.; Lentz, H.; Lüdemann, H.-D. *Biochem. Biophys. Res. Comm.* **1976**, 72, 1063-1070.
3. Hatcher, P.G.; VanderHart, D.L.; Earl, W.L. *Org. Geochem.* **1980**, 2, 87-92.
4. Wilson, M.A.; Pugmire, R.J.; Zilm, K.W.; Goh, K.M.; Heng, S.; Grant, D. *Nature*, **1981**, 294, 648-650.
5. Wilson, M.A. *NMR techniques and applications in geochemistry and soil chemistry*; Pergamon Press: Oxford, **1987**.
6. Thorn, K.A.; Weber, E.J.; Spidle, D.L.; Pettigrew, P.J. In *Humic substances in the global environment and implications in human health*; Senesi, N.; Miano, T.M., Ed.; 6th International Meeting International Humic Substances Society Abstracts Monopoli, Italy; Segreteria Organizzativa: Bari, Italy **1992**, pg 119.
7. Hatcher, P.G.; Bortiatynski, J.M.; Minard, R.D.; Dec, J.; Bollag, J.-M. *Environ. Sci. Technol.* **1993**, 27, 2098-2103.
8. Bortiatynski, J.M.; Hatcher, P.G.; Minard, R.D. In *NMR spectroscopy in environmental science and technology*; Nanny, M.A.; Minear, R.A.; Leenheer, J.A., Ed.; Oxford Universiy Press, in press.
9. Bortiatynski, J.M.; Hatcher, P.G.; Minard, R.D; Dec, J.; Bollag, J.M. In Humic substances in global environment and implications on human health; Senesi, N.; Miano, T.M., Eds.; Elsevier: Amsterdam, **1994**, pp. 1091-1010.
10. Preston, C.M. In *NMR of humic substances in coal*; Wershaw, R.L.; Mikita, M.A., Ed.; Lewis Publishers, Inc., Michigan, **1987**, p. 3-32.
11. Thorn, K.A.; Folan, D.W.; MacCarthy, P. *Characterization of the international humic substances society standard and reference fulvic and humic acids by solution state carbon-13 (^{13}C) and hydrogen-1 (^1H) nuclear magnetic resonance spectroscopy*; U.S. Geological Survey, Water Resource Investigations Report, Denver, Co, 1989; 89-4196, pp. 1-4.
12. Thorn, K.A. In *Humic substances in the Suwannee River, Georgia: Interactions, properties, and proposed structures*; Averett, R.C.; Leenheer, J.A.; McKnight, D.M.; Thorn, K.A., Eds.; U.S. Geological Survey, OFR, 87-557, **1989**, pp. 255-309.
13. Buddrus, J.; Burba, P.; Herzog, H.; Lambert, J. *Anal. Chem.* **1989**, 61, 628-631.
14. Lambert, J.; Burba, P.; Buddrus, J. *Mag. Res. Chem.* **1992**, 30, 221-227.
15. Hatcher, P.G. *Org. Geochem.*, **1987**. 11, 31-39.
16. Thorn, K.A.; Folan, D.W.; Arterbrun, J.B.; Mikita, M.A.; MacCarthy, P. *Sci. Tot. Environ.* **1989**, 81/82, 209-218.
17. Thorn, K.A.; Steelink, C.; Wershaw, R.L. *Org. Geochem.* **1987**, 11, 123-137.
18. Mikita, M.A.; Steelink, C. *Anal. Chem.* **1981**. 53, 1715-1717.
19. Preston, C.M.; Ripmester, J.A. *Can. J. Soil Sci.* **1983**, 63, 495-500.
20. Thorn, K.A.; Arterburn, J. B.; Mikita, M.A. *Environ. Sci. Technol.* **1992**, 26, 107-116.
21. Haider, K.; Spiteller, M.; Reichert, K.; Fild, M. *Intern. J. Environ.Anal. Chem.***1992**, 46, 201-211.
22. Grinwalla, A.S.; Mikita, M.A. *Environ. Sci. Technol.* **1992**, 26, 1148-1150.

23. Thorn, K.A.; Mikita, M.A. *Sci. Total Envion.* **1992**, 113, 67-87.
24. Vassallo, A.M.; Wilson, M.A.; Collin, P.J.; Oades, J.M.; Waters, A.G.; Malcolm, R.L. *Anal. Chem.* **1987**, 59, 558-562.
25. Fründ, R.; Lüdemann, H.-D. *Z. Naturforsch.* **1991**, 46c, 982-988.
26. Fründ, R.; Lüdemann, H.-D. *Sci. Total Environ.* **1989**, 81/82, 157-168.
27. Kinchesh, P.; Powlson, D.S.; Randall, E.W. *Europ. J. Soil Sci.* **1995**, 46, 125-138.
28. Hatcher, P.G.; Wilson, M. A. *Org. Geochem.* **1991**, 17, 293-299.
29. Baldock. J.A.; Oades, J.M.; Waters, A.G.; Peng, X.; Vassallo, A.M.; Wilson, M.A. *Biogeochem.* **1992**, 16, 1-42.
30. Wilson, M.A.; Vassallo, A. M.; Perdue, E.M.; Reuter, T.H. *Anal. Chem.* **1987**, 59, 551-558.
31. Hatcher, P.G.; Schnitzer, M. A.; Vassallo, M.; Wilson, M.A. *Geochim. Cosmochim. Acta* **1989**, 53, 125-130.
32. Preston, C.M.; Newman, R.H. *Can. J. Soil Sci.* **1992**, 72, 13-19.
33. Malcolm, R.L. *Anal. Chim. Acta* **1990**, 232, 19-30.
34. Hatcher, P.G.; Faulon, J.-L.; Clifford, D.A.; Mathews, J.P. In *Humic substances in global environment and implications on human health*; Senesi, N.; Miano, T. M., Eds.; Elsevier: Amsterdam, **1994**, p. 133-138.
35. Schulten, H.-R.; Schnitzer, M. A. *Naturwissenschaften* **1995**, 82, 487-498.
36. Kögel-Knabner, I.; Zech, W.; Hatcher, P.G. *Soil Sci. Am. J.* **1991**, 55, 241-247.
37. Inbar, Y.; Hadar, Y.; Chen, Y. *Sci. Tot. Environ.* **1992**, 113, 35-48.
38. deMontigny, L.E.; Preston, C.M.; Hatcher, P.G.; Kögel-Knabner, I. *Can. J. Soil Sci.* **1993**, 73, 9-25.
39. Haider, K. In *Microbial communities is soil*; Jensen et al. Eds.; Elsevier: London, **1986**, p. 133-147.
40. Wang, M.C.; Huang, P.M. *Sci. Tot. Environ.* **1992**, 113, 147-158.
41. Saiz-Jimenez, C.; de Leeuw, J.W. *J. Anal. Appl. Pyrol.* **1986**, 9, 99-119.
42. Schulten, H.-R.; Plage, B.; Schnitzer, M. *Naturwissenshchaften* **1991**, 78, 311-312.
43. Schulten, H.-R.; Schnitzer, M. *Soil Sci.* **1992**, 153, 205-224.
44. Saiz-Jimenez, C.; Hermosin, B.; Ortega-Calvo, J.J. *Water Res.* **1994**, in press.
45. Kögel-Knabner, I.; Hatcher, P.G. *Sci. Tot. Environ* **1989**, 81/82, 169-177.
46. Kögel-Knabner, I.; Hatcher, P.G.; Tegelaar, E.W.; de Leeuw, J.W. *Sci. Tot. Eniviron.* **1992**, 113, 89-106.
47. Baldock, J.A.; Oades, J.M.; Vassallo, A.M.; Wilson, M.A. *Aust. J. Soil Res.* **1989**, 27, 725-746.
48. Baldock, J.A.; Oades, J.M.; Vassallo, A.M.; Wilson, M.A. *Aust. J. Soil Res.* **1990**, 28, 193-212.
49. Zelibor, J.L.; Romankiw, L.; Hatcher, P.G.; Colwell, R.R. *Appl. Environ. Microbiol.* **1988**, 54, 1051-1060.
50. Schnitzer, M. In *Humic substances in soil sediments and water*; Aiken, G.R.; McKnight, D.M.; Wershaw, R.L.; MacCarthy, P., Eds.; John Wiley & Sons: New York, **1985**, p. 303-325.
51. Benzing-Purdie, L.M.; Cheshire, M.V.; Williams, B.L.; Sparling, G.P.; Ratcliffe, C.I.; Ripmeester, J.A. *J. Agric. Food Chem.* **1986**, 34, 170-176.
52. Benzing-Purdie, L.; Cheshire, M.V.; Williams, B.L. *J. Soil Sci.* **1992**, 43, 113-125.
53. Preston, C.M; Ripmeester, J.A.; Mathur, S.P.; Lévesque, M. *Can. J. Spectrosc.* **1986**, 31, 63-69.
54. Cheshire, M.V.; Williams, B.L.; Benzing-Purdie , L.M.; Ratcliffe, I.C.; Ripmeester, J.A. *Soil Use and Management* **1990**, 6, 90-92.

55. Almendros, G.; Fründ, R.; Gonzalez-Vila, F.J.; Haider, K.; Knicker, H.; Lüdemann, H.-D. *FEBS-Letters* **1991**, 282, 119-121.
56. Knicker, H.; Fründ, R.; Almendros, G.; Gonzalez-Vila, F.J.; Martin, F.; Lüdemann, H.-D. *Humus-uutiset* **1991**, 3, 313-315.
57. Zhuo S.; Wen, Q.; Du, S. *Chin. Sci. Bull.* **1992**, 37, 508-511.
58. Benzing-Purdie, L. M.; Ripmeester, J. A.; Preston, C.M. *Agric. Food Chem.* **1983**, 31, 913-915.
59. Preston, C.M.; Rauthan, B.S.; Rodger, C.; Ripmeester, J.A. *Soil Sci* **1982**, 134/5, 277-293.
60. Knicker, H.; Fründ, R.; Lüdemann, H.-D. *Naturwissenschaften* **1993**, 80, 219-221.
61. Knicker, H.; Fründ, R.; Lüdemann, H.-D. In *NMR spectroscopy in Environmental Science and Technology*; Nanny, M.A.; Minear, R.A. Leenheer, J.A., Eds.; Oxford: University Press, **1996**, (in press).
62. Knicker, H.; Scaroni, A.W.; Hatcher, P.G. *Org. Geochem.* **1996**, in press.
63. Knicker, H.; Hatcher, P.G.; Scaroni, A.W. *Energy & Fuels* **1995**, 9, 999-1002.
64. Knicker, H. *Quantitative ^{15}N- und ^{13}C-CPMAS-Festkörper- und ^{15}N-Flüssigkeits-NMR-Spektroskopie an Pflanzenkomposten und natürlichen Böden.*; Dissertation, Universität Regensburg: Regensburg, FRG, 1993.
65. Knicker, H.; Lüdemann, H.-D. *Org. Geochem.* **1995**, 23, 329-341.
66. Martin, G. J.; Martin, M.L.; Gouesnard, J.P. *^{15}N-NMR spectroscopy.* **1981**, Springer Verlag: Berlin.
67. Witanowski, M.; Stefaniak, L.; Webb, G. A. In *Annual reports on NMR spectroscopy*; Webb, G. A., Ed.; Academic Press: London, **1986**, vol. 18.
68. Witanowski, M.; Stefaniak, L.; Webb, G. A. In *Annual reports on NMR spectroscopy* Webb, G.A., Ed.; Academic Press: London, **1993**, vol. 25.
69. Schaefer, J.; Stejskal, E.O.; McKay, R.A. *Biochem. Biophys. Res. Commun.* **1979**, 88, 274-280.
70. Skokut, T.A.; Varner, J.E.; Schaefer, J.; Stejskal, E.O.; McKay, R.A. *Plant. Physiol.* **1982**, 69, 308-313.
71. Jacob, G.S.; Schaefer, J.; Stejskal, E.O.; McKay, R.A. *Biochem. Biophys. Res. Commun.* **1980**, 97, 1176-1182.
72. Anderson, H.A.; Bick, W.; Hepburn, A.; Stewart, M. In *Humic substances II*; Hayes, M.H.B.; MacCarthy, P.; Malcolm, R.L.; Swift, R.L., Eds.; John Wiley & sons: Chichester, **1989**, p. 223-253.
73. Schulten, H.-R.; Schnitzer, M. *Naturwissenschaften* **1993**, 80, 29-30.
74. Maillard, L.C. *C.R. Soc. Biol.* **1917**, 72, 599-601.
75. Knicker, H.; Hatcher, P.G. **1996**, Manuscript in preparation
76. Hedges, J.I.; Keil, R.G. *Mar. Chem.* **1995**, 49, 37-139.
77. Derenne, S.; Largeau, C.; Taulelle, F. *Geochim. Cosmochim. Acta* **1993**, 57, 851-857.
78. Clore, G. M.; Szabo, A.; Bax , A.; Kay, L. E.; Driscoll, P. C.; Gronenborn, A. M. *J. Am. Chem. Soc.* **1990**, 112, 4989-4991.
79. Harris, R. K. *Nuclear magnetic resonance spectroscopy*; Pitman: London, 1983.
80. von Philipsborn, W.; Müller, R. *Angewandte Chemie* **1986**, 25, 383-486.
81. Frye, J. S.; Bronnimann, C. E.; Maciel, G. E. In *NMR of humic substances and coal* ; Warshaw, R. L.; Mikita, M. A. Eds.; Lewis Publishers, Inc.: Chelsea, Michigan, 1987 pg. 42.

Chapter 6

Structural Characterization of Humic Substances Using Thermochemolysis with Tetramethylammonium Hydroxide

Jose Carlos del Rio[1] and Patrick G. Hatcher

Department of Geosciences, Pennsylvania State University, University Park, PA 16802

The focus of this paper centers on a review of the recent development of the TMAH thermochemolysis as a rapid, low-cost, and easily implemented technique for the structural analysis of humic substances. Pyrolysis, or rather chemolysis, in the presence of TMAH has been used for the structural characterization of humic substances from different origins. The procedure methylates carboxylic groups and hydroxyl groups, rendering the chemolytic products more amenable to chromatographic separation. The examination of humic substances with this technique reveals the presence of a series of benzenecarboxylic acid methyl esters as well as long-chain fatty acid methyl esters and dimethyl esters. The methylated structures produced by TMAH differ dramatically from those obtained by conventional pyrolysis, calling into question the recently proposed structure of humic acids which are based mostly on conventional pyrolysis. The procedure consists mainly of a thermally assisted chemolysis rather than true pyrolysis, and consequently is effective at subpyrolysis temperatures. This means that the procedure can also be easily implemented in sealed glass tubes. In general, the studies to date have demonstrated that this technique provides excellent preservation of the original structures containing carboxyl and hydroxyl groups in lignin monomers owing to protection of the functional groups from thermal reactions. The TMAH thermochemolysis also induces β-ether bond lysis in lignin. This is significant because the procedure shows potential in characterizing lignin-derived compounds in much the same way as the CuO oxidation procedure.

Characterization of complex organic matter like humic substances (HS) is a formidable task (1). A variety of destructive and non-destructive methods have been applied. Among the non-destructive methods, spectroscopic methods, such as NMR and FT-IR have proven to be very useful in providing information about the structure of these materials as a whole. Of these, NMR has proven to be the best method for bulk characterization, especially solid-state ^{13}C NMR, where the relative contribution of specific carbon types can be made (2). Different structural parameters (aromaticity,

[1]Current address: Instituto de Recursos Naturales y Agrobiologia de Sevilla, Consejo Superior de Investigaciones Cientificas, P.O. Box 1052, 41080–Seville, Spain

0097–6156/96/0651–0078$15.00/0

carboxyl content, etc.) could be determined directly. However, for more detailed information on the structure of HS at a molecular level, degradative methods need to be applied. Among them, wet chemical methods and pyrolytic methods have been widely used in the past years to get insight into the chemical constitution of these materials (*1,3,4*).

Wet chemical methods have the advantage of providing information on the molecular fragments comprising HS, but exhibit several disadvantages as they often require high quantities of sample, they interfere with the sample, and they sometimes are very drastic. Wet chemical degradations such as the CuO method of Hedges and Ertel (*5*) can identify components derived from lignin and other biopolymers (*3,6*).

Pyrolytic techniques coupled with mass spectrometry (*7*) and gas chromatography-mass spectrometry (*4,8*) have proved to be particularly useful in structural characterization of HS due to detailed information obtained at a molecular level and the ease of sample preparation. They usually require low amounts of sample and no chemical pretreatments are needed, avoiding possible contamination from laboratory manipulations. In many instances pyrolysis is very reproducible and the results can be interpreted both qualitatively and quantitatively, and finally, it is easily performed in a pyroprobe. The thermal degradative products can be then related to original moieties present in the structure of the HS. The pyrolysis of polar macromolecular materials is well known to produce volatile polar products which can be chromatographed; however, very polar products also produced are often so polar as to remain attached to the column and of insufficient volatility to be chromatographed. These products are simply not observed and remain unquantified. This is because the high polarity of products leads to further decomposition before volatilization, prevents transfer from the pyrolysis unit to the GC column, or makes it impossible to pass through a GC column. Also, some significant structural moieties can be heavily modified by unwanted thermal reactions which may lead to misinterpretation of the structure of the HS (*9-12*).

Recently, a new technique has been introduced for the characterization of polar macromolecules such as HS. This technique was originally introduced as pyrolysis with *in situ* methylation (*13*). The presence of a gaseous methylating agent, tetramethylammonium hydroxide (TMAH), at the point of pyrolysis greatly assists in converting polar products to less polar derivatives which are more amenable to chromatographic separation (*9,13-17*). This procedure avoids decarboxylation and produces the methyl esters of carboxylic acids and methyl ethers of hydroxyl groups, rendering many of the polar products volatile enough for gas chromatographic analysis. Thus, it is possible to separate and detect many more structurally significant products than that observed previously by conventional pyrolysis-GC-MS. This technique for the analysis of HS has greatly enhanced product yields and produces some products not observed by conventional pyrolysis (*9,11,15-17*).

Challinor (*13,18,19*) first introduced this technique for the simultaneous pyrolysis and methylation of phenolic polymers, and since then, it has been applied to different biopolymers (*14,20,21*), humic materials (*9-11,15-17,22-25*), asphaltenes and kerogens (*26,27*) and natural and fosil resins and resinites (*28,29*).

Several authors (*9,12,14,16,30*) have pointed out that the reaction involved in the TMAH/pyrolysis scheme is one of chemolysis rather than pyrolysis. Because the procedure is more likely a thermally-assisted chemolysis or "thermochemolysis", sub-pyrolysis temperatures of 300°C have also been found to effectivelly produce a suite of products similar to that observed at higher pyrolysis temperatures (*16,21*). Moreover, in a recent paper by McKinney *et al.* (*31*), a procedure was outlined for the characterization of lignin at subpyrolysis temperatures of 300°C in the presence of TMAH using sealed pyrex tubes. Since then, this technique has been extended to the characterization of lignin in fresh and degraded woods (*32*) and in coalified woods (*33*).

Application of the TMAH/Thermochemolysis Procedure to Humic Materials

As noted above, the use of TMAH/thermochemolysis for the structural characterization of different humic materials have been applied in two different forms: pyrolysis of the humic materials in the presence of TMAH or a confined reaction of the humic materials with the TMAH in a sealed glass ampoule at lower temperatures. Both methods release the same types of compunds.

Lignin is an important component in the structure of most humic materials and many lignin-derived compounds are released in the TMAH/thermochemolysis procedure. A model for a gymnosperm lignin is depicted in Figure 1 to help the understanding of much of the discussion refered to the lignin structure along the text. This structure is essentially a modified model of Adler (*34*) for spruce wood, and it is important to note several structural features for further reference. While the gymnosperm lignin is made up exclusively of guaiacyl units, angiosperm and grass lignins contain also syringyl and *p*-hydroxycoumaryl units. Most lignin units are linked to each other via β-O-4 linkages but other linkages included β-β, α-O-4, β-5 and 5-5' linkages using the standard notation for lignin units (Figure 1).

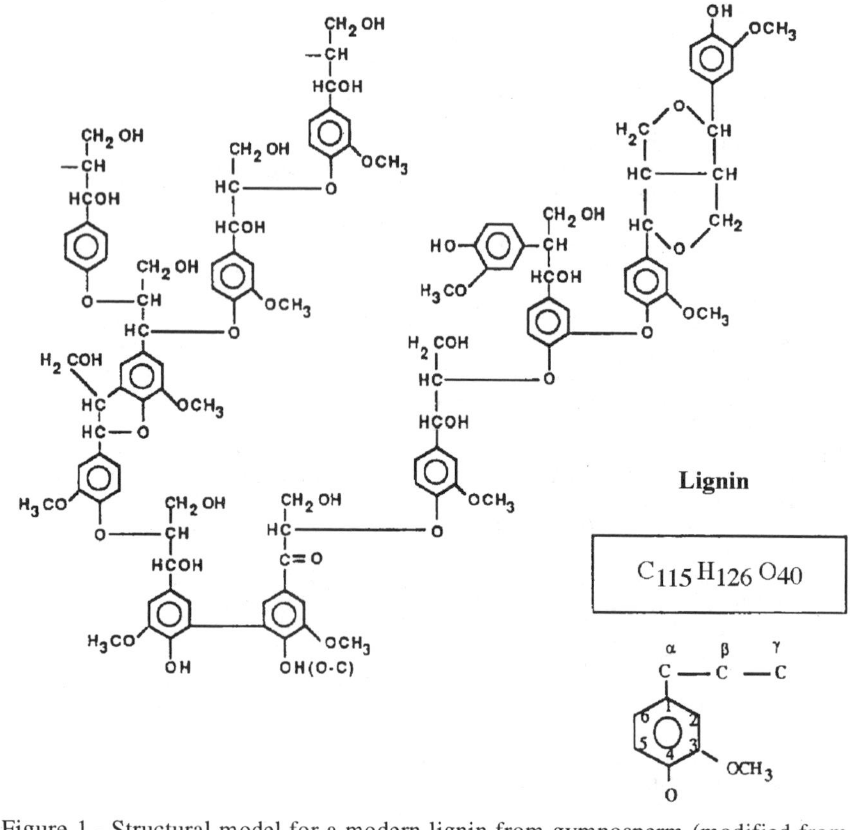

Figure 1.- Structural model for a modern lignin from gymnosperm (modified from Adler, *34*). The numbering convention is shown for a typical structural unit of lignin. Reproduced with permission from reference 34. Copyright 1990 Elsevier Science Ltd.—UK.

Pyrolysis in the Presence of TMAH. A wide set of Fulvic Acids (FA), Humic Acids (HA) and related materials from different origins have been thus far studied using the TMAH/thermochemolysis procedure (*9,15-17,22,24,25*). The TMAH procedure has mainly been performed in pyroprobe units, in much the same way as the conventional flash pyrolysis using either quartz tubes and/or a platinum coil, by mixing the humic material with a few drops of the reagent prior to heating.

Characterization of Fulvic Acids. Figure 2 shows the chromatogram of the products released after the pyrolysis of the FA isolated from a water-logged peatland, both in the absence and in the presence of TMAH. Peaks identifications are in Table I. The flash heating of the FA in the presence of TMAH yields mainly derivatives from polysaccharides and lignin which have become incorporated into the FA macromolecular structure. Methyl esters of aliphatic carboxylic acids were also released. This procedure yields products that are not released with conventional pyrolysis, because they are too polar or too reactive. The most striking feature is the identification of aromatic acids, as methyl esters, which have been previously reported as being building blocks of the FA structure in models based upon oxidative degradations (*1*). Saiz-Jimenez *et al.* (*15,22*) and Martin *et al.* (*9*) published the only studies thus far in which pyrolysis in the presence of TMAH is used for the structural characterization of FAs. They found that large quantities of aromatic acids were released from different soil and lake FAs, suggesting that these compounds might represent final steps in the oxidation of the side-chain during microbial degradation of lignins. Benzenecarboxylic acid moieties have not previously been released from humic substances upon conventional pyrolysis techniques due to decarboxylation

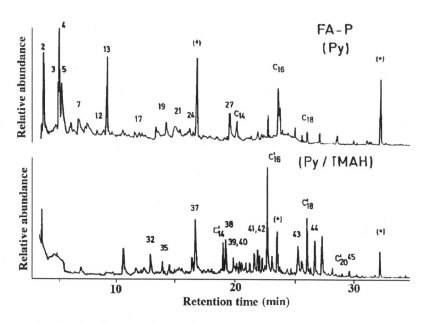

Figure 2.- Total Ion Chromatogram of the thermal degradative products obtained after pyrolysis of the FA isolated from a water-logged peatland, in the absence (Py) and in the presence of TMAH (Py/TMAH). For peak identifications refer to Table I. Contaminants are noted by (*) and the peaks labeled Cn are fatty acids and C'n are fatty acid methyl esters. Reproduced with permission from reference 9. Copyright 1994 Elsevier Science—Netherlands.

Table I. Main thermal degradative products identified in the pyrolysates of the fulvic acids isolated from a water-logged peatland Reproduced with permission from reference 9. Copyright 1994 Elsevier Science—Netherlands.

Conventional pyrolysis	
1	acetic acid
2	furfural
3	methylfurfural
4	dimethyltrisulphide
5	phenol
6	2-furancarboxylic acid
7	cresol
8	guaiacol
9	furanmethanol
10	benzenediol
11	2-methyl-1H-isoindol-1,3(2H)-dione
12	2-methyl-3-hydroxypyran-4-one
13	dimethyltetrasulphide
14	benzoic acid
15	2-butenoic acid
16	benzenediol
17	1,3-benzofurandione
18	2,2'-bifuran
19	methyl-1,3-benzofurandione
20	benzopyran-2-one
21	methoxy-1,3-benzofurandione
22	1H-isoindol-1,3-2(H)-dione
23	dibenzofuran
24	dimethyl-1,3-benzofurandione
25	4-oxopentenoic acid
26	acetovanillone
27	biphenol
28	dibenzofuranol
pyrolysis/TMAH	
29	phosphoric acid trimethyl ester
30	butanedioc acid dimethyl ester
31	permethylated monosaccharide
32	methoxybenzenecarboxylic acid methyl ester
33	dichlorobenzenecarboxylic acid methyl ester
34	permethylated monosaccharide
35	benzenedicarboxylic acid dimethyl ester
36	unknown
37	dimethoxybenzenecarboxylic acid methyl ester
38	trimethoxybenzenecarboxylic acid methyl ester
39	dimethoxydimethylbenzenecarboxylic acid methyl ester
40	methoxybenzenedicarboxylic acid dimethyl ester
41	dimethoxybenzenedicarboxylic acid dimethyl ester
42	benzenetricarboxylic acid trimethyl ester
43	methoxybenzenetricarboxylic acid trimethyl ester
44	benzenetetracarboxylic acid tetramethyl ester
45	methoxybenzenetetracarboxylic acid tetramethyl ester
46	benzenepentacarboxylic acid pentamethyl ester

process. The use of pyrolysis in the presence of TMAH avoids decarboxylation by protecting the carboxyl group, and releases them as their methyl esters. The aboveauthors agreed in that the benzenecarboxylic acids would represent structural constituents of FAs. In addition, these results are also consistent with the high carboxyl carbon contents reported in the structure of FAs as measured by non-degradative ^{13}C-NMR (*35*) and other methods (*1*).

Characterization of Humic Acids. Similar compounds are also released from soil HA and Figure 3 shows the chromatograms of the compounds released after pyrolysis in the presence of TMAH for two selected HA (*17*). High proportions of long-chain fatty acid methyl esters, as well as phenolic derivatives and aromatic acid methyl esters, were released. The released compounds might represent structural components of the humic macromolecule, as also suggested by different authors

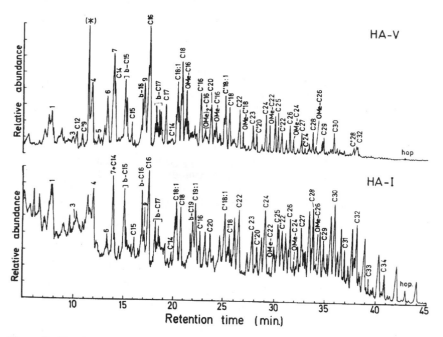

Figure 3.- Total Ion Chromatogram of the thermal degradative products obtained after pyrolysis of two HA (HA-V: *Typic Chemoxerert* and HA-I: *Histic Humaquept*) in the presence of TMAH. Key labels for aromatic compounds are:.(**1**) 4-methoxybenzenecarboxylic acid methyl ester, (**2**) 1,3,4-trimethoxybenzene, (**3**) benzenedicarboxylic acid dimethyl ester, (**4**) 3,4-dimethoxybenzenecarboxylic acid methyl ester, (**5**) methyl-3,4-dimethoxybenzenecarboxylic acid methyl ester, (**6**) 4-methoxybenzenepropenoic acid methyl ester, (**7**) 3,4,5-trimethoxybenzenecarboxylic acid methyl ester, (**8**) 2-tetradecylfuran, (**9**) 3,4-dimethoxybenzenepropenoic acid methyl ester. Key labels for aliphatic compounds are: (Cn) monocarboxylic acid methyl esters, (Cn:1) unsaturated monocarboxylic acid methyl esters, (b-Cn) branched monocarboxylic acid methyl esters, (C'n) dicarboxylic acid dimethyl esters, (C'n:1) unsaturated dicarboxylic acid dimethyl esters, (OMe-Cn) methoxymonocarboxylic acid methyl esters, (OMe2-Cn) dimethoxymonocarboxylic acid methyl esters, (OMe-C'n) methoxydicarboxylic acid dimethyl esters, (hop) hopanoids. Reproduced with permission from reference 17. Copyright 1995 Elsevier Science–Netherlands.

(*10,12,16,17*).The different series of aliphatic acids released after pyrolysis in the presence of TMAH consisted mainly of C_6-C_{34} mono- and dicarboxylic acids as well as their methoxylated counterparts. The different series of aliphatic acids may be chemically bound to the HA matrix in a form similar to that suggested by Schnitzer and Neyroud (*36*). The α,ω-alkanoic diacids may act as bridges in the macromolecule, their content being related to the cross-linking state of the macromolecular network. Homologous series of fatty acid methyl esters and a,w-alkanedioic acid methyl esters, and triterpenoid compounds with ursane, oleanane and hopane skeletons were also detected in the structure of different HAs upon pyrolysis in the presence of TMAH by different authors (*10,16,17,25*).

Different phenolic derivatives with lignin-related structures were also released. Among them, we note the presence of some benzenecarboxylic acids, some of them being lignin structures with *p*-hydroxycoumaryl, guaiacyl and syringyl skeletons, peaks (1) 4-methoxybenzenecarboxylic acid methyl ester, (4) 3,4-dimethoxybenzenecarboxylic acid methyl ester, (7) 3,4,5-trimethoxybenzenecarboxylic acid methyl ester respectively, as well as some benzenepropenoic acid methyl esters, peaks (6) 4-methoxybenzenepropenoic acid methyl ester and (9) 3,4-dimethoxybenzenepropenoic acid methyl ester. The presence of these compounds among the building blocks of the structure of HA has been previously undetected by conventional pyrolysis.

The aromatic acids released from different HA upon pyrolysis in the presence of TMAH probably represent original components of the HA structure released by the thermolytic action of TMAH (*10,12,16,17*). This observation is supported by the TMAH thermochemolysis data of Hatcher *et al.* (*23*) and Hatcher and Clifford (*16*) for a volcanic soil humic acid. In fact, the ^{13}C-NMR spectrum of this particular HA (shown in Figure 4) clearly indicates that it is composed of only aromatic and carboxyl carbons. Conventional pyrolysis of these HA produced trace quantities of volatile products without the release of any significant compounds while pyrolysis in the presence of TMAH yielded mainly benzenecarboxylic acid methyl esters (Figure 5), in accordance with the NMR data.

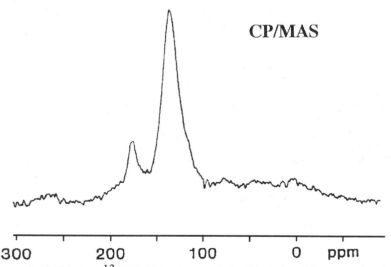

Figure 4.- Solid-state ^{13}C-NMR spectrum of the HA isolated from a volcanic soil Reproduced with permission from reference 16. Copyright 1994 Elsevier Science Ltd.—UK.

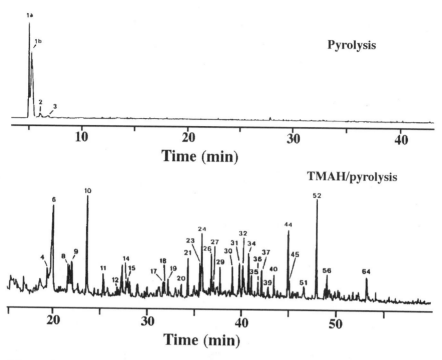

Figure 5.- Pyrolysis and TMAH/pyrolysis of the HA isolated from a volcanic soil. For peak identification refer to Table II. Reproduced with permission from reference 16. Copyright 1994 Elsevier Science Ltd.—UK.

All these studies seem to corroborate the presence of aromatic and aliphatic acids in the structure of HA, which become apparent in pyrolysis only after protection of carboxyl and phenolic groups. Taking into account all these data, and in agreement with other authors (*10,12,16*), it is apparent that the structural model for humic substances, based largely on only conventional pyrolysis studies (*37*), is inaccurate. In that paper (*37*) the "network" structure of HA is presented as mostly long-chain alkylaromatic compounds devoid of oxygen constituents, largely because no oxygen-containing compounds were reported among the pyrolysis products dominated by alkyl aromatic compounds. The elemental analysis, largely ignored, indicated the presence of 33-34% of oxygen in the HA investigated. The same authors have now revised their original structure, with the inclusion of oxygen in the form of carboxyls, phenolic and alcoholic hydroxyls, carboxyl esters and ethers (*38,39*) attached to their original "network". The model is now more consistent with the elemental data but is clearly inconsistent with the results derived from the pyrolysis data originally presented by these authors.

Pyrolysis in the presence of TMAH has also been applied to the structural characterization of HAs isolated from low-rank coals. Figure 6 shows the chromatogram of the compounds released after the TMAH/pyrolysis of the HA isolated from a humic coal from Konin (Poland). A large variety of components were released, the lignin-derived phenol derivatives and aliphatic acid methyl esters being the most prominent. A series of fatty acid methyl esters were identified in the range from C_{10} to C_{34}, with maxima at C_{16} and C_{18} showing an even-over-odd

Table II. Thermal degradative products released after the pyrolysis of the HA isolated from the volcanic ash. Reproduced with permission from reference 16. Copyright 1994 Elsevier Science Ltd.—UK.

Conventional pyrolysis

1a	chlorotrifluoroethane
1b	chloroethane
2	benzene
3	toluene

Pyrolysis/TMAH

4	phenol
6	2-butanedioc acid dimethyl ester
8	hexanoic acid methyl ester
9	butanedioc acid dimethyl ester
10	benzoic acid methyl ester
11	nonanoic acid methyl ester
12	methoxy methyl phenol
14	1-methyl-2,5-pyrrolidinedione
15-19	unidentified
20	C6H12)2 (t)
21	phosphoric acid dioctadecyl ester
23	2-propenoic acid-3-phenyl methyl ester
24	methoxybenzoic acid methyl ester
26	L-proline, 1-methyl-5-oxo, methyl ester (t)
27	dimethylundecanoic acid methyl ester (t)
29	1,3,5-triazine-2,4,6 (1H,3H,5H) trione,1,3,5-trimethyl (t)
30	ninhydrin
31	benzenedicarboxylic acid dimethyl ester
32	benzenedicarboxylic acid dimethyl ester
34	benzenedicarboxylic acid dimethyl ester
35	2,4 (1H,3H)-pyrimidinedione,1,3,5-trimethyl ester
36	methyl trimethoxy phenol (t)
37	tetradecanoic acid methyl ester
39	dimethoxybenzaldehyde
40	3,5-dimethoxybenzoic acid methyl ester
44	dimethoxybenzoic acid dimethyl ester
45	methyltetradecanoic acid methyl ester
51	unidentified
52	hexadecanoic acid methyl ester
56	trimethoxybenzoic acid methyl ester
64	unidentified

(t) tentatively identified compounds.

predominance. This series may be derived from ester bound long-chain carboxylic acids linked to the HA macromolecular network via ester bonds. A series of n-alkanes/n-alkenes, with chain lengths up to C_{33} and no even-over-odd predominance, was also released, indicating the presence of significant amounts of polyalkyl components in the structure of this HA. The fact that the series of n-alkanes/n-alkenes

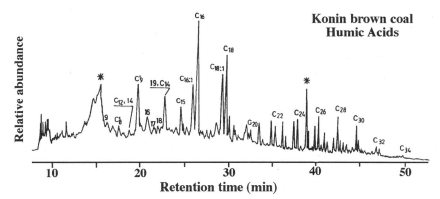

Figure 6.- Total Ion Chromatogram of the thermal degradative products obtained after pyrolysis of the HA isolated from the Konin (Poland) brown coal in the presence of TMAH. Key labels for aromatic compunds are: (9) 4-methoxybenzenecarboxylic acid methyl ester, (14) benzenedicarboxylic acid dimethyl ester, (16) 3,4-dimethoxybenzenecarboxylic acid methyl ester, (17) 3,4-dimethoxybenzeneacetic acid methyl ester, (18) 4-methoxycinnamic acid methyl ester, (19) 3,4,5-trimethoxy-1-ethylbenzene. Key labels for aliphatic compounds are: (Cn) monocarboxylic acid methyl esters, (Cn:1) unsaturated monocarboxylic acid methyl esters, (C'n) dicarboxylic acid dimethyl esters.

and long-chain carboxylic acid methyl esters are released together upon pyrolysis in the presence of TMAH seems to suggest that the long-chain carboxylic acids may arise from different aliphatic moieties than the *n*-alkanes/*n*-alkenes. This is also supported by the fact that there is no parallel between the chain length distribution of *n*-carboxylic acids and the distributions of hydrocarbons in the HA.

Again, a striking feature in this study of coal-derived HA is the release of lignin-derived phenol derivatives bearing carboxyl groups in the side-chain (benzenecarboxylic acid, benzeneacetic acid and benzenepropionic acid), as their methyl esters, as well as aliphatic acid methyl esters, which have not been observed previously by conventional pyrolysis. Del Rio *et al.* (*24*) also conducted TMAH/pyrolysis of HA isolated from a peat and a lignite. Figure 7 shows the chromatograms of the compounds released after TMAH/pyrolysis of the HA isolated from Padul peat and PGR lignite. An increasing amount of benzenecarboxylic acid methyl esters and benzylic ketones were observed in the TMAH thermochemolysis products of HA from coals of increasing rank. This fact is in agreement with the model proposed by Hatcher (*40*) for the transformation of lignin up to low-rank coal levels. According to this model, the content of carboxyl groups in the lignin macromolecule increases during the first stages of the coalification process, and reaches the maximum at the lignite stage. McKinney and Hatcher (*33*), in studies of TMAH/thermochemolysis of gymnosperm coalified woods of increasing rank, also observed that as the lignin biopolymer is altered and oxidized through time, carboxylic acid methyl ester functionalities increase in relative abundance. Oxidation of the C-3 side chain of the lignin structure would produce the benzenecarboxylic acids and the benzylic ketones identified.

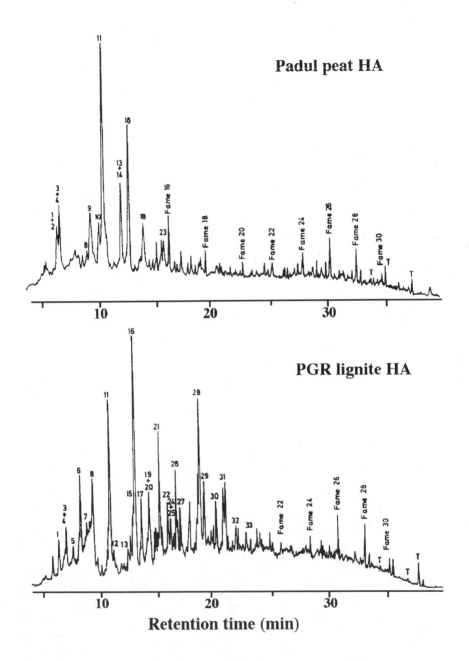

Figure 7.- Total Ion Chromatogram of the thermal degradative products obtained after pyrolysis of the HAs isolated from Padul peat and PGR lignite in the presence of TMAH. Peaks are identified in Table III. Reproduced with permission from reference 24. Copyright 1994 Elsevier Science Ltd.—UK.

Table III. Identification of the compounds released after pyrolysis-methylation of the Padul peat and PGR lignite HAs. Reproduced with permission from reference 24. Copyright 1994 Elsevier Science Ltd.—UK.

No.	Compound
1	3-methoxybenzenecarboxylic acid, methyl ester
2	4-methoxyacetophenone
3	2,5-furandicarboxylic acid, dimethyl ester
4	4-methoxybenzenecarboxylic acid, methyl ester
5	4-methoxybenzeneacetic acid, methyl ester
6	1,2-benzenedicarboxylic acid, dimethyl ester
7	1,3-benzenedicarboxylic acid, dimethyl ester
8	2,5-thiophenedicarboxylic acid, dimethyl ester
9	3,4-dimethoxybenzenemethanol, methyl ether
10	3,4-dimethoxyacetophenone
11	3,4-dimethoxybenzenecarboxylic acid, methyl ester
12	3,4-dimethoxybenzeneacetic acid, methyl ester
13	3-(4-methoxyphenyl)-2-propenoic acid, methyl ester
14	3,4,5-trimethoxyacetophenone
15	3-methoxybenzene-1,2-dicarboxylic acid, dimethyl ester (t)
16	3,4,5-trimethoxybenzenecarboxylic acid, methyl ester
17	4-methoxybenzene-1,3-dicarboxylic acid, dimethyl ester (t)
18	3,4,5-trimethoxybenzenemethanol, methyl ether
19	2,3,4-trimethoxybenzenecarboxylic acid, methyl ester
20	2-methoxybenzene-1,5-dicarboxylic acid, dimethyl ester (t)
21	1,2,4-benzenetricarboxylic acid, trimethyl ester (t)
22	3,4-dimethoxybenzene-1,2-dicarboxylic acid, dimethyl ester (t)
23	3-(3,4-dimethoxyphenyl)-2-propenoic acid, methyl ester
24	1,3,5-benzenetricarboxylic acid, trimethyl ester (t)
25	4,5-dimethoxybenzene-1,3-dicarboxylic acid, dimethyl ester (t)
26	4,5-dimethoxybenzene-1,2-dicarboxylic acid, dimethyl ester (t)
27	3-methoxybenzene-1,2,4-tricarboxylic acid, trimethyl ester (t)
28	2-methoxybenzene-1,3,5-tricarboxylic acid, trimethyl ester (t)
29	5-methoxybenzene-1,3,5-tricarboxylic acid, trimethyl ester (t)
30	1,2,3,4-benzenetetracarboxylic acid, tetramethyl ester (t)
31	1,2,4,5-benzenetetracarboxylic acid, tetramethyl ester (t)
32	5-methoxybenzene-1,2,3,4-tetracarboxylic acid, tetramethyl ester (t)
33	2-methoxybenzene-1,3,4,5-tetracarboxylic acid, tetramethyl ester (t)

(t) tentatively identified compounds.

It is likely that the benzenecarboxylic acids are derived from lignin units where the α-carbon of the side-chain has been oxidized to a carboxyl group (*32*), and different studies have interpreted the structures and distributions of the methyl esters of benzenecarboxylic acids as indicating that these structural units exist as either free or ester-bound structures in the HA (*10,16,17,24*). However, it has recently become clear that certain benzenecarboxylic acids are produced from unoxidized lignin moieties by the TMAH reagent in the course of the reaction at elevated temperatures (*41*). Examination of the products from the TMAH thermochemolysis of a model lignin dimer, free of any carboxylic functionality, showed the production of relatively large amounts of methylated benzenecarboxylic acid derivatives. These authors concluded that the release of benzenecarboxylic acids after TMAH/thermochemolysis of HA are only partially indicative of the presence of oxidized lignin units in the HA structure, and due caution is warranted in interpreting them as being derived exclusively from constituent structural entities.

TMAH/Thermochemolysis in Sealed Glass Tubes. It has recently been established that the TMAH procedure can also be conducted at sub-pyrolysis temperatures of 300°C (*16,21*). The same distribution of products are observed at pyrolysis and sub-pyrolysis temperatures, supporting the idea that the predominant effect of TMAH is a saponification/esterification reaction, as also pointed out by several other authors (*9,12,14,30*). Further investigations have prompted a study of the feasibility of carrying out the procedure at 300°C in a glass microreactor constructed of nothing more than a glass ampoule (*31*), without the need of specialized pyrolysis equipment. Therefore, the TMAH thermochemolysis procedure becomes a rapid, low-cost, and easily implemented technique for the analysis of HS. This procedure can be easily implemented in any laboratory having gas chromatographic capabilities. This is a potentially valuable advantage, because it makes the technique readily available to most geochemical laboratories.

The procedure of TMAH thermochemolysis in sealed glass ampoules has already been applied to different materials such as lignins, fresh and degraded wood samples and coalified woods (*31-33*). This procedure still induces cleavage of the β-O-4 bonds in lignin, releasing a distribution of products similar to that obtained from the CuO oxidation procedure, except for the fact that the CuO procedure mainly produces lignin monomers which are either the acid, aldehyde or methyl ketone derivatives. In the TMAH procedure, several phenol propanoids having one, two or three methoxyls at the propanoid side-chain are also released. Therefore, similar ratios as those calculated in the CuO oxidation procedure can be calculated here. In particular, this procedure appears to be more sensitive for calculating the acid/aldehyde (Ad/Al) ratios, displaying a larger dynamic range than that observed by CuO oxidation (*32*).

The TMAH thermochemolysis in sealed glass ampoules has been found to be an excellent approach for the characterization of other humic-like materials, such as dissolved organic matter (DOM). Several DOM samples, ranging from riverine systems to coastal oceans and oceanic samples have been recently studied using TMAH thermochemolysis procedure. Figure 8 shows the products released after the TMAH thermochemolysis of the Suwannee river DOM. Different lignin-derived phenol derivatives were identified among the released products, and these originate from the cleavage of the β-O-4 bonds from incorporated lignin-derived structures. The major products identified are the benzoic acid derivatives of the guaiacyl and syringyl structures, 3,4-dimethoxybenzenecarboxylic acid methyl ester (peak **37**) and 3,4,5-trimethoxybenzenecarboxylic acid methyl ester (peak **44**) respectively, with minor amounts of the cinnamyl derivative 4-methoxybenzenecarboxylic acid methyl ester (peak **17**). This means that this DOM is mainly composed of oxidized lignin moieties at the α-carbon of the side-chain. Benzeneacetic and benzenepropionic acid methyl esters were also found in the Suwannee river DOM. Benzoic acid derivatives were also identified as major compounds in the DOM from decomposing *Juncus effusus* upon pyrolysis in the presence of TMAH (*42*). A striking feature observed in the Suwannee river DOM is the identification, as major peaks, of 1,3,5-trimethoxybenzene (peak **21**) and 2,4,6-trimethoxytoluene (peak **25**), which are structures not related to lignin. Similar compounds, however, have been previously found as products from the TMAH thermochemolysis of cutan (*43*), the highly aliphatic and resistent biopolymer present in leaf cuticles from certain plants (*44*), which might indicate the contribution of this biopolymer to the structure of the Suwannee river DOM.

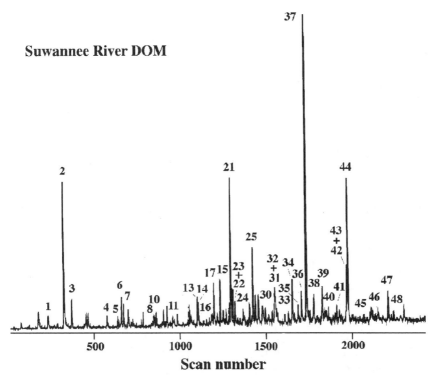

Figure 8.- Total Ion Chromatogram of the thermal degradative products obtained after TMAH/thermochemolysis in sealed glass tubes, at 250°C, of the dissolved organic matter (DOM) isolated from the Suwannee River. Peaks are identified in Table IV.

The Use of Tetrabutylammonium Hydroxide as an Alternative Alkylating Reagent

While the TMAH method is shown here to yield mono-, di-, and trimethoxyphenols derived from the three primary lignin units, it can also produce the same phenols from another component of vascular plant material, namely tannins. Tannins contain di- and trihydric phenols which can be expected to be methylated with TMAH. This is especially true with the hydrolyzable tannins which contain gallic acid esterified to sugars. The TMAH will clearly produce trimethoxy benzoic acid methyl ester, a major product also expected from TMAH/thermolysis of lignin from angiosperms. The non-hydrolyzable tannins contain dihydric phenols but these are carbon-bonded to the proanthocyanidin units and are unlikely to form products which can be similar to those from lignin.   Similarly, a very well-documented consequence of the coalification process, such as the demethylation of methoxyl groups of guaiacyl and syringyl structures of lignin and the formation of the corresponding catechols (*45,46*) can also be highly biased by the use of TMAH. This reagent will produce the methyl ethers of the free phenolic groups of both methoxyphenols and dihydric phenols; the latter cannot be differentiated from the original methoxy phenolic structures in lignin.

Table IV. Identities of the products relased after TMAH
thermochemolysis of the dissolved organic matter isolated from the
Suwanee river.

1	pentanoic acid methyl ester
2	butanedioic acid dimethyl ester
3	methylbutanedioic acid dimethyl ester
4	pentanedioic acid dimethyl ester
5	2-methylpentanediic acid dimethyl ester
6	1,2-dimethoxybenzene
7	1,4-dimethoxybenzene
8	3,4-dimethoxytoluene
9	3-methoxy-4-acetyl-2-5(H)-furanone
10	2,5-dimethoxytoluene
11	4-methoxybenzaldehyde
12	2-vinyl-5-methoxy-2,3-dihydrobenzofuran
13	1,2,3-trimethoxybenzene
14	3-methoxybenzoic acid, methyl ester
15	1,2,4-trimethoxybenzene
16	4-methoxyacetophenone
17	4-methoxybenzoic acid, methyl ester
18	benzeneacrylic acid, methyl ester (*cis* or *trans*)
19	3,4-dichlorobenzoic acid, methyl ester
20	1-(4-methoxyphenyl)-methoxypropane (isom.)
21	1,3,5-trimethoxybenzene
22	3,4,5-trimethoxytoluene
23	1-(4-methoxyphenyl)-methoxypropane (isom.)
24	4-methoxybenzeneacetic acid, methyl ester
25	2,4,6-trimethoxytoluene
26	3,4-dimethoxybenzenemethanol, methyl ether
27	Benzenedicarboxylic acid, dimethyl ester (3 isom.)
28	1-(3,4-dimethoxyphenyl)-1-methoxyethane
29	3-chloro-4-methoxybenzenecarboxylic acid, methyl ester
30	1-(3,4-dimethoxyphenyl)-1-propene
31	3,4-dimethoxybenzaldehyde
32	dimethoxybenzenemethanol, methyl ether
33	1-(3,4-dimethoxyphenyl)-2-methoxypropane
34	dimethoxybenzoic acid, methyl ester
35	dimethoxybenzenemethanol, methyl ether
36	3,4-dimethoxyacetophenone
37	3,4,dimethoxybenzoic acid, methyl ester
38	1-(4-methoxyphenyl)-1,2-dimethoxypropane
39	3,4-dimethoxybenzeneacetic acid, methyl ester
40	*trans*- 4-methoxyacrylic acid, methyl ester
41	3,4,5-trimethoxyacetophenone
42	2-methoxy-1,5-benzenedicarboxylic acid, methyl ester
43	3-(3,4-dimethoxyphenyl)propanoic acid, methyl ester
44	3,4,5-trimethoxybenzoic acid, methyl ester
45	1-(3,4-dimethoxyphenyl)-1,2-dimethoxypropane
46	*cis*- 1-(3,4-dimethoxyphenyl)-1,3-dimethoxy-1-propene
47	3,4,5-trimethoxybenzeneacetic acid, methyl ester
48	*trans*- 1-(3,4-dimethoxyphenyl)-1,3-dimethoxy-1-propene

Alternative reagents are used to assist in differentiating the originally free and methylated phenolic groups. Martin *et al.* (*20*) used tetrabutylammonium hydroxide (TBAH) for the structural characterization of lignins, and del Rio *et al.*, (*12*) used TBAH to study the formation of catechols during the coalification process. Pyrolysis in the presence of TBAH introduces a butyl moiety in the originally free hydroxyl group (forming an O-butyl ether) that can thus be distinguished from the original methoxyl groups in the structure of the lignin. Pyrolysis in the presence of TBAH only methylates phenolic groups, while the aliphatic hydroxyls remain unbutylated, probably due to their weakly acidic character. Pyrolysis in the presence of TBAH of the HA isolated from low-rank coals have shown the release of moieties with guaiacyl and syringyl structures having the OH groups at C-3 and C-5 either methylated or butylated. Larger amounts of compounds having the OH groups at C-3 and C-5 butylated were released from the lignite than from the peat, indicating that coalification induces the demethylation of the lignin structure to form catechols which become butylated with TBAH.

Main Advantages of the TMAH/Thermochemolysis Procedure in Comparison with Other Chemical Degradative and Pyrolysis Methods

The application of the TMAH/thermochemolysis procedure has several advantages over classical wet chemical and pyrolytic techniques. First, TMAH thermochemolysis produces more lignin-derived products with intact or partially altered side chains than other procedures such as the CuO oxidation procedure or conventional flash pyrolysis. This allows one to evaluate the degree of preservation of side-chain carbons in the original sample. Since these side-chain carbons are sensitive to environmental modification by biological processes, the TMAH procedure may eventually prove useful for assessing the rate of lignin modification during biological alteration. In comparison to conventional pyrolysis, this method also avoids decarboxylation of preexisting carboxylic moieties and releases aromatic and aliphatic acids as their methyl esters. Methylation also renders many of the polar products volatile enough for chromatographic analysis, which otherwise can be highly biased.

Second, the TMAH procedure methylates lignin phenols which have been microbially modified by demethylation reactions. Demethylation of lignin units by microbial enzymes to produce catechols is well known (*47*). Unfortunately, the CuO oxidation destroys catechols by ring opening, removing them from the analytical window. The TMAH/thermochemolysis, however, methylates catechols, preserving their analytical integrity. Partially demethylated lignin units in HS or other environmental samples will have their catechols preserved and analyzed as a lignin product. This means that the TMAH thermochemolysis may have a capability of detecting lignin-derived products with greater sensitivity and range, especially where microbial degradation is important.

Finally, and probably the most important advantage, the TMAH procedure is easily performed, carrying out the procedure at 250°C-300°C in a glass microreactor constructed of nothing more than a glass ampoule (*31*). This procedure has been applied to several samples such as lignins, fresh and degraded woods, coalified woods and dissolved organic matter (*31-33*), and the release of phenyl propane units having one, two or three methoxyl groups at the propanoid side-chain is an indication that the reactions conditions are sufficient for the hydrolysis of the β-O-4 bonds in the lignin. The presence of intense peaks for the vanillic acid and vanillin derivatives is an indication that this procedure will allow for calculation of acid/aldehyde ratios as is typically done for the CuO oxidation procedure. The ability to perform the TMAH thermochemolysis procedure in glass tubes allows for quantitative measurements by the addition of internal standards.

Conclusions

A new analytical procedure, termed TMAH thermochemolysis, has been used to assess the structural characterization of HS from a variety of samples. Although this procedure has mainly been performed in pyroprobe units at pyrolysis temperatures, it can also be conducted at subpyrolysis temperatures in sealed glass tubes. Therefore, this procedure can be easily implemented in any laboratory having gas chromatographic capabilities, in contrast to other chemolytic or pyrolytic procedures. This is a potentially valuable advantage, because it makes the technique readily available to most geochemical laboratories.

A significant feature of this method is that it may be able to trace lignin inputs where extensive degradation has occurred and resulted in sufficient alteration of lignin to render it undetected by the CuO or pyrolysis procedure. Microbial demethylation of lignin would not impact the TMAH thermochemolysis because the lignin ancestry is preserved in contrast to the CuO procedure which destroys demethylated lignin units. In essence, the TMAH procedure may be a more sensitive indicator of lignin-derived inputs than the more laborious CuO procedure or conventional pyrolysis.

In general, the data demonstrated that this technique provides relatively good preservation of the original carboxyl and hydroxyl groups in lignin phenols units present in the macromolecular structure of HS owing to protection of the functional groups from thermal reactions.

Literature Cited

(1)   Schnitzer, M.; Khan, S.U. *Humic Substances in the Environment*; Dekker: New York, 1972.
(2)   Wilson, M.A. *NMR Techniques and Applications in Geochemistry and Soil Chemistry*; Pergamon Press: Oxford, 1987.
(3)   Ertel, J.R.; Hedges, J.I. *Geochim. Cosmochim. Acta* **1985**, *49*, 2097.
(4)   Saiz-Jimenez, C.; de Leeuw, J.W. *J. Anal. Appl. Pyrol.* **1986**, *9*, 99.
(5)   Hedges, J.I.; Ertel, J.R. *Anal. Chem.* **1982**, *54*, 174.
(6)   Goñi, M.; Hedges, J.I. *Geochim. Cosmochim. Acta* **1990**, *54*, 3073.
(7)   Bracewell, J.M.; Haider, K.; Larter, S.R.; Schulten, H.-R. In *Humic Substances II. Search of Structure;* Hayes, M.H.B., MacCarthy, P., Malcolm, R.L., Swift, R.S., Eds.; Wiley: New York, 1989; pp 181-222.
(8)   Hempfling, R.; Schulten, H.-R. *Org. Geochem.* **1990**, *15*, 131.
(9)   Martin, F.; Gonzalez-Vila, F.J.; del Rio, J.C.; Verdejo, T. *J. Anal. Appl. Pyrol.* **1994**, *28*, 71.
(10)   Saiz-Jimenez, C. *Env. Sci. & Technol.* **1994**, *28*, 1773.
(11)   Saiz-Jimenez, C.; Ortega-Calvo, J.J.; Hermosin, B. *Naturwissenschaften* **1994**, *81*, 28.
(12)   del Rio, J.C.; Gonzalez-Vila, F.J.; Martin, F. *Trends Anal. Chem.* **1996**, *15*, 70.
(13)   Challinor, J.M. *J. Anal. Appl. Pyrol.* **1989**, *16*, 323.
(14)   de Leeuw, J.W.; Baas, W. *J. Anal. Appl. Pyrol.* **1993**, *26*, 175.
(15)   Saiz-Jimenez, C.; Hermosin, B.; Ortega-Calvo, J.J. *Water Res.* **1993**, *27*, 1693.
(16)   Hatcher, P.G.; Clifford, D.J. *Org. Geochem.* **1994**, *21*, 1081.
(17)   Martin, F.; del Rio, J.C.; Gonzalez-Vila, F.J.; Verdejo, T. *J. Anal. Appl. Pyrol.* **1995**, *31*, 75.
(18)   Challinor, J.M. *J. Anal. Appl. Pyrol.* **1991**, *18*, 233.
(19)   Challinor, J.M. *J. Anal. Appl. Pyrol.* **1991**, *20*, 15.
(20)   Martin, F.; del Rio, J.C.; Gonzalez-Vila, F.J.; Verdejo, T. *J. Anal. Appl. Pyrol.* **1995**, *35*, 1.

(21) Clifford, D.J.; Carson, D.M.; McKinney, D.E.; Bortiatynski ,J.M.; Hatcher, P.G. *Org. Geochem.* **1995**, *23*, 169.
(22) Saiz-Jimenez, C. *Environ. Sci Technol.* **1994**, *28*, 197.
(23) Hatcher, P.G.; Faulon, J.L.; Clifford, D.A.; Mathews, J.P. In *Humic Substances in the Global Environment and Implications on Human Health*; Senesi, N.; Miano, T.M., Eds.; Elsevier: 1994; pp 133-138.
(24) del Rio, J.C.; Gonzalez-Vila, F.J.; Martin, F.; Verdejo, T. *Org. Geochem.* **1994**, *22*, 885.
(25) Chiavari, G.; Torsi, G.; Fabbri, G.; Galletti, G.C. *Analyst* **1994**, *119*, 1141.
(26) Kralert, P.G.; Alexander, R.; Kagi, R.I. *Org. Geochem.* **1995**, *23*, 627.
(27) del Rio, J.C.; Gonzalez-Vila, F.J.; Martin, F.; Verdejo, T. *Org. Geochem.* **1996**, (in press).
(28) Anderson, K.B.; Winans, R.E. *Anal. Chem.* **1991**, *63*, 2901.
(29) Clifford, D.J.; Hatcher, P.G. In *Amber, Resinite and Fossil Resins*; Anderson, K.B.; Crelling, J.C., Eds.; ACS Symposium Series 617; American Chemical Society: Washington, D.C., 1995; pp 92-104.
(30) Challinor, J.M. *J. Anal. Appl. Pyrol.* **1995**, *29*, 223.
(31) McKinney, D.E.; Carson, D.M.; Clifford, D.J.; Minard, R.D.; Hatcher, P.G. *J. Anal. Appl. Pyrol.* **1995**, *34*, 41.
(32) Hatcher, P.G.; Nanny, M.A.; Minard, R.D.; Dible, S.C.; Carson, D.M. *Org. Geochem.* **1995**, *23*, 881.
(33) McKinney, D.E.; Hatcher, P.G. *Int. J. Coal Geol.* **1996**, (in press).
(34) Adler, E. *Wood Sci. Tecnol.* **1977**, *11*, 69.
(35) Hatcher, P.G.; Schnitzer, M.; Dennis, L.W.; Maciel, G.E. *Soil Sci. Soc. Am. J.* **1981**, *45*, 1089.
(36) Schnitzer, M.; Neyroud, J.A. *Fuel* **1975**, *54*, 17.
(37) Schulten, H.-R.; Plage, B.; Schnitzer, M. *Naturwissenschaften* **1991**, *78*, 311.
(38) Schulten, H.-R.; Schnitzer, M. *Naturwissenschaften* **1993**, *80*, 29.
(39) Schulten, H.-R.; Schnitzer, M. *Naturwissenschaften* **1995**, *82*, 487.
(40) Hatcher, P.G. *Org. Geochem.* **1990**, *16*, 959.
(41) Hatcher, P.G.; Minard, R.D. *Org. Geochem.* **1995**, *23*, 991.
(42) Wetzel, R.G.; Hatcher, P.G.; Bianchi, T.S. *Limnol. Oceanogr.* **1995**, *40*, 1369.
(43) McKinney, D.E.; Bortiatynsky, J.M.; Carson, D.M.; Clifford, D.J.; de Leeuw, J.W.; Hatcher, P.G. *Org. Geochem.* **1996**, (in press).
(44) Tegelaar, E.W.; de Leeuw, J.W.; Largeau, C.; Derenne, S.; Schulten, H.-R.; Muller, R.; Boon, J.J.; Nip, M.; Sprenkels, J.C.M. *J. Anal. Appl. Pyrol.* **1989**, *15*, 29.
(45) Stout, S.A.; Boon, J.J.; Spackman, W. *Geochim. Cosmochim. Acta.* **1988**, *52*, 405.
(46) Hatcher, P.G.; Lerch, H.E. III; Kotra, R.K.; Verheyen, T.V. *Fuel* **1988**, *67*, 1069.
(47) Crawford, R.L. *Lignin Bodegradation and Transformation;* Wiley: New York, 1981.

Chapter 7

Characterization of Aquatic Humic and Fulvic Materials by Cylindrical Internal Reflectance Infrared Spectroscopy

Nancy A. Marley, Jeffrey S. Gaffney, and Kent A. Orlandini

Environmental Research Division, Argonne National Laboratory, Building 203, 9700 Cass Avenue, Argonne, IL 60439

Cylindrical internal reflectance (CIR) techniques have been applied to humic and fulvic acids that were size fractionated by using hollow-fiber ultrafiltration methods with cutoffs of 0.1μm and 100,000, 30,000, 10,000, 3,000, and 500 molecular weight. The dissolved organic carbon and major cation contents were compared with the CIR spectra to estimate the active carboxylate units in each size fraction. Comparison of infrared spectra at various pH values for aquatic humics and for model polycarboxylate compounds (polymaleic acid and polyacrylic acid) indicated that the principal metal binding functionalities are carboxylate groups.

Humic and fulvic materials naturally present in ground and surface waters can act as strong complexing agents for metals and radionuclides and therefore can increase their migration and transport of these species in the geosphere (1-7). The distribution of bound metals and radionuclides in natural waters varies across the size distribution of humic materials. In most cases, the small organics (3,000 molecular weight) are the most active complexing agents (7-10). Evidence suggests that the carboxylate functional groups, which are mainly responsible for the aqueous solubility of these natural organics, are the most active complexing sites within the humic and fulvic acid molecules, with the smaller fulvic acids possessing the highest percentage of carboxylates (11-13).

Vibrational spectroscopy is the method of choice for the characterizing functional groups in complex organic molecules. Infrared transmission spectroscopy has been used on dried humics pressed into KBr pellets to determine the relative carboxylate content of humic materials (14-16). However, interferences arise from the presence of water bands and possible alterations of the samples under the high pressures used to form the pellets. Diffuse-reflectance techniques can avoid some of the difficulties associated with the KBr pressed-pellet method (9,17-18). To obtain a spectrum analogous to an absorption spectrum, the data are transformed from reflectance units to Kebulka-Munk (K-M) units. However, K-M units are related to

0097–6156/96/0651–0096$15.00/0

the molar absorption coefficient and the scattering coefficient, which vary with particle size and packing density of the sample. There is concern that the K-M transformation of data tends to amplify strong absorption bands (small relative reflectance) over weak ones (large relative reflectance) and may therefore bias any quantitative interpretation of results (*17*).

Raman spectroscopy has been used to characterize large organic molecules analogous to humic and fulvic acids in aqueous solution (*19*). However, humic materials absorb at the visible wavelengths used for laser Raman, and the resulting fluorescent background overwhelms any Raman signals. Fourier transform Raman, with near-infrared lasers, has been used successfully for other highly fluorescent macromolecules (*20*), but a recent attempt to characterize humic and fulvic acids by this technique yielded uncertain results (*21*).

Although the absorption of liquid water yields too strong a background signal to permit infrared spectroscopy of aqueous solutions by traditional techniques, the recent development of cylindrical internal reflectance (CIR) overcomes these limitations. This method permits the quantitative study of aqueous solutions by probing the interface between the water solution and an internal reflectance crystal, effectively providing a highly reproducible cell with an extremely short optical path length (*22*). The CIR techniques have been used to study the behavior of Aldrich humic acid in aqueous solution (*23*) and to quantitatively determine the carboxylate content of humic and fulvic materials obtained from surface waters (*7,24*).

For this study, humic and fulvic materials obtained from a small glaciated bog were separated into five size fractions by using hollow-fiber ultrafiltration techniques. The major cations associated with these organics are reported as a measure of the natural binding capacity of each size range. The structural characteristics and carboxylate content of each group were studied by CIR spectroscopy; the results are compared with those obtained by traditional Fourier transform infrared techniques. To aid in spectral interpretation, results were compared to those for selected model polyelectrolytes and simple acids.

Sampling and Isolation of Aquatic Humic and Fulvic Materials

Water samples were obtained from Volo Bog, located in an Illinois nature preserve northwest of Chicago. The small glaciated bog is surrounded by sedge peat and has no surface inlet or outlet. The water is of low nutrient content and has a pH of 4-5. Water samples were first passed through a 35 μm screen to remove large particulates and microorganisms, then prefiltered with a 0.45 μm Millipore filter to remove suspended solids. Humic materials were concentrated from 60 gallon water samples with hollow-fiber filter cartridges (Amicon Division, W.R. Grace and Co.) with effective size cutoff diameters of 0.1 μm and 100,000, 30,000, 10,000, and 3,000 nominal molecular weight. An additional flat-disk filter membrane in a vortex mixing stirred cell was used to separate the 500 nominal molecular weight species.

Experimental Methods

Dissolved organic carbon (DOC) measurements were obtained on the size-fraction ated concentrates with a PHOTOchem Organic Carbon Analyzer (Sybron, Model E3500) as an indication of the concentration of humic materials in each fraction. DOC values are reported as ppmC (mgL^{-1}) The major cations in each fraction were determined by inductively coupled plasma spectroscopy (Instruments SA, Model JY 86).

Detailed experimental procedures for obtaining infrared spectra on humic and fulvic acids have been reported previously (9,22,25-26) and will be briefly described here. Infrared spectra were taken on the size-fractionated samples by using a Fourier transform infrared spectrometer (Mattson, Polaris) with a cooled Hg/Cd/Te detector. Dried humic and fulvic materials were studied by diffuse reflectance infrared spectroscopy (Spectra Tech DRIFT accessory) and reported in K-M units, as well as by transmission absorbance in a KBr pellet. Infrared absorption spectra were obtained directly on the aqueous size-fractioned concentrates with CIR (Spectra Tech CIRCLE accessory). Raman spectra were taken by using an argon ion laser (Spectra-Physics Model 2025-05), a triple-grating monochromator (Spex Triplemate Model 1877), and a photodiode array detector system (Princeton Applied Research Model 1420). All Raman and infrared spectra were taken at 2 cm^{-1} resolution.

Characterization of Dissolved Organic Carbon and Trace Elements

The total concentration of dissolved organic carbon (DOC) found in the aqueous fractions from Volo Bog (0.45 μm) was 25 ppm. The DOC was distributed among the five size fractions as shown in Table I.

Table I   Dissolved Organic Carbon (DOC) and Major Elements in Size-fractionated Samples from Volo Bog.

Total (<0.4 μm) ppm		0.45μm-0.1 μm	0.1μm-100K	1-30K	30-3K	3-0.5K
			% of total			
DOC	25	7	16	6	44	27
Na	990	0	13	3	7	77
Ca,Mg,Ba	10650	3	5	19	21	52
Fe	340	9	44	9	38	0
Si	590	0	15	3	2	80

Most of the DOC was found in the two smallest fractions with 44% at 30,000-3,000 molecular weight and 27% at 3,000-500 molecular weight. The distribution was bimodal, like that for many surface waters. Most groundwaters have an even larger fraction of the DOC in the small size ranges, apparently because of the loss of material in the 0.45-0.1 μm and the 0.1-100K size fractions during the migration of surface waters into the aquifers. The small size fractions in the Volo Bog surface waters also had the largest fractions of major cations (Na, Ca, Mg, and Ba) associated with the DOC, as well as an appreciable fraction of the dissolved silica. The one exception was Fe, of which 44% was found between 0.1μm and 100,000 molecular weight, although 47% of the Fe was found in the three smallest size fractions. These results are in agreement with previous work showing the smaller size fractions to be the most active in binding metals and radionuclides (*7-10*).

Infrared Characterization of Unfractionated Organics

Figure 1 presents the infrared spectra of Volo Bog humic materials in the carboxylate region as traditionally studied for the total DOC fraction smaller than 0.45 μm for dried samples obtained by using the KBr pellet (spectrum A, Figure 1) and by the diffuse-reflectance (spectrum B, Figure 1) techniques. Because hollow-fiber ultrafiltration does not rely upon chemical separation, the sample is in its natural state and is not stripped of metals or other cations. Thus, the carboxyl groups are in the form of acid salts with all major counter ions present.

Both spectra A and B in Figure 1 show characteristically broad bands, centered at 1590 and 1415 cm^{-1}. The major contribution to these bands is generally accepted to arise from the asymmetrical and symmetrical stretching vibrations of the carboxylate anion (*9,12,14*). However, the bands have also been attributed to the C=C stretch of aromatic rings (*16,18*), the amide N-H stretch at 1610 cm^{-1} (*17*), the C-H deformation of CH_3 and CH_2 groups and the OH deformation of phenolic groups at 1400 cm^{-1} (*15*). Another broad band is centered at 1107 cm^{-1}. Previous assignments attributed this band to the C-OH stretching vibration of aliphatic alcohols (*12,17*), although the C-C stretch can also occur in this region.

Spectra A and B in Figure 1 can be compared to the aqueous phase spectrum taken on the same sample by CIR spectroscopy (spectrum C, Figure 1). The solution-phase spectrum has narrower bands in this region than were obtained with either of the solid-phase methods. Band widths at half-maximum are 72 and 89 cm^{-1} for the 1635 and 1390 cm^{-1} bands, respectively; widths of 190-215 and 160 cm^{-1} were obtained similarly from the KBr pellet and diffuse-reflectance bands. In addition, both major bands in spectrum C appear as doublets. The carboxylate asymmetric stretch at 1590 cm^{-1} in the dry samples appears at 1635 and 1565 cm^{-1} in the aqueous phase. The symmetric stretch is shifted from 1415 cm^{-1} in the dry samples to 1395 and 1370 cm^{-1} in the aqueous phase. In addition, two weaker bands which appear at 1103 and 937 cm^{-1}.

Figure 2 shows similar spectra taken of the sodium salt of polymaleic acid, a polyelectrolyte commonly used as a model compound for humic carboxylate behavior. The solid spectra taken by KBr pellet and diffuse-reflectance techniques are shown in spectra A and B (Figure 2), as before. The aqueous-phase spectra collected by CIR are shown in Figure 2 for pH values of 3 (spectra C) and 8 (spectra D). All carboxylate bands were split in the KBr pellet technique (spectrum A, Figure 2) confirming the presence of strong hydrogen bonding within the pressed pellet (*17*). This binding

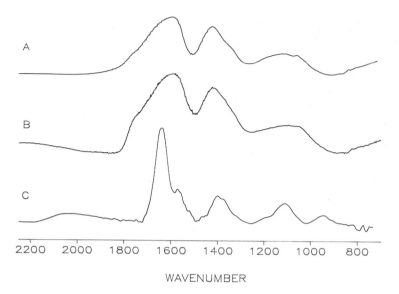

Figure 1. Infrared spectra of natural organic materials isolated from Volo Bog obtained on solid samples by A) the KBr pellet technique and B) the diffuse reflectance method and C) in aqueous solution by CIR.

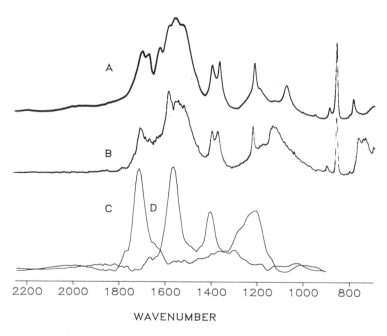

Figure 2. Infrared spectra of the sodium salt of polymaleic acid obtained on solid samples by A) the KBr pellet technique and B) the diffuse reflectance method . Aqueous solution spectra were obtained using CIR at C) pH=2 and D) pH=8.

could be responsible for the band broadening in solid phase humic acid spectra. Band splitting, although present, is not as severe in the diffuse-reflectance spectrum (spectrum B, Figure 2) because the sample was prepared without the high pressures applied using the pellet method. The symmetric carboxylate stretch near 1556 and 1543 cm^{-1} for the KBr pellet and the diffuse-reflectance spectra appears at 1560 cm^{-1} in the pH 8 aqueous-phase spectrum (spectrum D) and the symmetric carboxylate stretch at 1399/1364 and 1396/1371 cm^{-1} in the solid-phase spectra is at 1400 in the aqueous-spectrum at pH 8. The band at 1714 cm^{-1} in the aqueous-phase at pH 2 arises from the asymmetric C=O stretch of hydrogen bonded COOH, and that at 1210 cm^{-1} arises from the C-O stretch.

The Raman spectra of polymaleic acid are shown in Figure 3 for pH 8 (spectra A), pH 5 (spectra B), and pH 2 (spectra C). Changing the pH causes less change in these spectra than for the infrared spectra. The symmetric C=O stretch at 1615 cm^{-1} in acidic solution corresponds to the asymmetric stretch in the infrared spectrum at 1714 cm^{-1}. The existence of this pair of bands in the infrared and Raman is unambiguous evidence for dimerization of the carboxyl (27), indicating internal hydrogen bonding between the maleic acid units within the polymer. The symmetric COO- stretch appears at 1388 cm^{-1} and shifts to 1380 in the protonated acid form.

Infrared Characterization of Fractionated Organics

Infrared CIR spectra of size fractionated organic materials from Volo Bog are shown in Figure 4 (spectrum pairs A-D). The band assignments for these spectra are given in Table II. The top spectrum of each pair in Figure 4 was taken at pH 8 while the bottom spectrum was taken at pH 2.5. The C=O asymmetric stretch appears in acidic solution at about 1705 cm^{-1}, a lower frequency than expected for protonated carboxyl monomers, indicating the possibility of internal hydrogen bonding between carboxyls. Examination of Fourier transform Raman spectra for the companion band could confirm this possibility.

Table II. Infrared Absorption Bands for Size-factionated Aqueous Organics Isolated from Volo Bog.[a]

	0.45µm-0.1 µm	0.1µm-100K	100-30K	30-3K	3-0.5K
C=O asym str	1705	1708	1705	--	1701
COO-M asym str	1645	1625	1620	1635	1649
COO⁻ asym str	1563	1562	1565	--	1560
COO⁻ sym str	1390	1395	1395	1396	1400
COO-M sym str	1360sh	1340sh	1340sh	--	1340sh
C-O str (COOH)	1235	1235	1228	--	1236
C-C or C-OH str	1100	1100	1100	--	1103

[a] sym=symmetric, asym=asymmetric, str=stretch, sh=shoulder

The asymmetric carboxylate stretch appears at about 1560 cm^{-1} in basic solution (top spectra, Figure 4). The C-O stretch of COOH can be seen at about 1230

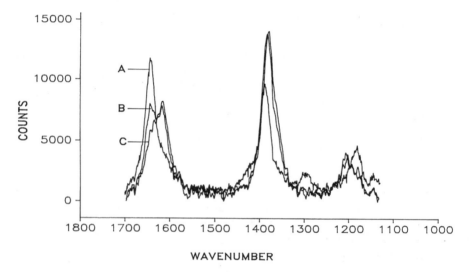

Figure 3. Raman spectra of polymaleic acid in aqueous solution at A) pH=8, B) pH=5 and C) pH=2.

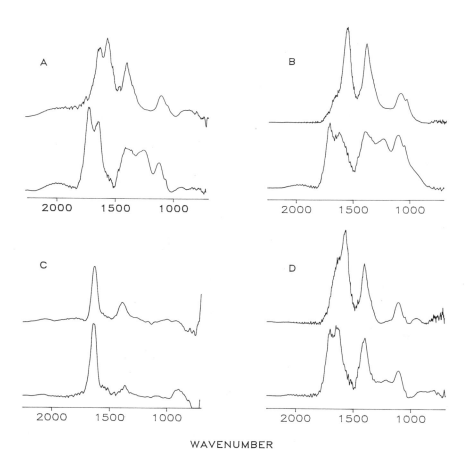

WAVENUMBER

Figure 4. CIR spectra of size-fractionated organics from Volo Bog at pH=8 (top spectra) and pH=2.5 (bottom spectra). Size ranges are A) 0.1μm to 100,000 nominal molecular weight, B) 100,000-30,000 nominal molecular weight, C) 30,000-3,000 nominal molecular weight, and D) 3,000-500 nominal molecular weight.

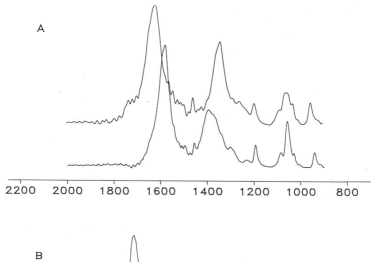

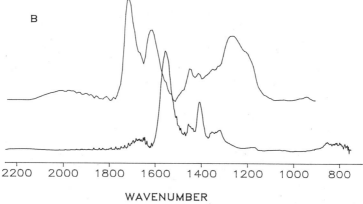

WAVENUMBER

Figure 5. CIR spectra of A) the sodium salt of mandelic acid with added iron (top) and without iron (bottom) and B) the sodium salt of polyacrylic acid with added zinc (top) and without zinc (bottom).

cm^{-1} in acidic solution (bottom spectra); it is most prominent in the larger size fractions. The bands near 1630 and 1390-1340 cm^{-1} are present at all pH levels and are the only bands present in the size range 3,000-30,000 molecular weight. (Figure 4, spectra pair C). We propose that these bands are due to the asymmetric and symmetric stretches of carboxylates complexed with metal cations. If the metal is more strongly associated with one oxygen of the COO$^-$ group than the other, that oxygen possess more single-bond character. This difference tends to shift the out-of-phase band higher and the in-phase band lower in frequency than for a sodium salt, where both bonds are equal (*27*). This interpretation is supported by the major cation analyis performed on the samples (see Table I), because the 3-30K size fraction contains 38% of the iron from the sample. The general trend in the strengths of these bands observed in the size-fractionated samples also correlates with the levels of total dissolved cations.

Figure 5 shows the CIR spectra of the sodium salt of mandelic acid (spectra pair A) and polyacrylic acid (spectra pair B). The bottom spectrum in each pair is the acid alone; the top spectrum is the acid with metal cation added. For both the simple acid and the polymer, the asymmetric COO$^-$ stretch is shifted to higher frequencies (38 and 63 cm^{-1}). In pair A (Figure 5), the symmetric stretch is shifted 56 cm^{-1} lower. In pair B (Figure 5) this band is obscured by the broad C-O stretch of the partially protonated acid. The asymmetric COO- band shifts for the humic samples range from 89 cm^{-1} for the smallest size range to 63 cm^{-1} for the range 0.1 µm to 100,000 molecular weight. Similar shifts for the symmetric stretch would place the bonded carboxylate at approximately 1340 cm^{-1}. In each spectrum, a shoulder appearing in this region does not disappear at acidic pH.

Conclusions

Cylindrical internal reflectance spectra were obtained on size-fractionated materials isolated from a small glaciated bog. Band assignments were made on the basis of their behavior under changing pH and comparison to model compounds. The bands appearing at about 1630 and 1390-1340 cm^{-1} were assigned to the asymmetric and symmetric stretches of carboxylates bound to metal cations. The increased intensities for these bands in the smaller size fractions were found to be strongly correlated with increased cation content, consistent with this interpretation of the infrared spectra.

The appearance of the C=O asymmetric stretch at very low frequency (1705 cm^{-1}) is an indication of internal hydrogen bonding between carboxylate units within the molecule. The appearance of a companion band near 1620 cm^{-1} in the Raman spectrum would confirm this association. Although humic materials cannot be studied by conventional Raman spectroscopy because of the high-fluorescence background, Fourier transform Raman spectroscopy using YAG-laser excitation in the near-infrared could be used to study internal hydrogen bonding and conformational changes in humic and fulvic acids under varying conditions.

Cylindrical internal reflectance infrared spectroscopy presents many advantages over conventional infrared techniques for the study of aquatic humic and fulvic materials. Samples can be studied in their natural state and in the aqueous environment from which they are isolated. Sample alterations due to drying and exposure to high pressures in the pellet forming process are avoided. In addition, smaller band widths and the ability to manipulate the sample by changing pH or cation concentrations, aid in band identification, allowing changes in ionization, metal bonding, and conformational effects to be studied under varying environmental conditions.

Acknowledgments

This work was performed at Argonne National Laboratory and was supported by the United States Department of Energy, Office of Energy Research, Office of Health and Environmental Research, under Contract W-31-109-ENG-38.

Literature Cited

1. Moulin, V.; Ouzounian, G. *Appl. Geochem.* **1992,** *Suppl. Issue 1,* 179-186.
2. Dearlove, J.P.L.; Longworth, G.; Ivanovich, M.; Kim, J.I.
   Delakowitz, B.; Zeh, P. *Radiochim. Acta* **1991,** *52/53,* 83-89.
3. Kim, J.I.; Zeh, P.; Delakowitz, B. *Radiochim. Acta,* **1992,** *58/59,* 147-154.
4. Choppin, G.R.; Clark, S.B. *Marine Chem.* **1991,** *36,* 27-38.
5. Choppin, G.R. *Radiochim. Acta* **1992,** 58/59, 113-120.
6. Rao, L.; Choppin, G.R.; Clark, S. *Radiochim. Acta* 1994, 66/67, 141-147.
7. Marley, N.A.; Gaffney, J.S.; Orlandini, K.A.; Cunningham, M.M. *Environ. Sci. Technol.* **1993,** *27,* 2456-2461.
8. Marley, N.M.; Gaffney, J.S.; Orlandini, K.A.; Dugue, C.P. *Hydrol. Proc.* **1991,** *5,* 291-299.
9. Marley, N.A.; Gaffney, J.S.; Orlandini, K.A.; Picel, K.A.; Choppin, G. *Sci. Total Environ.* **1992,** *113,* 159-177.
10. Gaffney, J.S.; Marley, N.A.; Orlandini, K.A. *Environ. Sci. Technol.* **1992,** *26,* 1248-1250.
11. Schnitzer, M. *Soil Sci* **1991,** *151,* 41-58.
12. Ricca, G.; Severini, F. *Geoderma* **1993,** *58,* 233-244.
13. Machado, A.S.C.; Joaquim, C.G. *Lab. Info. Manage.* **1992,** *17,* 249-258.
14. Byler, M.D.; Gerasimowicz, W.V.; Susi, H.; Snitzer, M. *Appl. Spectrosc.* **1987,** *41,* 1482-1430.
15. Yonebayashi, K.; Hattori, T. *Soil Sci. Plant Nutr* **1989,** *3,* 383-392.
16. Alberts, J.J.; Flip, Z.; Herkorn, N. *Cont. Hydrol.* **1992,** *11,* 317-330.
17. Baes, A.U.; Bloom, P.R. *Soil Sci. Soc. Am.* **1989,** *53,* 695-700.
18. Niemeyer, J.; Chen, Y.; Bollag, J.-M. *Soil Sci. Soc. Am.* **1992,** *56,* 135- 140.
19. Lewis, E.N.; Kalasinsky, V.F.; Levin, I.W. *Appl.Spectrosc.* **1988** *42,* 88-1193.
20. Levin, I.W.; Lewis, E.N. *Anal. Chem.* **1990,** *62,* 1101-1111.
21. Yang, Y.-H.; Li, B.-N.; Tao, Z.-Y. Spectrosc. Lett. **1994,** *27,* 649-660.
22. Marley, N.A.; Gaffney, J.S.; Cunningham, M.M. *Spectroscopy* **1992,** *7,* 44-53.
23. Morra, M.J.; Marshall, D.B.; Lee, C.M. Commun. *Soil Sci. Plant Anal.* **1989,** *20,* 851-867.
24. Cabaniss, S.E. *Anal. Chim. Acta* **1991,** *255,* 23-30.
25. Marley, N.A.; Ott, M.; Feary, B.L.; Benjamin, T.M.; Rogers, P.S.Z.; Gaffney, J.S. *Rev. Sci. Instrum.* **1989,** *59,* 2247-2253.

26. Marley, N.A.; Gaffney, J.S. *Appl. Spectrosc.* **1990**, *44*, 469-476.
27. Colthup, N.B.; Daly, L.H.; Wiberley, S.E. In *Introduction to Infrared and Raman Spectroscopy*; Third Edition; Academic Press, Inc.: San Diego, CA, **1990**; pp 317-318.

Chapter 8

Data Treatments for Relating Metal-Ion Binding to Fulvic Acid as Measured by Fluorescence Spectroscopy

Michael D. Hays[1], David K. Ryan[1], Stephen Pennell[2], and Lisa Ventry Milenkovic[3]

[1]Department of Chemistry and [2]Department of Mathematics, University of Massachusetts, Lowell, MA 01854
[3]Quality Automation Sciences, Inc., 750 Verona Lake Drive, Fort Lauderdale, FL 33326

Intensity changes in the natural fluorescence of fulvic acid (FA) caused by the binding of metal ions have been well documented. Various quantitative models have been developed relating the measured fluorescence signal to the amount of metal ion bound to fulvic acid. Stern-Volmer, linear, and nonlinear models developed for 1:1 binding between metal ions and fulvic acid ligand sites have been used to calculate concentrations of FA binding sites (C_L), and conditional stability constants (K). However, the ability of these models to describe metal complexation by the polydispersed fulvic acid system is somewhat limited. The presence of at least two fluorophores, and possibly a third, associated with metal ion binding in fulvic acid strongly suggests the need for multiple binding site models. Existing linear and nonlinear models will be reviewed for both fluorescence quenching and enhancement. A new modified 1:1 Stern - Volmer model will be introduced as well as two site and multiple site models. Application of the models to Cu^{2+} binding by fulvic acid and certain well defined model systems are discussed.

Microbial reactions occurring in soil, resulting in the breakdown of plant and animal tissue consisting of proteins, cellulose and lignin are the main production pathway of organic material or humic matter (*1*). Fulvic acid is a class of humic substances consisting of heterogeneous macromolecules found naturally in soil and aquatic environments. Once formed, FA molecules exhibit a variety of structural characteristics that influence their chemical reactivity. These include polyfunctional and polyelectrolytic qualities that are central to effecting interactions between the dissolved organic matter and metal ions (*2-5*). Environments surrounding metal ions in FA systems depend on concentration of involved constituents and multiple type ligand affinities for metal ion; this is termed metal ion speciation. Investigation of metal ion speciation in the aquatic environment is useful because free metal ions, such as Al^{3+} and

0097–6156/96/0651–0108$15.00/0
© 1996 American Chemical Society

Cu^{2+}, have shown toxic effects on a diverse assortment of aquatic biota (*6-8*). The very same metal ions, on the other hand, portray a reduction or complete eradication of toxic effects when complexed with natural organic matter. Fate and transport of metal ions in the environment are also governed by associations with fulvic acid material. Therefore, determination of stability constants between FA ligand sites and potentially hazardous metal ions should be considered fundamentally important.

Molecular fluorescence spectroscopy is a commonly employed analytical method that is sensitive to certain chemical properties of FA (*9-13*). Fulvic acid's molecular fluorescence is principally due to conjugated unsaturated segments and aromatic moieties present in the macromolecule (*14*). Several types of fluorescence spectra can be measured, including an excitation emission matrix or total luminescence spectrum, constant offset synchronous fluorescence, excitation spectra, and emission spectra, furnishing the researcher with useful data. The ability to resolve and select multiple fluorescent species makes these approaches extremely useful for studying FA relative to its chemical reactivity.

Fluorescence emission scans exhibit a notable decrease in fluorescence intensity when paramagnetic metal ions, such as Mn^{2+}, Co^{2+}, and Cu^{2+} become bound to ligand sites on FA; this phenomenon is referred to as fluorescence quenching (*15*). A second phenomenon of importance with respect to metal ion FA fluorescence investigations is the occurrence of fluorescence enhancement. Fulvic acid fluorescence emission scans have shown enhancement effects upon interaction with both Al^{3+} and Be^{2+} ions in solution (*16-20*). This chapter will primarily focus on data treatments describing fluorescence quenching of fixed wavelength measurements or emission scans due to Cu^{2+} ion binding to the FA ligand. However, new data treatment techniques specifically describing enhancement relationships will also be investigated. FA metal ion titration data as well as hypothetical curves representing both fluorescence quenching and enhancement phenomena will be presented.

Fluorescence Quenching of Soil Fulvic Acid Emission Scans

Typical fluorescence spectra of 15 mg/L FA at pH 6.00 and 0.1 molar ionic strength are shown in figure 1. These spectra were obtained at a excitation wavelength of 335 nm (λ_{ex}), and emission wavelengths were scanned from 300 nm to 600 nm. A broad featureless peak is commonly seen for FA emission spectra, positioned between approximately 390 nm to 550 nm. The maximum of the FA fluorescence emission peak is located at approximately 450 nm. In figure 1, the sharp intense peak at 335 nm is the Rayleigh scattering peak. The small shoulder on the FA fluorescence band, present at 375 nm, is the Raman water band peak. Figure 1 includes five separate FA emission spectra with each successive spectra exhibiting a decrease in overall fluorescence intensity or quenching due to incremental additions of Cu^{2+} to the FA solution. The Cu^{2+} binds to the FA material directly linking metal ion complexation to the observed decrease in analytical signal (*21*).

Nonlinear Fluorescence Quenching Model

The original quantitative data treatment relating FA fluorescence intensity changes to metal ion binding was developed by Ryan and Weber (*22*). The nonlinear model uses

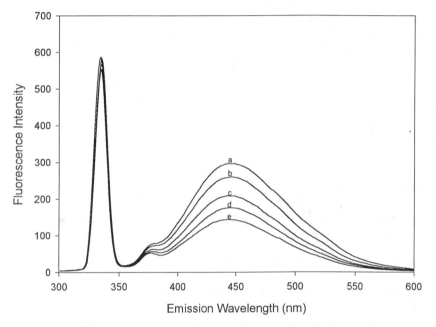

Figure 1. Fluorescence quenching of emission spectra due to Cu(II) ion titration of 15 ppm soil fulvic acid at 0.1 M ionic strength and pH 5. The FA sample was excited at 335 nm and emission wavelengths were scanned from 300 to 600 nm. Cu(II) concentrations of each emission spectra are: (a) 0 uM, (b) 4.9 uM, (c) 9.8 uM, (d) 14.7 uM, and (e) 22 uM.

data in the form of a spectrophotometric titration at fixed pH and relates the quenched fluorescence to a conditional stability constant (K) between the ligand and metal ion and the total concentration of FA ligand sites on the molecule (C_L). This elementary data treatment was validated using the amino acid L-tyrosine as a model compound. The model compound also exhibits fluorescence quenching when bound to Cu^{2+}, and therefore K and C_L could be calculated from Cu^{2+} titrations of L-tyrosine monitored by fluorescence. Data treatment results showed good reproducibility between runs and produced close agreement between the computed K and C_L compared to the model compound's theoretical stability constant and experimentally fixed concentration. Finally, application of the nonlinear data treatment model was attempted on FA fluorescence quenching curves. Results showed conditional stability constant and ligand site concentration values similar to results generated for other speciation methods, such as anodic stripping voltammetry, ion selective electrode, and equilibrium dialysis.

Development of the Nonlinear Fluorescence Quenching Model. The main function of the nonlinear model is to relate aquatic equilibrium considerations based on conventional solution thermodynamics to observed fluorescence intensity changes occurring as metal ion is added at fixed pH. The fundamental relationship equating fluorescence quenching to complexation is present in equation 1.

$$[ML]/C_L = (I_0-I)/(I_0-I_{RES}) \qquad (1)$$

In equation 1 [ML] is the concentration of metal-ligand complex and C_L is the total concentration of ligand present in solution. I_0 is the intensity of ligand at the initial point in the titration and is the intensity due to free, unquenched ligand material. Very often intensity data is placed on a relative scale and I_0 is equal to 100. I in equation 1 represents the overall intensity at each incremental titration point. I_{RES} is the residual fluorescence intensity due to metal ion-FA complex fluorescence and is the FA fluorescence signal left unquenched by metal ion at the titration endpoint.

Once fluorescence intensity changes are related to metal ion speciation in solution, a conditional formation constant and mass balance equations may be employed to define a final nonlinear relationship between K, C_L and fluorescence intensity signals.

$$M + L \leftrightharpoons ML \qquad (2)$$

$$K=[ML]/[M][L] \qquad (3)$$

$$C_M=[M]+[ML] \qquad (4)$$

$$C_L=[L]+[ML] \qquad (5)$$

Equation 2 is the 1:1 equilibrium reaction between a metal ion, M, and ligand, L, resulting in the formation of the complex, ML, with charges omitted for simplicity. Equation 3 is the expression for the conditional stability constant, K, the free metal ion concentration, [M], free ligand concentration, [L] (all forms of ligand not associated

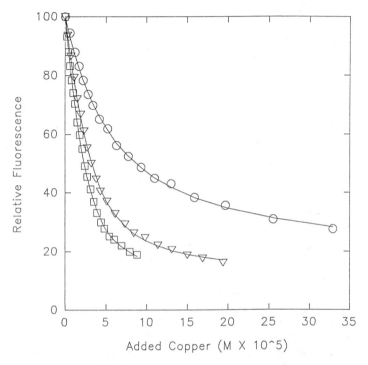

Figure 2. Fluorescence quenching of 15 ppm soil fulvic acid at 0.1 molar ionic strength titrated with Cu(II) ion (O) pH 5, (▽) pH 6, and (□) pH 7. The solid lines (—) illustrate the calculated intensity values from the nonlinear data treatment approach.

with metal), and [ML], the concentration of complex. The experimentally controlled total metal concentration C_M and the unknown total ligand concentration C_L, are equated to concentrations of free and complexed species in the mass balance equations 4 and 5. Equations 1 and 3-5 are algebraically manipulated and used in conjunction with the quadratic formula to derive the final nonlinear expression given in equation 6.

$$I = ((I_{RES}-I_O)/2KC_L)\{(KC_L+KC_M+1)-[(KC_L+KC_M+1)^2-4K^2C_LC_M]^{0.5}\}+I_O \quad (6)$$

The final nonlinear equation, containing K, C_L, and I_{RES} as parameters to be determined, requires data input from measured FA fluorescence intensities (I and I_O), and total metal ion additions at each titration interval (C_M). Equation 6 may be solved by nonlinear regression. FA fluorescence quenching curves for titrations with Cu^{2+} at pH 5, 6, and 7 are shown in figure 2. The lines through the data points represent the best fit curve of equation 6 to each set of data.

The Modified Stern Volmer Equation

A second nonlinear fluorescence quenching data treatment method developed by Ventry and Ryan may also be used to extract conditional stability constants, and ligand concentrations from titrations of FA with Cu^{2+} (*23*). The model designed is a modification of the original Stern-Volmer theory defined by equation 7 which accounts for either static or dynamic quenching of fluorescent species.

$$I_O/I = 1 + K[Q] \quad (7)$$

The variable [Q] is the concentration of free quencher existing in the system. The K value, as defined previously, is the conditional stability constant when the observed quenching is static (i.e. not collisional). I_O is the intensity of the unquenched sample when Q=0 and I represents fluorescence intensities at various values of Q.

Derivation of the Modified Stern-Volmer Model for Metal Complexation. The transformation from the original Stern-Volmer equation to the modified Stern-Volmer equation is based on allowing for a quantum efficiency change of free ligand fluorescence upon interaction with complexed forms (e) and the assumption of a significant quantum efficiency for the complex (d) (*24*). The Stern-Volmer relationship may be altered to include e and d and [M] substituted for [Q] for the chemical system being described .

$$(I_O-I)/I = [(1-e)+K[M](1-d)]/[e+K[M]d] \quad (8)$$

A simplified version of equation 8 was created by recognizing that [M]=0 at the titration beginning, and $I=I_O$. Consequently, e is equivalent to one, and the reduced version of equation 8 is:

$$(I_O-I)/I =[K[M](1-d)]/[1+K[M]d] \quad (9)$$

After determining the simplified equation 9, Ventry (23) postulates that $I_{RES}=I_Od$ where d is related to the residual and initial fluorescence intensity. Manipulation of mass balance and stepwise formation constant relationships, and application of a similar derivation procedure used in the nonlinear model, yields equation 10. Equation 10, like the nonlinear model equation, relates observed changes in FA fluorescence intensity I, to total metal, C_M, with a conditional stability constant (for the metal ion and FA) and the degree of complexation of the FA. The modified Stern-Volmer equation is:

$$I=[200+2KI_{RES}C_M-I_{RES}[(KC_L+KC_M+1)-((KC_L+KC_M+1)^2-4K^2C_MC_L)^{0.5}]]$$
$$/[2+2KC_M-[(KC_L+KC_M+1)-((KC_L+KC_M+1)^2-4K^2C_LC_M)^{0.5}]] \qquad (10)$$

Equation 10 may be solved for K, C_L, and I_{RES} by way of nonlinear regression analysis. Careful examination reveals that the modified Stern-Volmer equation is mathematically identical to the original nonlinear model developed by Ryan and Weber (22). Fluorescence quenching curves for Cu^{2+}-FA and application of the modified Stern-Volmer data treatment to the experimental information are shown in figure 2. Since the nonlinear data treatment and the modified Stern-Volmer equations are algebraically identical, their ability to fit experimental data and provide meaningful parameters is the same.

Linear Model Based on 1:1 Complex Formation

A linear data treatment model using an iterative calculation procedure allowing the computation of conditional stability constants between Cu^{2+} and FA, and the concentration of ligand sites has also been developed (23, 25). In the linear model, fluorescence intensity and total added metal ion concentration data are manipulated such that a linear relationship of the form y=mx+b is derived. Overall, this simplified data handling strategy is underutilized: most researchers examining metal ion complexation with FA have chosen to employ nonlinear approaches when estimating conditional stability constants and total ligand concentrations (16, 26-28). However, when studying results generated from both nonlinear and linear data treatment methods, deviations between the parameters generated by the two procedures are minimal. The linear treatment not only gives essentially the same results as nonlinear methods, but has also been used to generate conditionally stability constants and ligand concentration values for the model compound salicylic acid. The close agreement of these results to literature data validates the ability of the linear data treatment to calculate important equilibrium parameters (29).

A second linear model that will not be covered in detail was introduced by Frimmel and Hopp. This model allows for the determination of stability constants and specific ligand concentrations by simultaneously adding both metal and FA to solution, but keeping intensity ratios (I/I_O) equal. Stability constants for complexes formed relative to increasing ligand and metal concentrations are produced and compared with literature values (29).

Mathematical Derivation of the Linear Data Treatment Model. Formation of the linear model is based on the Scatchard model which states:

$$v/[L] = -Kv + K \qquad (11)$$

In equation 11, $v=[ML]/C_L$, and all other symbols used are the same as defined previously for nonlinear and modified Stern-Volmer models. Theoretically, a plot of v versus $v/[L]$ should yield a straight line with K as the y intercept and -K as the slope. Algebraic manipulation of C_L and C_M mass balance relationships (equations 4 and 5) and substitution into equation 11, yields equation 12.

$$v/(([L]/C_L)C_M) = -KC_L(v/C_M) + K \qquad (12)$$

The $[L]/C_L$ ratio in equation 12 is equivalent to (1-v) based on the mass balance equation for the ligand and the definition of v.

Linear regression analysis on the straight line relationship described by $v/(1-v)C_M$ versus v/C_M should provide a straight line with a slope of $-KC_L$ and a y intercept of K. The I_{RES} value, representing residual intensity present at the end of the titration, found within the v term, is continually altered by an iterative approximation procedure. The approximation method uses the correlation coefficient (r^2) of the line to determine the optimum value of I_{RES}. The maximum r^2 value constitutes the best fit for a particular set of data with the best I_{RES}. A typical graph exhibiting manipulated data and the best fit straight line through the results is shown in figure 3.

Fluorescence Enhancement Data Treatment Model

Recently, Al^{3+} and Be^{2+} interactions with FA have been examined in detail by using fluorescence titration techniques (*16-20*). Fluorescence enhancement effects were observed for the reaction of FA with Al^{3+}, but attempts to quantify Al^{3+} associations with the FA ligand were limited at best. Upon close investigation of enhancement phenomena it is apparent that Ryan and Weber's nonlinear model (*22*) may also be applied to enhancement data. The main difference between fluorescence enhancement and quenching during a spectrophotometric titration is photophysical. For enhancement the quantum yield of the metal-FA complex is greater than the quantum efficiency of free ligand the opposite effect is true for quenching. In the original nonlinear derivation, quantum yield values for the complex and free ligand are mathematically eliminated. As a result, quantum yields are rendered meaningless when treating fluorescence titration data. Consequently, the final enhancement equation is shown as equation 6. Hypothetical fluorescence enhancement curves produced by using equation 6 are shown in figure 4. Each curve represents a different K value, while I_{RES} and C_L remain constant.

Multisite Fluorescence Quenching

Fluorescence evidence for multiple functional groups on fulvic acid molecules has been well documented (*14, 30*). Only recently has fluorescence data linked the associations

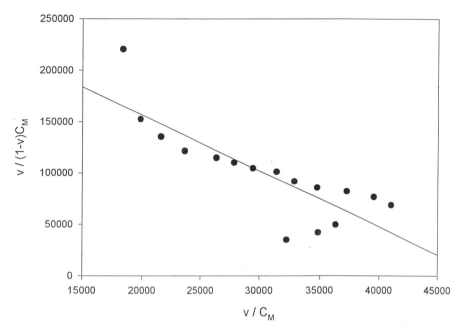

Figure 3. Linear data transformation of 15 ppm FA at 0.1 molar ionic strength and pH 5 titrated with Cu(II) ion at pH 6. (—) - Linear data treatment result.

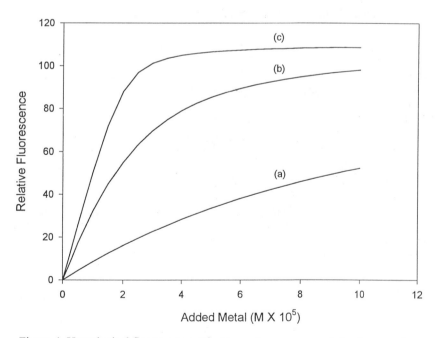

Figure 4. Hypothetical fluorescence enhancement curves generated using equation 6 with C_L=20 uM, I_{RES}=110 and K values of (a) 10^4, (b) 10^5, and (c) 10^6.

between metal ions and FA to the presence of several types of binding sites, each distinctly affecting fluorescence (*16, 27, 28*). Now multisite metal ion binding by FA may be observed by utilizing constant offset synchronous fluorescence spectroscopy. This scanning method shows variable quenching rates at discrete spectral locations for FA. This observed effect indicates separate binding sites associated with the measured fluorescence signal. Fluorescence lifetime measurements have also exhibited the possibility of differentiating between separate metal ion binding sites at constant emission and excitation wavelengths (*31*). Lifetime studies show clearly that at fixed wavelengths, an FA fluorescence emission spectrum contains three distinguishable fluorophores that may be associated with different type ligand sites for metal ion binding. Thus, multiple independent sites may simultaneously modify fluorescence intensity upon addition of quencher.

Modeling of Multisite Fluorescence Quenching. A traditional multisite fluorescence quenching model using a Stern-Volmer approach has been developed and applied to quenching curves involving residual protein fluorescence (*32*). More recently, however, the multiple site model has been used to describe structural characteristics in diverse polymer environments (*33-35*). The multisite Stern-Volmer model shown in equation 13 may be used to define multiple fluorescent binding sites present under one emission peak

$$I_L/I = 1/\{ f_1/(1+K_1[M]) + f_2/(1+K_2[M]) + ... + f_n/(1+K_n[M])\} \qquad (13)$$

For the purposes of this discussion only two fluorescent sites will be used. In equation 13, f_1 and f_2 are the fraction of total emission from each fluorescent site under unquenched conditions. K_1 and K_2 are conditional stability constants for each fluorescent ligand site available to bind metal, and may be represented as:

$$K_1 = [ML1]/[M][L1] \qquad (14)$$

$$K_2 = [ML2]/[M][L2] \qquad (15)$$

Equations 14 and 15 equate [ML1] and [ML2] to concentrations of metal bound at FA site one and site two respectively, and as before, [M] is equal to the free metal concentration in solution during the metal ion titration. Values of [L1] and [L2] are the unbound ligand concentrations or the concentrations of each fluorescent site capable of binding metal.

In order to fully define the titration system it is essential to account for the total metal ion added to solution, C_M, and the total ligand concentrations, C_{L1} and C_{L2}, available to complex metal. This is done via mass balance equations shown in equations 16 thru 18.

$$C_M = [M] + [ML1] + [ML2] \qquad (16)$$

$$C_{L1} = [ML1] + [L1] \qquad (17)$$

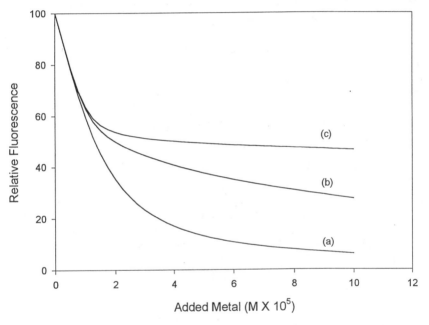

Figure 5. Hypothetical multisite fluorescence quenching curves generated using equations 13 and 22 with $C_{L1}=C_{L2}=10$ uM, $f_1=f_2=0.5$, and (a) $K_1=10^6$, $K_2=10^5$ (b) $K_1=10^6$, $K_2=10^4$ and (c) $K_1=10^6$, $K_2=10^3$.

$$C_{L2} = [ML2] + [L2] \qquad (18)$$

Equation 16 relates the total concentration of metal added to solution to the concentration of both complexes and to the free metal ion concentration. This equation may be redefined to describe the actual number of binding sites involved in quenching for a particular system (i.e. three or more sites). Equations 17 and 18 are mass balance equations characterizing the total ligand concentration for each fluorescent site.

After defining aquatic equilibrium equations for the two component system, the total concentration of metal ion added to solution and the fluorescence intensity modifications must be related to the conditional stability constants (K_1 and K_2) and the concentrations of metal ion binding FA sites (C_{L1} and C_{L2}). Using equations 14 and 17 it is possible to relate the concentration of complex at site one to its conditional stability constant, site concentration and to the free metal ion existing in solution. This relationship is shown in equation 19.

$$[ML1] = K_1 C_{L1}[M]/K_1[M] + 1 \qquad (19)$$

A similar statement may be derived for the second site in an identical manner and is given in equation 20. In order to relate the unknown parameters to the total metal concentration

$$[ML2] = K_2 C_{L2}[M]/K_2[M] + 1 \qquad (20)$$

equation 21 is developed by incorporating equations 19 and 20 into equation 16. Finally equation 21 is multiplied through by $(K_1[M] + 1)(K_2[M] + 1)$ and a third degree

$$C_M = [M] + (K_1 C_{L1}[M]/K_1[M] + 1) + (K_2 C_{L2}[M]/K_2[M] + 1) + ...$$
$$+ K_n C_{Ln}[M]/K_n[M] + 1 \qquad (21)$$

polynomial is formed as shown in equation 22.

$$K_1 K_2[M]^3 + \{K_1 K_2(C_{L1}+C_{L2}-C_M) + K_1 + K_2\}[M]^2 +$$
$$\{C_{L1}K_1+K_2 C_{L2}-C_M(K_1 + K_2+1)\}[M] - C_M = 0 \qquad (22)$$

For given values of K_1, K_2, C_{L1}, C_{L2} and C_M equation 22 may be solved numerically for [M]. It can be shown that only one positive solution for equation 22 exists, assuming that K_1, K_2, C_{L1}, C_{L2} and C_M are all positive. While numerically solving for [M] in equation 22, K_1, K_2, C_{L1}, C_{L2} and C_M are treated as unknown parameters. The solution is reached by employing nonlinear least squares regression. Ultimately, experimental fluorescence intensity data is compared with predicted intensity values generated from equation 13 to determine the best K_1, K_2, C_{L1}, C_{L2} values. Hypothetical fluorescence quenching curves for the two site system are shown in figure 5. Application of the multisite data treatment to a model chemical system consisting of a mixture of two fluorescent amino acids (L-tryptophan and glycyltryptophan) results in the verification of the multisite model strategy. Direct treatment of the FA experimental fluorescence data also indicates the presence of two metal binding fluorophores present under the one emission peak.

Multisite Fluorescent Enhancement Modeling

Fluorescence enhancement due to metal binding by a variety of functional groups present on multiple FA fluorophores is also very likely to occur. Like fluorescence quenching, contributions from diverse fluorophores to observed enhancement phenomena may be modeled mathematically. The creation of a two component fluorescence enhancement model has been attempted by isolating FA from a forest soil, fractionating the material by ultrafiltration methods, and titrating the ultrafiltered portions (16). After addition of Al^{3+} to solution, fluorescence enhancement was measured by synchronous scan spectroscopy. Work done by Lakshman and coworkers (16) represents the first known attempt at quantifying a two component fluorescent enhancement system, but, a close examination of their derivation seems necessary. In the proposed two site enhancement model the formation of two separate classes of metal-FA complex is postulated. If this hypothesis is valid and each class site consumes metal ion simultaneously during a titration by drawing from the same pool of metal, the total metal concentration, C_M, is no longer given by equation 4, but is reasonably given by equation 16. The derivation proposed by Ryan and Weber (22) and used by researchers in reference 16 is not likely valid in this instance.

A new approach to modeling multisite fluorescent enhancement data similar to the multisite quenching data treatment is proposed here. Incorporating fluorescence intensity changes and multicomponent mass balance equations describing the multiple site fluorescent ligand system is necessary. The new model uses the expression in equation 23 to describe

$$I = I_1 + I_2 \tag{23}$$

the maximum intensity signal (I) due to two fluorescent ligand sites (I_1 and I_2) present under one peak at any point in the titration. Intensity enhancement due to metal complexation with ligand for a one site model is proportional to the product from the fraction of total ligand bound and residual intensity when I_0 is zero as is shown in equation 24. In equation 24 I_{RES} is the intensity at the end of the titration where no

$$I = ([ML]/ C_L)I_{RES} \tag{24}$$

more complex may form. For two fluorescent sites the intensity due to each site is $I_1 = ([ML1]/C_L)I_{RES}$ and $I_2 = ([ML2]/C_L)I_{RES}$ where C_L is the sum of C_{L1} and C_{L2}. The other symbols present were defined previously for the two site fluorescence quenching model. By using definitions of I_1 and I_2 and setting $\alpha_1 = [ML1]/C_L$ and $\alpha_2 = [ML2]/C_L$

we can state overall intensity found in equation 24 as equation 25. Equation 25 relates total intensity under one emission peak to fraction of metal complexation by two fluorescent

$$I = (\alpha_1 + \alpha_2)I_{RES} \tag{25}$$

sites and to the residual intensity. Equation 25 may be further altered by substituting in equations 19 and 20 and by assuming that $C_L = C_{L1} + C_{L2}$. By using this approach both fluorescent binding sites are accounted for by using mass balance relationships. This preliminary multisite enhancement model remains untested, but seems to be a reasonable approach in attempting to solve this somewhat more complex problem.

Application of Fluorescence Models on Two Site Systems. Recently, both one site and two site binding models have been applied to FA fluorescence spectra (*16, 27, 28*). Machado and coworkers observed multiwavelength variations in synchronous fluorescence spectra of FA, and by using a self modeling mixture analysis technique (SIMPLISMA) detected two components upon varying Cu^{2+} ion concentration in solution. Concentration profiles of the two ligand sites were obtained at variable pH. Ryan and Weber's (*22*) nonlinear model was applied to the fluorescence variation for each site versus the total added copper concentration in a similar manner to the work of Lakshman and coworkers (*16*). When examining Machado and coworkers' (*27,28*) results, clearly the two observed FA metal ion binding sites were simultaneously quenched over similar concentration ranges of added copper ion. The simultaneous quenching of the two sites during one titration indicates copper ion consumption by both sites at the same total added metal ion concentration. The application of a one site titration model for each of the two sites would be inappropriate because each one site model does not account for the metal ion taken up at the parallel site. In other words, the fundamental assumption of a one site mass balance, even when used twice, is not consistent with a two site system.

Simultaneous, variable rate quenching occurring at discrete spectral locations clearly shows that the total concentration of metal ion added to solution during the titration procedure is equivalent to the sum of free metal and of metal tied up in complexes. If each separate spectrum is treated with the one component model while actually two components exist, equation 4 from the one site nonlinear model will exclude the concentration of a second FA binding site. Consequently, the total metal concentration term located in the one binding component model will be incorrectly balanced. The elimination of the second binding site concentration term increases the free metal ion term in equation 4, thus leading to underestimation of the conditional stability constant at the site being studied. A model that accounts for changes in intensity at both affected sites and correctly applies aquatic equilibrium principles is necessary for treating this type of data. We have developed such a model and validated it with known chemical systems as well as applied it to FA. These results will be published elsewhere.

Problems Encountered when Applying Fluorescence Models

At pH values of 5 and 6 the total copper concentration added to solution exists almost entirely in its free form (Cu^{2+}). However, at pH 7 the total copper concentration no longer represents only free Cu^{2+}. Various hydroxy species such as $CuOH^+$, $Cu(OH)_2$, and $Cu_2(OH)_2^{2+}$ are found in solution. Interactions between these various hydroxy species and FA molecules are possible, but highly unlikely. Nevertheless, the described fluorescence modelling employs a total added metal concentration term which assumes all metal added is in its free form. This assumption is fundamentally incorrect for copper solutions near pH 7 and above, and depending on which metal is used and at what equilibrium pH the assumption could adversely effect data treatment results. Reactions involving hydrolysis may be accounted for using a side reaction coefficient (*36*). The side reaction coefficient can estimate the extent of hydrolysis reactions occurring in solution. By using overall formation constants (β_1, β_2,....β_n), and equilibrium pH it is possible to determine the side reaction coefficient. The side reaction coefficient value remains constant throughout the titration, and may be incorporated into all previously introduced fluorescence data treatments.

Precipitation Effects and Fluorescence Modelling. Metal ions may cause precipitation of fulvic acid under certain equilibrium conditions (*21, 22*). Ryan and Weber (*22*) developed a Rayleigh scattering method for detecting precipitation in solutions containing FA and metal ions which is based on standard nephelometry. Observed increases in Rayleigh scattering intensity indicate increasing fulvate precipitate formation. Increases in solid FA forms in suspension could cause dramatic problems when treating fluorescence data since adsorption reactions occurring on surfaces of precipitates are not accounted for by fluorescence measurements. Ryan and Weber (*22*) eliminated fluorescence quenching data whose Rayleigh scattering intensity exceeded twice the original Rayleigh scattering intensity at the titrations beginning. By following this procedure these researchers eliminated potential problems and errors associated with the effect of precipitation on the solution phase equilibria being modelled.

Summary

The mathematical modeling of FA metal ion interactions using fluorescence spectra has been explored in detail. The ability of nonlinear and linear one component models in determining conditional stability constant and ligand concentrations has been reviewed. The one site fluorescence enhancement data treatment was shown to be identical to the original nonlinear quenching data treatment. Consequently, application of the nonlinear one site model is appropriate for treating quenching and enhancement of FA fluorescence due to metal ion interactions. Preliminary transformation of one component fluorescence models into multiple component fluorescence models has been shown. Theory linking both types of models is based on conventional thermodynamic principles. Application of the two site models shows preliminary evidence of success in detecting multiple conditional stability constants and multiple ligand concentrations. Discussion describing precipitation and hydrolysis reaction problems during FA titrations was presented. Further development of model application to particular fluorescent systems is necessary and should be the subject of future research efforts.

Literature Cited

(1) Haider, K. *Humic Substances in the Global Environment and Implications on Human Health*; Elsevier Science: **1994**; pp 91-107.

(2) Tipping, E. *Environ. Sci. Technol.* **1993**, *27*, 520-529.

(3) Westall, J. C. Jones, J. D.; Turner, G. D.; Zachara, J. M. *Envir. Sci. Technol.* **1995**, *29*, 951-959.

(4) Green, S. A.; Morel, F. M. M.; Blough, N. V. *Envir. Sci. Technol.* **1992**, *26*, 294-301.

(5) Ephraim, J. H.; Allard, B. *Environ. Int.* **1994**, *20*, 89-95.

(6) Karlsson-Norrgren, L.; Bjorklund, I.; Ljungberg, O.; Runn, P. *Journal of Fish Diseases* **1986**, *9*, 11-25.

(7) Neville, C. M. *Can. J. Fish. Aquat. Sci.* **1985**, *42*, 2004-2018.

(8) Nor, M. Y. *Humic Substances in the Global Environment and Implications on Human Health*; Elsevier Science B. V. **1994**; pp 1055-1062.

(9) Laane, R. W. P. M. *Marine Chemistry* **1982**, *11*, 395-401.

(10) Miano, T. M.; Sposito, G.; Martin, J. P. *Soil Sci. Soc. of Am. J.* **1988**, *52*, 1016-1019.

(11) Coble, P. G.; Green, S. A.; Blough, N. V.; Gagosian, R. B. *Nature* **1990**, *348*, 432-435.

(12) Hayase, K.; Tsubota, H. *Geo. Cosmo. Acta.* **1985**, *49*, 159-163.

(13) Marino, D. F.; Ingle, D. F. *J. Anal. Chim. Acta.* **1980**, 23-30.

(14) Datta, C.; Ghosh, K.; Mukherjee, S. K. *Journ. Indian Chem. Soc.* **1971**, *48*, 279-287.

(15) Ryan, D. K.; Thompson, C. P.; Weber, J. H. *Can. J. Chem.* **1983**, *61*, 1505-1509.

(16) Lakshmann, S.; Mills, R.; Patterson, H. *Anal. Chim. Acta.* **1993**, *282*, pp. 101-108.

(17) Esteves da Silva, J. C. G.; Machado, A. A. S. C.; Garcia, T. M. O. *Appl. Spec.* **1995**, *49*, 1500-1506.

(18) Shotyk, W.; Sposito, G. *Soil Sci. Soc. of Am. J.* **1988**, *52*, 1293-1297.

(19) Blaser, P.; Sposito, G. *Soil Sci. Soc. of Am. J.* **1987**, *51*, 612-619.

(20) Silva, C. S. P. C. O.; Esteves da Silva, J. C. G.; Machado, A. A. S. C. *Appl. Spec.* **1994**, *48*, 363-372.

(21) Saar, R. A.; Weber, J. H. *Anal. Chem.* **1980**, *52*, 2095-2100.

(22) Ryan, D. K.; Weber, J. H. *Anal. Chem.* **1982**, *54*, 986-990.

(23) Ventry, L. S. Metal Ion Binding by Humic Materials: An Integrated High Performance Liquid Chromatography and Fluorescence Quenching Study. **1989**, Northeastern University. PhD.

(24) Patonay, G.; Shapira, A.; Diamond, P.; Warner, I. M. *J. Phys. Chem.* **1986**, *90*, 1963-1966.

(25) Ventry, L. S.; Ryan, D. K.; Gilbert, T. R. *Microchemical Journal.* **1991**, *44*, 201-214.

(26) Seritti, A.; Morelli, E.; Nannicini, L.; Giambelluca, A.; Scarno, G. *The Sci. Tot. Environ.* **1994**, *148*, 73-81.

(27) Machado, A. A. S. C.; Esteves da Silva, J. C. G.; Maia, J. A. C. *Anal. Chim. Acta.* **1994**, *292*, 121-132.

(28) Esteves da Silva, J. C. G.; Machado, A. A. S. C. *Chem. Intel. Lab Sys.* **1995**, *27*, 115-128.

(29) Frimmel, F. H.; Hopp, W. *Fresenius Z. Anal. Chem.* **1986**, *325*, 68-72.

(30) Senesi, N.; Sakelariadou, F. *Environ. Int.* **1994**, *20*, 3-9.

(31) Cook, R. L.; Langford, C. H. *Anal. Chem.* **1995**, *67*, 174-180.

(32) Ho, C. N.; Patonay, G.; Warner, I. M. *Trends in Anal. Chem.* **1986**, *5*, 37-43.

(33) Xu, W.; McDonough, R. C.,III; Langsdorf, B.; Demas, J. N.; Degraff, B. A. *Anal. Chem.* **1994**, *66*, 4133-4141.

(34) Carraway, E. R.; Demas, J. N.; Degraff, B. A. *Anal. Chem.* **1991**, *63*, 337-342.

(35) Demas, J. N.; Degraff, B. A.; Xu, W. *Anal. Chem.* **1995**, *67*, 1377-1380.

(36) Ryan, D. K.; Weber, J. H. *Environ. Sci. Technol.* **1982**, *16*, 866-872.

Chapter 9

Fluorescence Spectroscopic Studies of Al–Fulvic Acid Complexation in Acidic Solutions

David K. Ryan[1], Chih-Ping Shia[1], and Donald V. O'Connor[2]

[1]Department of Chemistry, University of Massachusetts,
Lowell, MA 01854
[2]Center for Fast Kinetics Research, University of Texas,
Austin, TX 78712

The complexation interaction between a soil fulvic acid (SFA) and aluminum (III) ion has been studied using several different types of fluorescence measurements. Excitation and emission scans have been employed along with total luminescence, fluorescence lifetime and fixed wavelength experiments. Titrations of SFA with Al^{3+} at fixed pH values of 4.00 and 5.00 as well as pH titrations were used to examine the unique fluorescence behavior of humic material complexes of aluminum. SFA was found to exhibit both fluorescence quenching and enhancement effects upon complexation of Al^{3+} depending on the pH and wavelengths employed. A pronounced shift in the fluorescence spectral maximum to longer excitation wavelengths and shorter emission wavelengths was observed for the Al-SFA complex. Striking similarities between the fluorescence behavior of the model compound salicylic acid and SFA in the presence of Al^{3+} were clearly evident.

Molecular fluorescence has been shown to be a valuable tool for studying the binding of certain metal ions to both isolated humic materials (1-4) and natural organic matter present in samples collected from lakes, rivers and other bodies of water (5,6). Paramagnetic metal ions tend to reduce or quench fluorescence which has been demonstrated for Cu^{2+} (1-5), Co^{2+} (4,6), Mn^{2+} (4), Ni^{2+} (1), Fe^{3+} and Fe^{2+} (7) with various samples of humic materials. Diamagnetic metal ions, on the other hand, may quench, show no effect or even enhance humic material fluorescence depending on the metal, the source of the humic material and other experimental factors (1,8-13)

A review of the published literature for Al^{3+}-humic interactions reflects variations in fluorescence behavior (8-12,14,15). Sposito and co-workers (11,12) concluded that there was a quenching effect for aluminum ion complexed with a chestnut leaf litter extract (LLE). Philpot et al. (14) also reported fluorescence quenching of humic acid (Aldrich) upon complexation with aluminum. However, unlike these studies, Plankey and Patterson (8-10) found that the fluorescence intensity

0097–6156/96/0651–0125$15.00/0

of fulvic acid was increased by the addition of aluminum ions. According to recent work by Luster et al. (15), natural organic matter (LLE) complexed with Al^{3+} can either decrease or increase the relative fluorescence intensity depending on the wavelength at which the measurements are made.

Humic materials have various binding sites with different complexation properties. The actual complexation interaction between Al^{3+} and humic material depends on the structural and conformational chemistry of the individual molecule, and the arrangement of functional groups for each site. Complexation with a metal ion changes the electronic polarization of both the metal ion and the binding site. This change could result in a fluorescence intensity increase or decrease at a specific emission wavelength, which is dependant upon the fluorescence properties of that particular binding site. The fluorescence spectrum is an overall result of the combination of intensities of all fluorophores versus the wavelength. Fluorophores associated with binding sites will also be affected by the presence of a metal ion, the properties of that metal ion, pH and other factors.

Therefore, it is possible to observe the fluorescence intensity increase at one emission wavelength and decrease at another wavelength when the humic material is bound to Al^{3+} or possibly other metal ions. Both kinds of fluorescence effects may be useful in calculating conditional stability constants (K) and the concentration of aluminum binding sites (C_L) on the humic material (11). The rigidity of the molecular structure may increase due to a small, highly charged cation like Al^{3+} binding with humic material. This may possibly increase the fluorescence quantum yield by reducing other possible nonradiative transitions (16).

According to the initial results reported here, a fluorescence enhancement effect for humic material binding with aluminum can be observed, while under different conditions a quenching of fluorescence predominated. Another very interesting fluorescence phenomenon described here is that after complexation with Al^{3+}, the maximum intensity of the humic material fluorescence shifts to longer excitation wavelengths and to shorter emission wavelengths in an excitation emission matrix (EEM).

This work is a preliminary investigation into the factors that influence changes in the fluorescence of a particular soil-derived fulvic acid (SFA) upon complexation with Al^{3+}. Conditions that produce quenching, enhancement and shifting of the fluorescence are examined. Fluorescence lifetime measurements of SFA and Al-SFA were also conducted and compared to previously reported results. In addition, Al^{3+} complexation with the model compound salicylic acid (SA) is examined. Significant evidence indicates that SA type sites are important for metal complexation in humic materials (17). Interestingly, SA has similar fluorescence properties to humic materials and undergoes fluorescence enhancement and a wavelength shift upon complexation with Al^{3+}.

Experimental

Apparatus. A Mark 1 Spectrofluorometer (Farrand Optical Co., Inc.) was utilized to collect fluorescence data for the experiments involving fluorescence titrations of SA or SFA and EEMs of SFA. Data was collected and processed with Lab Calc and Grams 386 software packages (Galactic Industries Corp., Salem, NH) running on IBM compatible personal computers. All fluorescence measurements were performed in standard 10 mm quartz cells at room temperature except for the titrations of salicylic acid and soil fulvic acid which were done at 25 °C with a constant temperature water bath (model 1150, VWR Scientific, Boston, MA) and a jacketed titration cell (EG&G Princeton Applied Research, Princeton, NJ). An Orion 960 Autochemistry System (Orion Research, Boston, MA) coupled with an Orion 91-02 glass combination pH electrode was used to measure the pH of all solutions.

The instrument used for both SFA pH titrations and SA EEM experiments was a Perkin-Elmer model MPF-44B spectrofluorometer (Perkin-Elmer Corp., Norwalk, CT). The excitation and emission spectra were obtained by using an SLM-Aminco 48000 phase shift spectrofluorometer (Milton Roy Co., Urbana, IL).

Fluorescence lifetimes were measured by time-correlated single photon counting using a mode-locked, synchronously pumped, cavity-dumped pyridine I dye laser (343 nm) or Rhodamine 6G dye laser (290 nm). Emissive photons were collected at 90° with respect to the excitation beam and passed through a monochromator to a Hamamatsu Model R2809U microchannel plate. Data analysis was made after deconvolution (18) of the instrument response function (FWHM ~ 80 ps).

Reagents. Solutions of salicylic acid (Fisher Scientific Co., Pittsburgh, PA) were prepared by dissolving the powdered reagent in dilute NaOH followed by neutralization of the excess base with $HClO_4$. Salicylic acid stock solutions were 1 mM and freshly made for each series of experiments. The soil fulvic acid used in these studies was obtained from Dr. James H. Weber, Department of Chemistry, University of New Hampshire. The isolation and characterization of SFA (19-21) and its metal ion binding and fluorescence properties (1-4) have been reported previously. Aluminum solutions were prepared from two sources; dissolving Al metal wire (J.T. Baker Co., Phillipsburgh, NJ) in $HClO_4$ or dissolving aluminum perchlorate (Alfa/Johnson Mathey, Ward Hill, MA) in deionized water. Sodium hydroxide and perchloric acid were purchased from Fisher Scientific Company. The ionic strength of the solutions in both EEM and fluorescence titration experiments for SA and SFA was made up with 0.1 M $NaClO_4$ (Aldrich Chemical Co., Milwaukee, WI). Both SFA and SA solutions were filtered through 0.4 μm pore size filters after adjustment to the pH of the experiment.

Results and Discussion

Monitoring the fluorescence of a 15 mg/L soil fulvic acid solution while adding aliquots of an Al^{3+} solution at pH 4.00 gives rise to the data shown in Figure 1. A distinct enhancement of fluorescence occurs from the initial value, set at 100%, to a maximum of more than 180% (Figure 1). Using non linear regression to fit a one site binding model to the data as discussed in a previous chapter (22) provides a conditional stability constant (K) of 6.2 x 10^5 (log K=5.79) and a concentration of

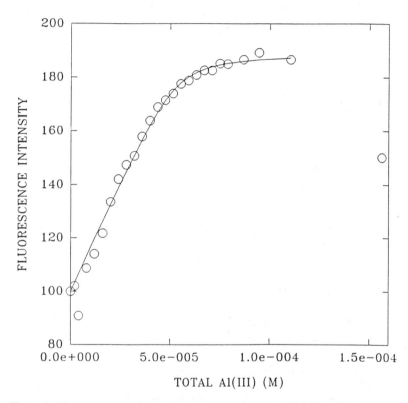

Figure 1. Fluorescence enhancement titration curve for 15.0 mg/L soil fulvic acid in 0.1 M NaClO$_4$ at pH 4.00 and 25 °C; (◯) measured fluorescence. Line is SigmaPlot fitted curve excluding last titration point. The fluorescence is monitored at an excitation wavelength of 360 nm and an emission wavelength of 420 nm.

SFA ligands sites of 50.6 µM. The solid line in Figure 1 shows the best fit of the model to the data.

SFA and Al-SFA pH Titrations. Changing the pH of the SFA solution to 5.00 and the excitation wavelength to 335 nm causes fluorescence quenching to occur when Al^{3+} is added as illustrated in Figure 2. The fluorescence emission spectrum (Figure 2) exhibits reduced fluorescence at nearly all wavelengths when Al^{3+} is present. Similar results are observed at higher pH as well. These results indicate that both pH and fluorescence wavelengths may be important in determining whether fluorescence enhancement or quenching are observed.

In order to study the effect of pH on the fluorescence of SFA and Al-SFA complexes, pH titrations were performed. Figure 3 shows the fluorescence intensity of two solutions containing of 15 mg/L SFA or 15 mg/L SFA with 200 µM Al^{3+} as the pH is varied from 2 to 10. The solution of SFA alone shows a modest increase in fluorescence from pH 2 to a maximum around pH 5 and then a decrease as pH is increased further (1). The solution containing Al^{3+} and SFA exhibits a slight increase from pH 2 to 3 and then a dramatic decrease in fluorescence (i.e., quenching) as pH is increased from 3 to 5.

This behavior is very likely caused by increased complexation of Al^{3+} as pH goes up. The Al-SFA curve in Figure 3 levels off between pH 5 and 7 where strong complexation of Al^{3+} by SFA produces a maximum fluorescence quenching or minimum fluorescence emission at these wavelengths.

As the pH is further increased above pH 7, fluorescence increases and approaches nearly the same intensity value observed at low pH (no Al^{3+} binding) which is the same intensity in the absence of Al^{3+} (Figure 3). The reason for this increase in fluorescence at high pH is probably the hydrolysis of Al^{3+} to various aluminum hydroxides. As the Al hydroxides form, Al-SFA complexes are broken up and little or no free Al^{3+} is available at pH 10 to bind with SFA and cause fluorescence quenching.

Fluorescence Peak Maxima. Figure 4 shows a total luminescence spectrum or excitation emission matrix (EEM) for 15 mg/L SFA in the form of a contour plot. Contour lines give the fluorescence intensity at essentially all values of excitation wavelength from 290 nm to 390 nm and all values of emission wavelength from 391 nm to 491 nm. The peak maximum is observed at an excitation wavelength of 335 nm and an emission wavelength of 446 nm. When 2×10^{-4} M Al^{3+} is added at pH 4 the intensity increases and the peak maximum moves to an excitation wavelength of 350 nm and an emission wavelength of 436 nm as shown in Figure 5.

The observed changes in the SFA peak maximum of 15 nm for excitation wavelength and 10 nm in emission wavelength are clearly significant. Although the cause of this change is not understood at this time, it's existence may have an effect on whether quenching or enhancement is observed. If fluorescence measurements are made at wavelengths near the 350 nm excitation and 436 nm emission maximum observed in the presence of Al^{3+}, then enhancement should be observed. Measurements made at other wavelengths may show lesser enhancement or even quenching.

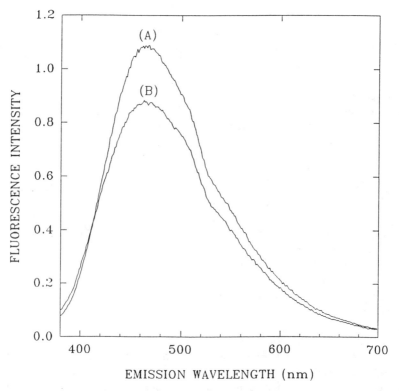

Figure 2. Fluorescence emission spectra of 100 mg/L soil fulvic acid at pH 5.00 with (A) no aluminum and (B) 100 μM Al^{3+}. The excitation wavelength is 335 nm.

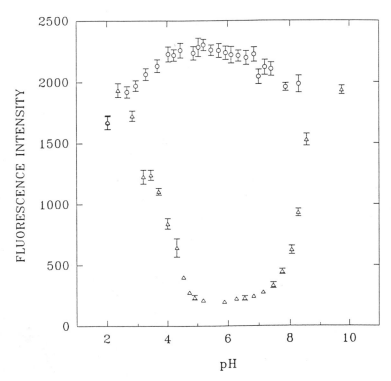

Figure 3. pH titrations of soil fulvic acid solutions with fluorescence monitored at an excitation wavelength of 340 nm and an emission wavelength of 455 nm. The samples are: (○) 15 mg/L soil fulvic acid; (▵) 15 mg/L soil fulvic acid with 200 μM Al(III) with error bars.

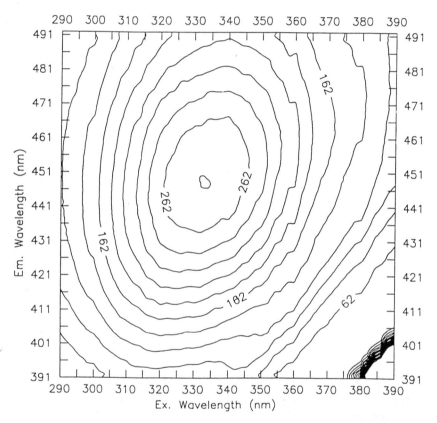

Figure 4. Excitation emission matrix contour plot of 15.0 mg/L soil fulvic acid in 0.1 M NaClO₄ at pH 4.00. The contour lines give fluorescence intensity in arbitrary units.

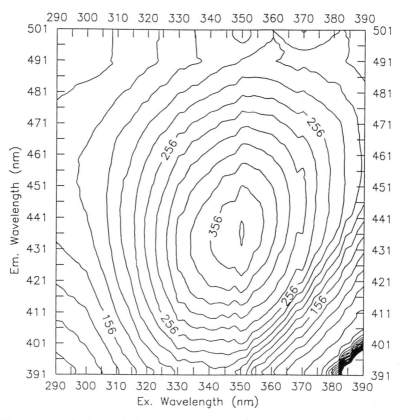

Figure 5. Excitation emission matrix contour plot of 15.0 mg/L soil fulvic acid with 2.00 x 10⁻⁴ M Al(III) in 0.1 M NaClO₄ at pH 4.00.

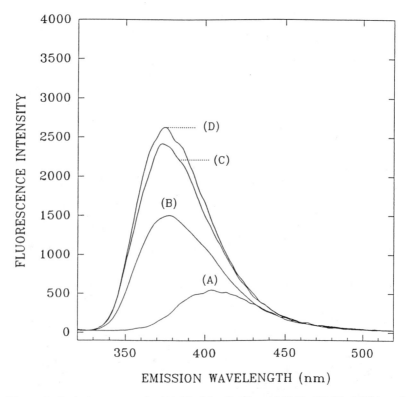

Figure 6. Emission spectra for (A) 10 µM salicylic acid (SA), (B) 10 µM SA and 20 µM Al(III), (C) 10 µM and 60 µM Al(III), and (D) 10 µM SA and 100 µM Al(III) at pH 4.00. The excitation wavelength is 317 nm.

Extrapolation of these results to other humic materials should be done with caution as they may or may not exhibit analogous behavior. However, experiments with the model compound salicylic acid under similar conditions showed nearly identical results. Enhancement of fluorescence and peak shifting are both exhibited in the salicylic acid spectrum show in Figure 6. At a fixed excitation wavelength of 317 nm, curve A (Figure 6) gives the emission spectrum of SA alone prior to any wavelength shift. The wavelength maximum for curve A is very close to 405 nm. When Al^{3+} is added at a concentration of 20 µM and the Al-SA complex forms, the peak shown in curve B of Figure 6 is not only increased approximately threefold, but its maximum also moves to less than 380 nm. Curve B also exhibits a very small shoulder near 400 nm resulting from the uncomplexed SA still remaining in solution. Further additions of Al^{3+} giving concentrations of 60 µM (curve C) and 100 µM (curve D) cause additional enhancement and clearly establish the wavelength maximum at 375 nm (Figure 6).

The overall shift in the fluorescence maximum of SA upon complexation of Al^{3+} is best seen by comparing the EEMs in figures 7 and 8. Figure 7 shows a SA fluorescence peak maximum at pH 4.00 of 301 nm for the excitation wavelength and 408 nm for the emission wavelength. When 60 µM Al^{3+} is added at pH 4.00, the resulting Al-SFA complexes give rise to a dramatically shifted peak maximum at 312 nm excitation and 381 nm emission wavelengths (Figure 8). This wavelength change is more than 10 nm in excitation wavelength and in excess of 25 nm in emission wavelength. Any quantitative work relying on fluorescence measurements of SA in the presence of Al^{3+} must carefully take into account this change in fluorescence wavelengths upon complexation. Researchers designing fluorescence experiments to measure the complexation of Al^{3+} by SA or SFA must also be cognizant of this phenomenon and make careful selection of monitoring wavelengths in order to measure the desired effect.

Fluorescence Lifetime Measurements of SFA. The time dependent decay, S(t), of SFA fluorescence can be described by three exponential terms, and the following equation fit to the data

$$S(t) = A_1 e^{-t/\tau_1} + A_2 e^{-t/\tau_2} + A_3 e^{-t/\tau_3}$$

where τ_1, τ_2 and τ_3 are the fluorescence lifetimes and A_1, A_2 and A_3 are the relative preexponential terms for the three components. The results obtained in this study are compare to the work of others in Table I. As stated by Cook and Langford (23), direct comparison of these three lifetimes should be done cautiously since they are a minimum set of parameters to fit the decay curves within experimental error. The lifetimes of SFA may be better described as a distribution of lifetimes.

All three component lifetimes became longer after SFA complexed Al^{3+}, but the first component (shortest lifetime) almost tripled in length due to the binding. The first component also had a smaller preexponential term, whereas the exponential terms of the other two components became larger for the Al-SFA complex. This indicates that the aluminum binding is similar for those fluorophores represented by compo-

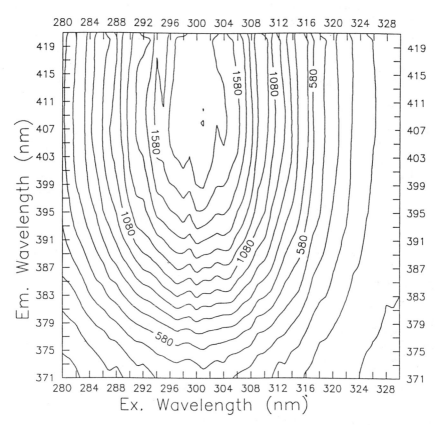

Figure 7. Excitation emission matrix contour plot of 10.0 μM salicylic acid in 0.1
M NaClO$_4$ at pH 4.00.

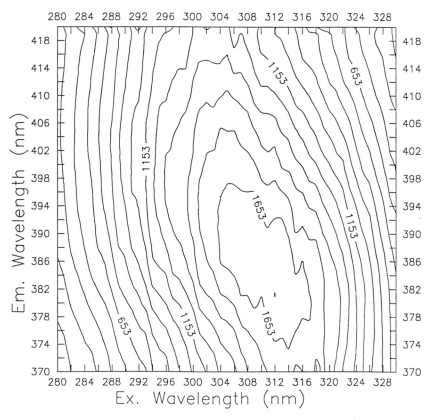

Figure 8. Excitation emission matrix contour plot of 10.0 μM salicylic acid with 60 μM Al(III) in 0.1 M NaClO$_4$ at pH 4.00.

nents 2 and 3, and the fluorophores of component 1 have distinguishably different binding properties.

Fulvic acids from different sources show a similar trend of having three components present in the fluorescence decay curves (Table I). The SFA used for this study and the Armdale FA have similar lifetime distributions. Laurentian FA, on the other hand, shows rather unique lifetime properties.

Table I. Fluorescence lifetime measurements of three fulvic acids (FA) and Al-soil fulvic acid (SFA).

Lifetimes (ns) and Preexponential Terms

Sample	First Component		Second Component		Third Component		Reference
	Life-time	Pre-exponent	Life-time	Pre-exponent	Life-time	Pre-exponent	
SFA	.205	.59	1.63	.30	8.17	.11	this study
Al-SFA	.593	.49	2.32	.37	8.49	.14	this study
Armdale FA	.2	.86	2.00	.11	7.00	.03	24
Laurentian FA	.05	.66	.430	.20	4.20	.14	23

Lifetime measurements of SA and the Al-SA complex gave essentially the same values of 5.4 and 5.3 ns, respectively, with an excitation wavelength of 290 nm and an emission wavelength of 400 nm. The errors on the fluorescence lifetime values given in Table I are ± 0.1 ns for the first and shortest lifetime, ± 0.2 ns for the second and ± 0.5 ns for the third.

Conclusion

The fluorescence intensity of SFA in the presence of Al^{3+} shows both quenching and enhancement behavior depending primarily on the wavelength used and to some extent the pH. The model compound salicylic acid exhibits very similar behavior showing wavelength shifts of more than 10 nm for excitation and 25 nm for emission upon complexation of Al^{3+}. Fluorescence lifetime measurements also show changes between complexed and uncomplexed SFA with the most significant change observed for the shortest lived fluorescence component.

One possible explanation of the data presented here is that fluorescence normally observed in the absence of aluminum is quenched as Al^{3+} complexes are formed. Simultaneously a new fluorescence peak for the complex may appear at slightly shifted wavelengths. The complex may have a higher quantum efficiency and therefore show greatly enhanced fluorescence. This hypothesis may be useful in explaining the fluorescence behavior of both SFA and SA in the experiments described here.

Literature Cited

1. Saar, R.A., & Weber, J.H. *Anal. Chem.* **1980**, *52*, 2095-2100.
2. Ryan, D.K., & Weber, J.H. *Anal. Chem.* **1982a**, *54*, 986-990.
3. Ryan, D.K., Thompson, C.P., & Weber, J.H. *Can. J. Chem.* **1983**, *61*, 1505-1509.
4. Ventry, L.S., Ryan, D.K., & Gilbert, T.R. *Microchem. J.* **1991**,*44*, 201-214.
5. Ryan, D.K., & Weber, J.H. C *Environ. Sci. Technol.* **1982b**, *16*, 866-872.
6. Cline, J.T. and Holland, J.F. ERDA Symposium Series, **1977**, *42*, 264-279.
7. Waite, T.D. and Morel, F.M.M. *Anal. Chim. Acta* **1984**, *162*, 263-274.
8. Plankey, B.J., & Patterson, H.H. *Environ. Sci. Technol.* **1987**, *21*, 595-601.
9. Plankey, B.J., Patterson, H.H., & Cronan, C.S. *Anal. Chim. Acta*, **1995**, *300*, 227-236.
10. Plankey, B. J. and Patterson, H.H. *Environ. Sci. Technol.* **1988**, *22*, 1454-1459.
11. Shotyk, W., & Sposito, G. *Soil Sci. Soc. of Am. J.* **1988**, *52*, 1293-1297.
12. Blaser, P. and Sposito, G. *Soil Sci. Soc. Am. J.* **1987**, *51*, 612-619.
13. Esteves da Silva, J.C.G., Machado, A.A.S.C., & Garcia, T.M.O. *Appl. Spec.* **1995**, *49*, 1500-1506.
14. Philpot, W.D., Bisogni, J.J. and Vodacek, A. U.S. Geological Survey, **1984**, USGS/G859(05).
15. Luster, J., Lloyd, T. and Sposito, G. in Humic Substances in the Global Environment and Implications on Human Health, N. Senesi an T.M. Miano, eds., Elsevier Science. **1994** pp. 1019-1024.
16. Cabaniss, S.E. *Environ. Sci. Technol.* **1992**, *26*, 1133-1139.
17. Morel, F.M.M. and Hering, J.G. Principles and Applications of Aquatic Chemistry. John Wiley & Sons, Inc. **1993**, p. 377.
18. O'Conner, D.V. and Phillips, D. Time Correlated Single Photon Counting. Academic Press, London. **1984**
19. Weber, J.H. and Wilson, S.A. *Water Res.* **1975**, *9*, 1070-1084.
20. Wilson, S.A. and Weber, J.H. *Chem. Geol.* **1977**, *19*, 285-293.
21. Wilson, S.A. and Weber, J.H. *Chem. Geol.* **1979**, *19*, 345-351
22. Hays, M.D., Ryan, D.K., Pennell, S. and Ventry-Milenkovic, L. **1996** Data treatments for relating metal ion binding to fulvic acid as measured by fluorescence spectroscopy. Preceding paper in this volume.
23. Cook, R.L., & Langford, C.H. *Anal. Chem.* **1995**, *67*, 174-180.
24. Power, J.F., LeSage, R., Sharma, D.K. and Langford, C.H. *Environ. Technol. Lett.* **1986**, *7*, 425-430.

Chapter 10

Investigation of Fulvic Acid–Cu^{2+} Complexation by Ion-Pair Reversed-Phase High-Performance Liquid Chromatography with Post-Column Fluorescence Quenching Titration

G. Christopher Butler and David K. Ryan

Department of Chemistry, University of Massachusetts,
Lowell, MA 01854

An ion-pair reversed-phase high performance liquid chromatography (IP-RP-HPLC) separation with post-column addition of Cu^{2+} and simultaneous UV and fluorescence detection is used to investigate the interactions between Cu^{2+} and separated fractions of soil and water fulvic acids. In previous steady-state fluorescence experiments, binding characteristics for humic materials have been determined based on fluorescence quenching data from titrating the sample with paramagnetic metal ions. In the present study, the humic material fractions from the IP-RP-HPLC separation are "titrated on-line" with incrementally increasing concentrations of Cu^{2+} and the extent of quenching of the fluorescence peaks is measured. A titration curve of percent fluorescence vs. metal concentration is plotted for each fraction and the ligand concentrations (C_L) and conditional stability constants (K) are calculated from nonlinear regression of a one-site model applied to the data.

Humic substances such as humic and fulvic acid have been shown to affect transport, bioavailability, and toxicity of metals in natural waters (1). Fluorescence quenching measurements have been used to determine binding characteristics of these naturally occurring organic substances under conditions typically found in aquatic environments (2-4). These studies have employed titrations of humic materials with metal ion, yielding results which describe the overall binding properties of the humic substance. However, since humic substances are polydisperse, comprising a wide range of compounds with differing structures, molecular weights, spectroscopic characteristics, and chemical and complexing properties, the resulting binding parameters, K and C_L, are average values for the many fractions in the material being tested.

0097–6156/96/0651–0140$15.00/0
© 1996 American Chemical Society

In order to investigate the properties of individual fractions of humic substances, various modes of high performance liquid chromatography (HPLC) have been employed. Hydrophobic interaction chromatography (5) has proved to be an effective separation technique, resulting in five distinct humic fractions from one sample. Structural analysis of these fractions was subsequently performed by infrared and ^{13}C nuclear magnetic resonance spectroscopy, and molecular weight distribution was also measured.

Size exclusion chromatography (SEC) has been used to measure molecular weight (MW) distribution of humic substances (3, 6-9). Coupled with detection methods such as molecular fluorescence spectroscopy and dissolved organic carbon analysis (7), electrochemical detection (9), and atomic emission spectroscopy (3), SEC has been used extensively to study humic-metal complexes. A major disadvantage of SEC is that it does not provide adequate resolution for separating humic materials as they do not appear to be made up of distinct fractions with large differences in MW.

Naturally-occurring humic-metal complexes have been isolated from estuarine systems and seawater using solid phase extraction (SPE) onto a C_{18} HPLC column to preconcentrate the sample (10-12). Samples were subsequently eluted from the SPE column at a much higher concentration and injected onto another HPLC column and detected by UV absorbance and a metal-sensitive detector, such as atomic fluorescence spectroscopy. The concentration of metal-humic complexes in natural aquatic environments was then calculated. However, there was some evidence of competitive binding of the metal ion between the organic matter and free silanol groups in the stationary phase resulting in a loss of metal in the column and erroneously low metal values (10).

More recent developments in the study of humic substances by RP-HPLC include the use of stepwise gradients (13-15) and fluorescence emission and synchronous spectra (15, 16). Separations using stepwise gradients resulted in improved resolution and a greater number of fractions when compared with previous RP-HPLC experiments. The use of fluorescence emission (15, 16) and synchronous spectra (16) have lead to better understanding of the differences in spectroscopic characteristics of humic fractions. In one such study (15) it was observed that for humic-like marine dissolved organic matter fluorescing material was mainly present in the fractions of intermediate polarity, while non-fluorescing material was predominantly present in the more non-polar fractions.

Studies have been performed utilizing HPLC coupled with fluorescence quenching to examine the metal-binding characteristics of humic fractions separated by size-exclusion chromatography (3) and IP-RP-HPLC (17) in an effort to determine if fractions had different binding and/or fluorescing characteristics. In these experiments, the modified Hummel-Dreyer mode of HPLC (18) was utilized, in which a constant concentration of the metal of interest is introduced in the mobile phase to prevent dissociation of the humic-metal complexes (9, 18). In HPLC-fluorescence quenching experiments, UV and fluorescence detectors were used to simultaneously collect chromatographic data. Fluorescence detection was utilized to measure free ligand since complexation of humic materials with paramagnetic transition metal ions has been shown to quench fluorescence. "Total" ligand was measured by UV detection, since complexation of the metal by the humic substance is not known to effect the absorbance of the molecule in the UV.

In a study utilizing SEC and fluorescence, retention time increased with increased concentration of Cu^{2+} which in this mode of HPLC would indicate that there had been a decrease in molecular size (*3*). It was also hypothesized that ionic interactions between the humic-metal complex and charged surfaces such as free silanol groups in the column packing may have also contributed to the increased retention time. In addition to changes in retention time, there was a decrease in peak area with increased copper concentration for both the UV and fluorescence chromatograms. This may have been due to an irreversible binding (in the time-scale of the separation) of copper-humic complexes to the stationary phase in the presence of increased Cu^{2+}.

In the case of IP-RP-HPLC (*17*) marked improvements in the separation of distinct fractions of fulvic acid were observed when no metals were present. However, when metal-ion containing mobile phase was introduced into the system, drastic changes occurred to both the UV and fluorescence chromatograms showing peaks merging or disappearing altogether and changing retention times. Since similar changes were shown for both the UV and fluorescence chromatograms, it was difficult to determine if decreases in the fluorescence peaks were due to quenching or to loss of the humic metal complexes to the column.

It is clear from the above studies that coupling HPLC and fluorescence using the Hummel-Dreyer mode leads to confusing results due to changes in the retention characteristics of the humic molecules in the column in the presence of metal ion. Ideally, the study of fluorescence quenching of fractions of humic materials on-line using HPLC and fluorescence detection should include a mode of HPLC which provides adequate ability to resolve fractions of humic materials as well as a means of introducing metals into the system in such a way that the chromatography of the humic sample is not disrupted. In the present study, humic samples are first separated by IP-RP-HPLC run in the isocratic mode. The column effluent is then mixed on-line with a copper reagent solution via a post-column mixing T connected to a reaction coil. In this way, eluting fractions of the humic sample are exposed to metal ion only after being separated, thereby eliminating the possibility that addition of metal will disrupt chromatographic performance.

Experimental

In the current study, IP-RP-HPLC was performed using a metal-free Dionex Gradient Pump Module (GPM) connected to a glass-lined C_{18} HPLC column (Scientific Glass Engineering model 250 GL4-ODS2-30/5) and a model PF-1 photodiode array (PDA) detector (Groton Technology, Inc.) and a dual monochromator fluorescence detector (model FD-300, Groton Technology, Inc.). All tubing, fittings, and detector flow cells were metal-free to minimize metal contamination. The post-column reagent system consisted of a second GPM pump capable of delivering known concentrations of Cu^{2+} by mixing predetermined proportions of the two reagent solutions. The outlet of the column and the reagent pump were connected by tubing to a mixing T followed by a reaction coil 12 inches in length. The reaction coil allowed mixing of the eluent and metal reagent streams with minimal band-broadening as well as providing time for complexation between the sample and metal ions.

The mobile phase was 37% acetonitrile in H_2O with 6.0 mM tetra-butylammonium perchlorate and 0.05 M N-morpholino-ethanesulfonic acid (MES) with a pH of 6.00 ± 0.05. The reagent solutions were, A) 0.05 M MES, pH = 6.00 ± 0.05 and B) 2.0 mM copper perchlorate and 0.05 M MES, pH = 6.00 ± 0.05. The HPLC was run isocratically with an eluent flow rate of 0.7 mL/min and a post-column reagent flow rate of 0.1 mL/min. The monitor wavelength of the PDA was set at 254 nm and the fluorescence excitation and emission wavelengths were set at 332 nm and 442 nm, respectively. The concentration of Cu^{2+} in the post-column reagent was set by entering the percentage of solution A and percentage of solution B on the reagent pump prior to beginning the chromatographic run and was held constant throughout. Subsequent sample runs were performed with the concentration of Cu^{2+} in the post-column reagent increased in increments of 0.1 mM, up to a maximum of 2.0 mM Cu^{2+} (100% reagent B). Humic samples were dissolved in a solution of 0.05 M MES and adjusted to pH = 6.00 ± 0.05 for a final concentration of 500 mg/L. The injection volume was 80 μL.

Results and Discussion

Samples used to demonstrate this method were soil fulvic acid (SFA) and water fulvic acid (WFA), both well-characterized materials obtained from Dr. James H. Weber at the University of New Hampshire (*19*). Figure 1 shows that the chromatographic method resulted in four fractions separated for the SFA. The use of IP-RP-HPLC with the biological buffer MES resulted in sufficient separation to eliminate the need for gradients as have been used in previous studies (*13-17*). Simultaneous collection of UV (254 nm) and fluorescence ($\lambda_{excitation}$ = 332 nm and $\lambda_{emission}$ = 442 nm) data showed similar chromatograms with peaks at the same retention times except in the case of the more non-polar (later-eluting) fractions which did not exhibit measurable fluorescence. This result is similar to that reported by Lombardi et al. (*15*) for marine DOM. Figure 2 shows a very similar separation for WFA.

An example of the quenching of fluorescence chromatograms of SFA appears in figure 3. All chromatograms are shown on the same scale with the concentration of Cu^{2+} in the post-column reagent given above the respective chromatogram. Peak heights and areas decrease with increased Cu^{2+} concentration as is expected, while the retention times and relative positions of the peaks remain unchanged (figure 3). However, a maximum decrease of approximately 10 % in UV peak area for each peak was observed at the higher Cu^{2+} concentrations. This may be due to precipitation of SFA in the presence of high concentrations of Cu^{2+} or other loss of SFA in the post-column system.

Fluorescence chromatograms for each titration experiment were integrated

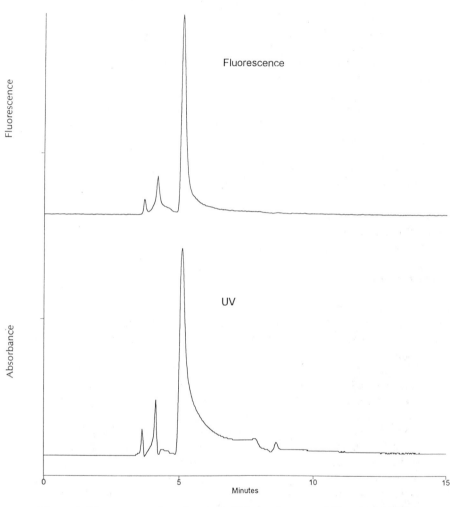

Figure 1. Fluorescence (top, $\lambda_{excitation}$ = 332 nm, $\lambda_{emission}$ = 442 nm) and UV (bottom) (254 mn) chromatograms of SFA (500 mg/L) with an injection volume of 80 µL. Eluent and post-column reagent (0.0 mM Cu^{2+}) flow rates were 0.7 mL/min. and 0.1 mL/min., respectively.

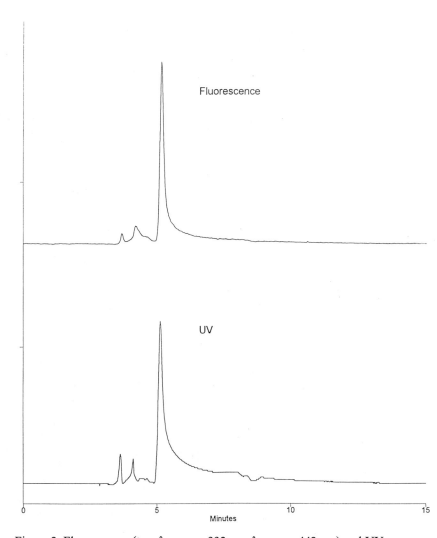

Figure 2. Fluorescence (top, $\lambda_{excitation}$ = 332 nm, $\lambda_{emission}$ = 442 nm) and UV (bottom) (254 mn) chromatograms of WFA (499 mg/L) with an injection volume of 80 µL. Eluent and post-column reagent (0.0 mM Cu^{2+}) flow rates were 0.7 mL/min. and 0.1 mL/min., respectively.

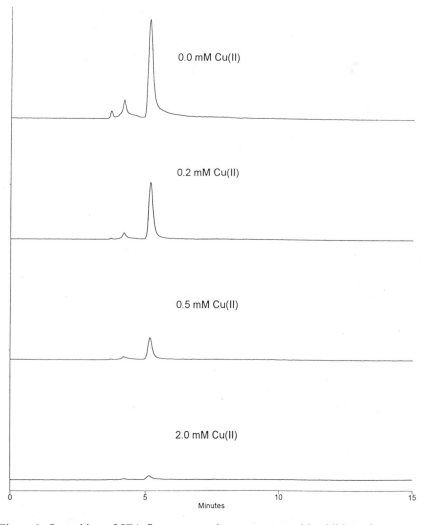

Figure 3. Quenching of SFA fluorescence chromatograms with addition of Cu^{2+}. Post-column reagent concentration of Cu^{2+} is listed above corresponding chromatogram. Conditions are the same as listed in figure 1.

and the areas of each individual peak, expressed as percent relative fluorescence for each experiment, were plotted against the corresponding concentration of Cu^{2+} in the reagent. The resulting quenching data was subjected to non-linear regression using the data treatment of Ryan and Weber (*2*) resulting in best-fit curves for each set of quenching data based on calculated values of I_{res}, the residual fluorescence, C_L, the concentration of ligand, and K, the 1:1 conditional stability constant. The best-fit quenching curves for SFA in figure 4 shows similar quenching for the three peaks, although peak 1 quenches more steeply relative to Cu^{2+} concentration than peaks 2 and 3. The quenching of the three fluorescence peaks separated from WFA shown in figure 5 differs more widely in terms of steepness and residual fluorescence. Peak 1 of the WFA shows a similar shaped quenching curve compared with peak 1 of SFA while curves for peaks 2 and 3 of the WFA are more broad and shallow showing less quenching for a given Cu^{2+} concentration. This may be due to the less polar fractions being weaker binders of Cu^{2+}. Similar quenching trends have been observed for Suwannee Stream fulvic acid and Aldrich humic acid.

Table I lists binding parameters for SFA based on total Cu^{2+} concentration, correcting for the dilution of the Cu^{2+} ion when post-column reagent stream was mixed with the column eluent. While these values show that this data treatment works well for SFA, binding parameters for WFA showed negative values for C_L and I_{res}, indicating that the quenching behavior of WFA is not fit by the model on which the data treatment is based. The fact that quenching of the WFA is not modelled well may be a result of the fundamental assumption of solution equilibrium not being met (*2*).

Table I. Fitted Binding Parameters for HPLC Post-Column Fluorescence Titration of SFA with Cu²⁺

Sample	C_L μM (std dev)	K (x 10⁻⁴) (std dev)	I_{RES} (std dev)
SFA			
Peak 1	22.5 ($\pm$ 2.3)	101.2 ($\pm$ 55.8)	5.6 ($\pm$ 1.1)
Peak 2	33.3 ($\pm$ 6.2)	39.9 ($\pm$ 32.8)	10.2 ($\pm$ 2.2)
Peak 3	46.2 ($\pm$ 5.3)	26.07 ($\pm$ 13.0)	6.6 ($\pm$ 1.9)

Since the quenching of the fluorescence of fractions eluting from the chromatographic column occurs as Cu^{2+} is added on-line, there may not be sufficient time for equilibrium to occur. This may also explain why data points in this study show greater deviation from the best-fit curves than has been observed in experiments performed under equilibrium conditions (*2*). However, based on a stability constant of 1.08×10^5 for complexation at pH 6 (*2*), a rate constant in the

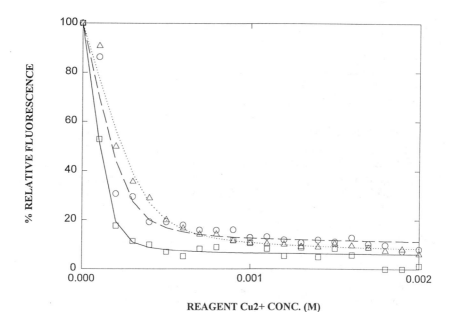

Figure 4. Fluorescence peak area quenching ($\lambda_{\text{excitation}}$ = 332 nm, $\lambda_{\text{emission}}$ = 442 nm) of SFA chromatographic peak 1 (△), peak 2 (O), and peak 3 (□), and best-fit quenching curves for peak 1 (-), peak 2 (---) and peak 3 (...).

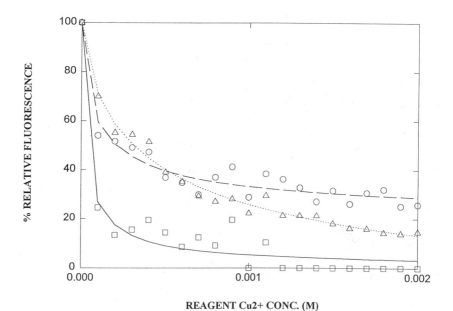

Figure 5. Fluorescence peak area quenching ($\lambda_{\text{excitation}}$ = 332 nm, $\lambda_{\text{emission}}$ = 442 nm) of WFA chromatographic peak 1 (△), peak 2 (O), and peak 3 (□), and best-fit quenching curves for peak 1 (-), peak 2 (---) and peak 3 (...).

range of 311.0 to 2948.4 s^{-1} (*20*) and a mixing time in the reaction coil and tubing of approximately 2.3 seconds, greater than 99 % complexation should occur before the humic material reaches the fluorescence detector, assuming a first-order rate constant and 1:1 humic:metal binding. Although it is evident from the quenching of fluorescence that binding does occur, it is possible in the time-scale of this experiment that some secondary effects of the binding on the humic molecules, such as intramolecular hydrogen bonding and conformational changes may not have time to occur.

The complicated nature of the instrumentation required for these experiments may also result in artifacts which may cause the fluorescence quenching to not conform to the model. Fluctuations in the proportioning of solutions A and B in the post-column reagent pump may result in error in the total Cu^{2+} concentration reported as well as variation in concentration of Cu^{2+} in the post-column system throughout an experiment. Likewise, fluctuations in pump flow rate of both the eluent and post-column reagent pumps may also effect the total Cu^{2+} concentration in the post-column system. Since each experiment was conducted over nine to twelve hours, temperature of the system and the fluorescence detector lamp output may also have contributed to variations in fluorescence quenching. A decreased run time for the chromatographic method and a decrease in the number of sample runs for each experiment may help to reduce this artifact.

Summary

A new method for combining HPLC with fluorescence quenching titrations has been demonstrated and provides information on the binding characteristics of individual fractions of humic materials. By mixing the metal titrant with the column effluent post-column, the chromatographic separation of the sample is not disrupted by the addition of metal, as shown by the fact that the retention time and relative positions of the peaks do not change with increasing Cu^{2+} concentration in the solution. However, some small decreases in UV peak height and area were observed with increased Cu^{2+} concentration. This method yields values for K, C_L, and I_{RES} similar to those reported in previous studies, but also provides information about the metal binding behavior of the individual fractions which make up the humic material. This method may also be adapted to other modes of HPLC, resulting in greater understanding of the effects of metal-complexation on other properties of humic substances.

Literature Cited

1. Paulauskis, J. D.; Winer, R. W. *Aquat. Toxicol.* **1988**, *12*, 273.
2. Ryan, D. K.; Weber, J. H. *Anal. Chem.* **1982**, *54*, 986.
3. Ventry, L. S. K. Ph.D. Thesis, Northeastern University, **1989**.
4. Ventry, L. S. K.; Ryan, D. K.; Gilbert, T. R. *Microchem. J.* **1991**, *44*, 201.

5.   Kalinowski, E.; Blondeau, R. *Mar. Chem.* **1988**, *24*, 29.
6.   Knuutinen, J.; Virkki, L.; Mannila, P.; Mikkelson, P.; Paasivirta, J.;
      Herve, S. *Water Res.* **1988**, *22*, 985.
7.   Frimmel, F. H.; Gremm, T.; Huber, S. *Sci. Total Environ.* **1992**, *117-118*,
      197.
8.   Chin, Y.; Aiken, G.; O'Loughlin, E. *Environ. Sci. Technol.* **1994**, *28*, 1853.
9.   Adamic, M. L.; Bartak, D. E. *Anal. Chem.* **1985**, *57*, 279.
10.   Mackey, D. J. *Mar. Chem.* **1985**, *16*, 105.
11.   Mills, G. L.; McFadden, E.; Quinn, J. G. *Mar. Chem.* **1987**, *20*, 313.
12.   Mackey, D. J.; Higgins, H. W. *J. Chromatogr.* **1988**, *436*, 243.
13.   Saleh, F. Y.; Chang, D. Y. *Sci. Total. Environ.* **1987**, *62*, 67.
14.   Lombardi, A. T.; Morelli, E.; Balestreri, E.; Seritti, A. *Environ. Technol.*
      1992, *13*, 1013.
15.   Lombardi, A. T.; Seritti, A.; Morelli, E. in *Humic Substances in the Global
      Environment and Implications on Human Health*; Senesi, N.; Miano, T.
      M., Eds., Elsevier: Amsterdam, **1994**, 851.
16.   Morelli, E.; Puntoni, F.; Seritti, A. *Environ. Technol.* **1993**, *14*, 941.
17.   Liu, X. M.S. Thesis, University of Massachusetts Lowell, **1993**.
18.   Williams, R. F.; Aivaliotis, M. J.; Barnes, L. D.; Robinson, A. K. *J.
      Chromatogr.* **1983**, *266*, 141.
19.   Weber, J. H.; Wilson, S. A. *Water Res.* **1975**, *9*, 1079.
20.   Rate, A. W.; McLaren, R. G.; Swift, R. S. *Environ. Sci. Technol.* **1993**, *27*,
      1408.

Chapter 11

Organic Geochemistry and Sources of Natural Aquatic Foams

Margaret S. Mills[1,4], E. M. Thurman[1], John Ertel[2], and Kevin A. Thorn[3]

[1]Water Resources Division, U.S. Geological Survey, 4821 Quail Crest Place, Lawrence, KS 66049
[2]Skidaway Institute of Oceanography, 10 Ocean Science Circle, Savannah, GA 31411
[3]U.S. Geological Survey, Mail Stop 408, 5293 Ward Road, Arvada, CO 80002

Aquatic foams and stream-water samples were collected from two pristine sites for humic substances isolation and characterization. Biomarker compounds identified in foam and stream humic substances included phospholipid fatty acids, steroids, and lignin. Results showed that foams had a 10 to 20 fold greater DOC concentration and were enriched in humic substances (90% by weight of DOC) that showed increased hydrophobicity, aliphatic character, and compositional complexity compared to host stream humic substances (55 to 81% by weight of DOC). Foam humic substances also were enriched in humic acid (36 to 83% by weight) compared to host stream humic substances (10 to 14% by weight). Biomarkers, which contributed less than 5% by weight to the DOC pool, indicated higher plants, bacteria, algae, fungi, and diatoms as DOC sources. It is proposed that aquatic foams may be important media for the concentration and transport of organic substances in the aquatic environment.

Foams are ubiquitous in the environment, commonly seen as discolored patches on streams, rivers, lakes, and sea water. They often are assumed to be anthropogenic in origin as they are aesthetically unpleasant, yet they frequently appear in pristine environments indicating a natural origin. The chemical nature of foam, however, is ill defined, and geomorphological and geochemical constraints on natural foam formation are not well characterized.

Foams form where surface-active chemicals or "surfactants" are present. Surfactants also are called amphipathic agents and are characterized by the presence of both hydrophilic and hydrophobic functional groups (1). Surface-active agents that may be present in natural waters include fulvic and humic acids, collectively termed humic substances, (2,3), fatty acids and lipids (4), and proteins (5), and all have been identified as contributing surfactants in the few characterization studies that have been conducted on natural foams in freshwater environments (5-10) and marine surface layers (11-16).

[4]Current address: 32 Bloswood Lane, Whitchurch, Hampshire RG28 7BJ, United Kingdom

As an approach to investigating the complex chemistry of natural foams, humic substances (compounds sufficiently nonpolar at pH 2.0 to be isolated by reverse phase on XAD-8 and recovered in 0.1N sodium hydroxide) were isolated from aquatic foam and associated stream water for chemical characterization and investigations into surfactant behavior. Humic substances were chosen because they represent natural organic compounds present in natural waters that are sufficiently nonpolar at pH 2.0 to be isolated by XAD-8 adsorption. As surfactants also possess moderately nonpolar characteristics it follows that humic substances may contain a significant surfactant component. We hypothesized that foam would be enriched in humic substances compared to stream samples and would show increased hydrophobicity, aliphaticity, and decreased carboxylation in order to sustain surface-active behavior.

The objectives of this research were threefold. Our first objective was to determine if the origin of freshwater foams in pristine environments was anthropogenic or natural. This question was of interest to commercial surfactant manufacturers, the U.S. Department of Interior and the National Park service, and the general public to negate their fears of man-made contamination of pristine areas. Our second objective was to obtain detailed chemistry of the compounds present in the foam. Such information enables identification of the sources of the foam, such as bacteria, plants, algae, etc., by identifying source-specific "biomarkers." The biomarkers used in this research included phospholipid fatty acids, steroids, and lignin compounds. Chemical characterization would also indicate chemical complexity of natural foams to provide some insight into how and why the foams exist. Our third objective was to quantify natural foam as it represents an abundant source of nonpolar organic compounds in the aquatic environment. It is, therefore, an extremely important medium in understanding how nonpolar, low solubility compounds exist in natural waters, how they are transported, and their lability and environmental fate. Understanding this behavior may provide fundamental information on the mechanisms of how highly insoluble compounds, including insoluble manmade contaminants, exist and are transported in the aqueous environment rather than partitioning directly onto sediments.

Foams were collected for characterization from two pristine but geographically contrasting areas -- Como Creek, a pristine mountain stream near Ward, Colorado, receiving discharge from direct snowmelt, and the Suwannee River draining the organic rich Okefenokee swamp in Georgia which receives 85% of its water from direct precipitation.

Materials and Methods

Equipment. Silver filters of 0.45-μm pore size and 47-mm and 142-mm diameter were obtained from Osmonics, Inc., Minnetonka, MN. Stainless-steel pressure filtration equipment (47-mm and 142-mm plate filtration units) was obtained from Millipore (Bedford, MA). XAD-8 resin was purchased from Supelco, Inc. (Bellefonte, PA) and cation-exchange resin was purchased from Fisher Scientific (Ag-MP-5, Springfield, NJ). A Millipore XX80 peristaltic pump was used in the isolation of humic substances (Bedford, MA), and the freeze drying unit was a FTS Systems Flex Dry freeze drier, an FTS Systems shell freezer, and an FTS model VP62A vacuum pump (Stone Ridge, NY). The specific conductance meter (model 604) was obtained from Amber Science, Inc., Eugene, OR.

Foam and Stream-Water Collection. Samples from Como Creek, Colorado, were collected daily during a 3-week period in late May and early June 1993, which coincided with peak snowmelt and discharge. Forty liters of collapsed foam (approximately 150 L of aerated foam) were collected from quiet pools in a floating PVC U-shaped foam trap positioned with the opening facing upstream. The U-shaped trap was assembled using PVC drainpipe, elbow joints and caps and anchored using plastic cords attached to sunken concrete blocks. The foam collection was allowed to collapse overnight at 4 °C before combining as one sample. The foam was filtered through a 0.45-µm silver filter onsite prior to storing at 4 °C for processing. A 180-L stream-water sample was collected in 4-L baked glass bottles during high discharge at the beginning of June and stored at 4 °C until processed.

Samples from the Suwannee River, Georgia, were collected from below the spillway draining the swamp during a 4-day period in early January 1994, which coincided with peak rainfall and high swamp water levels. Approximately 40 L of collapsed foam (approximately 150 L of aerated foam) were collected from quiet pools, allowed to collapse, and then combined into one container before returning to the laboratory for filtration and processing. A 100-L stream-water sample was collected in 4-L baked-glass bottles from the headwaters prior to departure from the site.

Sample Preparation of Foam and Water Samples and Humic Substances Isolation. All foam and water samples were filtered through 0.45-µm silver filter using stainless-steel filtration units. Silver filtration of Como Creek and Suwannee River foam samples resulted in build up of a brown extract on the filter paper, which was readily solubilized in 0.1N sodium hydroxide. This extract was refiltered through silver filters as a sodium hydroxide solution. This fraction was believed to be colloidal in nature and was treated as a separate humic fraction, called the "foam-extract" fraction. A part of the filtered foam was freeze dried directly and considered "raw" foam. Fulvic and humic acids were isolated from foam and stream-water samples via the XAD-8 adsorption technique developed by Thurman and Malcolm (*17*), freeze dried, and weighed. To obtain a sufficient mass of humic substances, each entire sample was used for one extraction. As multiple samples were not extracted, calculation of the error associated with humic substances isolation cannot be made, and the contributions of humic substances to the DOC content must be regarded as estimates.

Physical Measurements.

Surface Tension. Surface-tension measurements were made in triplicate using a Fisher Autotensiomat equipped with a 4 mm-diameter platinum ring (Fisher Scientific, Springfield, NJ). The instrument was calibrated using distilled water, methanol, and a solution of sodium dodecyl sulfate. The surface tension measurements for SDS were lower than published values, indicating that the standard was not completely pure (*18*). However, the SDS measurement was still used as a reference point for all samples. Fifty-milliliter (50-mL) solutions of freeze-dried raw foam and humic and fulvic acids from foam and stream-water samples were made up by stock dilution ranging in concentration from 10 to 5,000 mg/L depending on solubility constraints for surface-tension measurements at pH 7.0. pH adjustments were performed using 0.1N NaOH.

Foaming Ability and Foam Stability. Foaming ability and foam stability were determined in triplicate on 100-mg/mL solutions of freeze-dried raw foam and humic substances from stream, foam, and foam-extract samples. Measurements were made at 25 °C at pH 3.0, 7.0, and 10.0, with sample pH adjusted using 0.1N sodium hydroxide using the Ross Miles method (D1173-53 ASTM standards, 1991). Briefly, a 200-mL aqueous sample was poured into a glass pipette with a 2.9-mm inside-diameter orifice and allowed to fall 90 cm into 50 mL of the same solution contained in a glass cylindrical vessel (*19*). The initial foam height was measured immediately after all of the sample solution had drained from the upper pipette into the foaming vessel. The height of the foam remaining after 5 minutes was defined as the stable foam height (*19*).

^{13}C NMR. Quantitative liquid-state carbon-13 nuclear magnetic resonance (^{13}C NMR) spectra were recorded for humic and fulvic acid from Como Creek foam and for stream and foam fulvic- and humic- acid samples from the Suwannee River at the U.S. Geological Survey, laboratory in Arvada, CO. ^{13}C NMR could not be performed on other humic substances due to insufficient sample or instrument availability. The acquisition parameters used were as follows: ^{13}C NMR spectra were recorded on a Varian XL-300 NMR spectrometer at 75 MHz. Each sample (200 mg of freeze-dried material) was dissolved in deuterated water and deuterated sodium hydroxide was added to ensure solution a total solution volume of approximately 6 to 7 mL. Spectra were recorded using a 30,000 Hz spectral window, a 45° pulse width, a 0.199 second acquisition time, and a pulse delay of 10 seconds for quantitative spectra. The number of transients was 10,000, and line broadening was 50 Hz.

Chemical Analyses.

Dissolved Organic Carbon (DOC) Analysis. Dissolved organic carbon (DOC) was measured by the persulfate-ultraviolet oxidation method using a Dohrman DC-80 total organic carbon analyzer (Xertex Dohrman Inc., Santa Clara, CA). Potassium hydrogen phthalate solution was used to calibrate the instrument over a 1 to 50-mg/L standard curve. Sample injection volume was 1.0 mL, which gave an operating range of 0.1 to 160 mg/L. Inorganic carbon measurements were made by acidification to pH 2.0 using 85% phosphoric acid and sparging of carbon dioxide by bubbling nitrogen through the sample for approximately 5 minutes.

Elemental Analysis. Carbon, hydrogen, nitrogen, oxygen, sulfur, phosphorous, and ash determinations were made on freeze-dried samples of raw foam, and humic and fulvic acids isolated from Como Creek and Suwannee River foam and water samples by Huffman Laboratories (Golden, CO). Trace-metal analysis was performed on freeze-dried Como Creek foam by Huffman Laboratories (Golden, CO).

Titration Measurements. The carboxyl and phenolic hydroxyl content of humic and fulvic acids from stream, foam, and foam-extract samples was determined by titration using carefully calibrated 0.1N sodium hydroxide. Briefly, 20 mg of freeze-dried sample were dissolved in 10 mL of distilled water, then titrated to pH 8.0 and pH 10.0 with 50-μL increments of 0.1N sodium hydroxide, standardized against potassium hydrogen phthalate. pH measurements were performed using an Orion Research Ionalyzer (910) and pH electrode (Orion Research, Cambridge, MA). Carboxyl content of the sample was calculated from the amount of sodium hydroxide required to raise the pH to 8.0. Phenolic hydroxyl content was calculated as twice the amount of sodium hydroxide required to raise the pH from 8.0 to 10.0.

Phospholipid Fatty Acid Analysis. Phospholipid fatty acid analysis was carried out on selected humic and fulvic freeze-dried samples of stream, foam, and foam extract from the Suwannee River by Microbial Insights, Inc., (Knoxville, TN). Briefly, lipids were extracted by liquid extraction (chloroform:methanol, 2:1, v/v) with phosphate buffer (*20*). The total lipid extract then was separated into three lipid classes by silicic acid chromatography (*21,22*). Fatty acids esterified to the phospholipids were methylated to their methyl ester derivatives by mild alkaline methanolysis of the polar lipid fraction by direct transesterification, followed by their separation and gas chromatography/mass spectrometry (GC/MS) analysis. For identification of monounsaturation position and geometry the fatty acids were derivatized further at the double bond using a dimethyl disulfide derivatization, which then becomes favored location for fragmentation.

Steroid Analysis. Steroid analysis was carried out on selected humic and fulvic freeze-dried samples of stream, foam, and foam extract from the Suwannee River by Microbial Insights, Inc., (Knoxville, TN). Tri-methyl silyl derivatives of 3 β-ol sterols were prepared from either the neutral lipid or total lipid fraction by alkaline saponification as described by Nichols et al. (*23*).

Lignin Analysis. Lignin present in fulvic and humic acids from stream, foam, and foam-extract samples from Como Creek and the Suwannee River were oxidized to phenolic oxidation products by copper oxide oxidation via the method described by Hedges and Ertel (*24*). These phenolic oxidation products then were derivatized to trimethyl silyl derivatives for GC/MS analysis.

Results and Discussion

Chemical Characterization. Measurements of specific conductance, pH, and dissolved organic carbon (DOC) content were made on collapsed foam and stream-water samples from Como Creek and the Suwannee River (Table I). The chemistry of the stream water from each site was distinct, reflecting the different geochemical settings. Como Creek stream samples (n = 38) contained a low calculated dissolved solids (from a specific conductance average of 21 μS/cm), an average pH of 6.6, and an average DOC concentration of 6.5 mg/L, reflecting the chemistry of almost pure snow, which is the source for most of the stream water, combined with sparse surrounding vegetation. In contrast, Suwannee River samples contained a low calculated dissolved solids (from a specific conductance of 41 μS/cm), but had an acidic pH of 3.9 and a DOC concentration of 37.7 mg/L. The low calculated dissolved-solids content stems from the low concentrations of dissolved solids found in rainwater the main source of water for the swamp -- combined with the high cation-exchange capacity of underlying peat in the swamp, which exchanges cations for hydrogen ions (*24*). The low pH is due to the release of hydrogen ions from the peat combined with large concentrations of organic acids leached from the abundant surrounding vegetation. The leaching of organic matter from decaying swamp vegetation also results in the high DOC concentrations measured, as well as imparting a dark-brown discoloration to the swamp water.

Specific conductance, pH, and DOC concentrations also were distinctly different between stream-water and collapsed-foam samples from each site. Foams possessed a 10 to 20-fold increase in DOC content compared to stream-water samples combined with a higher calculated dissolved solids content and lower pH than their host stream. This indicates that the foams are apparently a concentration medium for both inorganic matter and organic acids.

A summary of the percentage contribution by weight of humic substances to DOC, and the contributions of humic versus fulvic acid is presented in Table II. Typically, humic substances are reported to make up 50 to 60% of the DOC pool in streams and rivers and up to 75% of the DOC pool in streams draining wetland areas (*25,26*).

Table I. Average values for specific conductance, pH, and dissolved organic carbon (DOC) (± 1 standard deviation) for silver-filtered stream and collapsed foam samples from Como Creek and the Suwannee River.

Sample	Number of samples (n)	Specific conductance (μS/cm)	pH (standard units)	DOC (mg/L)
		Como Creek, late May and early June 1993		
Stream water	38	21 (± 2.59)	6.6 (± 0.12)	6.5 (±1.42)
Collapsed foam	1	35	6.2	58.7
		Suwannee River, early January 1994		
Stream water	6	41 (± 3.74)	3.9 (±0.29)	37.7 (± 1.59)
Collapsed foam	1	117	3.3	625.0

Humic substances isolated from stream samples from each site fell within these common ranges, with Como Creek humic substances accounting for 55% of the DOC whereas the Suwannee River humic substances accounted for 81% of the DOC. In contrast, foam humic substances accounted for 93% of the DOC from Como Creek foam and for 89% of the DOC from the Suwannee River foam. Organic carbon determinations were not made on the foam-extract fraction, and the percentage contribution by weight of humic substances to this material could not be determined.

Table II. Percentage contributions (by weight) of humic substances to DOC and the percentages of fulvic and humic acids contributing to the humic substances.

Sample	Percentage contributions (by weight) of humic substances in DOC	% Fulvic acid (by weight) in humic substances	% Humic acid (by weight) in humic substances
Como Creek, late May and early June 1993			
Stream water	55	86	14
Foam	93	45	55
Foam extract	--	39	61
Suwannee River, early January 1994			
Stream water	81	90	10
Foam	89	64	36
Foam extract	--	17	83

The contribution of fulvic acid versus humic acid commonly in streams ranges from 80 to 95% fulvic acid for stream and blackwater samples and as much as 95% for ground-water samples (*25*). The high concentrations of fulvic versus humic acid are thought to reflect the higher solubility of fulvic acids and the limited solubility of humic acids in natural waters due to the latter's hydrophobic nature and stronger affinity for sorption on surfaces (*25*). In this study fulvic acid contributions from Como Creek stream humic substances (86%) and the Suwannee River (90%) fell within these ranges, and Suwannee River yields were comparable to previously recorded yields (*26*) for that river. In contrast, fulvic acid contributions in foam humic substances were significantly less, contributing only 45% for Como Creek foam and 64% for Suwannee River foam, with a corresponding increase in the humic-acid contribution. The "foam-extract" humic substances were enriched further in humic acid, with fulvic acid contributing only 39% to the Como Creek foam extract humic substances, and 17% to the Suwannee River foam extract humic substances. Thus, both foam and foam-extract samples from both sites were enriched in humic acid relative to stream humic substances. Differences between the sites exist in the contribution by weight of the foam-extract fraction to the foam, which was much smaller for Como Creek (11% by weight) than for Suwannee River (58% by weight). In comparison, therefore, Como Creek foam contains a smaller humic-acid content than the Suwannee River foam and also a smaller contribution of the humic-acid rich foam-extract fraction.

Table III. Elemental data (reported as percentage contribution by weight)[1] for raw foams and humic substances isolated from stream, foam, and foam-extract samples from Como Creek and the Suwannee River.

Sample	Percentage contribution by weight (%)						
	Carbon	Hydrogen	Oxygen	Nitrogen	Phosphorous	Sulfur	Ash
Como Creek, late May and early June 1993							
Stream fulvic acid	49.96	5.32	43.61	0.76	0.01	0.50	0.74
Stream humic acid	*Insufficient sample*						
Foam fulvic acid	53.28	5.80	38.88	.99	.04	.40	1.88
Foam humic acid	53.20	6.35	34.15	1.96	.10	.50	4.85
Foam-extract fulvic acid	52.27	5.97	37.19	1.24	.07	1.08	3.83
Foam-extract humic acid	55.66	5.18	31.08	2.20	.09	.83	1.80
Raw foam	42.40	5.64	34.20	1.25	.09	1.18	17.33
Suwannee River, early January 1994							
Stream fulvic acid	49.89	4.62	44.90	0.81	0.004	0.53	0.20
Stream humic acid	51.81	4.46	42.49	.91	.009	.32	.05
Foam fulvic acid	55.56	6.09	37.10	.69	.005	.26	.54
Foam humic acid	54.72	5.57	36.80	.90	.006	.31	.40
Foam-extract fulvic acid	54.20	5.75	38.37	.84	.009	.93	1.18
Foam-extract humic acid	54.06	5.67	34.31	1.16	.008	.32	6.10
Typical values reported in the literature (25)							
Stream fulvic acid	51.90	5.00	40.30	1.10	0.60	0.2	1.50
Stream humic acid	50.50	4.70	39.60	2.00	--	--	5.00

[1] Percentages may not add up to 100 due to rounding

Elemental analyses were performed on humic and fulvic acids from stream, foam, and foam-extract samples from Como Creek and the Suwannee River and on raw foam from Como Creek, (Table III). The elemental composition of humic substances from Como Creek and Suwannee River water samples are comparable to typical values reported in the literature (*25,27*) (Table III). However, a general trend of increasing carbon (from 52 to 55%), and hydrogen (from 5 to 6%), with a corresponding decrease in oxygen content (from 40 to 34-37%) can be seen from fulvic to humic acids and from stream, to foam, to foam-extract samples from both sites. The lower oxygen content indicates less oxygen-containing functional groups are present, which combined with the increase in carbon and hydrogen content indicate an overall increase in the hydrophobic nature of the foam fraction relative to stream humic substances. Furthermore, the contribution of carbon may actually be larger than that reported in samples with a high ash contribution (samples with > 2% ash).

A ratio commonly used as an indication of the aliphaticity of a sample is the atomic hydrogen to carbon ratio (H:C). An increase in the ratio reflects an increased aliphatic content (-CH2-) over aromatic carbon (-C=C-). The values of the atomic hydrogen to carbon ratios for each fulvic and humic acid and for typical humic substances are shown in Table IV (*25*). The H:C ratios for stream and foam samples follow a trend of increasing from stream (1.06 to 1.3) to foam and foam-extract fractions (1.1 to 1.4) for both Como Creek and the Suwannee River, indicating an increase in aliphatic character from stream to foam humic substances.

The carboxyl and phenolic hydroxyl contents of stream and foam humic substances were determined by direct titration with 0.1N sodium hydroxide and are reported in Table V with typical published values for comparison. Como Creek and the Suwannee River stream fulvic and humic acids all fall within typical values and are comparable with published values (*26*) (Table V). Foam and foam-extract samples from Como Creek and the Suwannee River, however, possess a lower number of carboxyl moieties per gram of organic carbon compared to stream humic substances and lower carboxyl contents for humic versus fulvic acids. Fulvic acids from Como Creek foam and foam extract have a ratio of one carboxyl group for every 10.6 to 10.7 carbon atoms compared to 7.7 for Como Creek stream fulvic acid. The humic acids have a much higher ratio of one carboxyl group for every 17.7 and 18.8 carbon atoms, respectively. Suwannee River foam samples showed similar trends of decreasing carboxyl content from stream, to foam, to foam extract, and from fulvic to humic acids as did Como Creek samples, but the differences between the different groups were not as large. The carboxyl to carbon ratio ranged only from 8.7 to 13.2 for fulvic acid and humic acid foam and foam-extract samples. To explain this smaller difference it is hypothesized that at the acidic pH of the Suwannee River (pH 3.8), more than 50% of the organic acids in solution are in protonated form as the average pKa is approximately 4.2 (*25*) (The pKa of an acid is the negative log (base 10) of the dissociation constant, Ka (pKa = log 10 Ka) (*28*)). This would increase overall hydrophobicity of humic and fulvic acid mixtures in solution, enhancing their ability to act as a partition medium for other hydrophobic species. Less distinction may exist between fulvic and humic acid as partitioning media because of the increase in hydrophobicity of the fulvic-acid fraction due to protonation. Hydrophobic compounds may partition, therefore, into fulvic-acid mixtures, lowering the overall carboxyl content and leading eventually to a smaller difference in overall carboxyl content between fulvic and humic acids. In contrast, the higher pH of Como Creek results in all carboxyl moieties existing in full dissociated form. This leads to a greater difference in the solubility, hydrophobicity, and partitioning capacity of fulvic versus humic acids, and therefore a greater difference in their relative nonpolar chemistry and carboxyl contents.

It is interesting to note that the carboxyl contents of foam and foam-extract samples from Suwannee foam are comparable, even though the foam-extract is more hydrophobic than foam according to elemental analyses. This indicates that the foam-extract fraction is not an insoluble precipitate on the 0.45-μm filter paper as it has a

Table IV. Atomic hydrogen to carbon ratios (H:C) for raw foams and humic substances isolated from stream, foam, and foam-extract samples from Como Creek and the Suwannee River.

Sample	H:C
Como Creek, late May and early June 1993	
Stream fulvic acid	1.3
Stream humic acid	*Insufficient sample*
Foam fulvic acid	1.3
Foam humic acid	1.4
Foam-extract fulvic acid	1.4
Foam-extract humic acid	1.1
Raw foam	1.6
Suwannee River, early January 1994	
Stream fulvic acid	1.1
Stream humic acid	1.03
Foam fulvic acid	1.3
Foam humic acid	1.2
Foam-extract fulvic acid	1.3
Foam-extract humic acid	1.3
Typical ratios (25)	
Stream fulvic acid	1.16
Stream humic acid	1.11
Wetland fulvic acid	1.06
Wetland humic acid	1.03

Table V. Carboxyl and phenolic content (meq/g) and carboxyl:carbon ratio of humic substances from stream, foam, and foam-extract samples from Como Creek and the Suwannee River.

Sample	Carboxyl content (meq/g)	Phenolic content (meq/g)	Carboxyl:carbon ratio
Como Creek, *late May and early June 1993*			
Stream fulvic acid	5.38	1.96	7.7
Stream humic acid	*Insufficient Sample*		
Foam fulvic acid	4.16	1.27	10.7
Foam humic acid	2.45	1.56	17.7
Foam-extract fulvic acid	4.40	2.15	10.6
Foam-extract humic acid	2.44	1.96	18.8
Suwannee River, *early January 1994*			
Stream fulvic acid	5.87	1.47	7.0
Stream humic acid	5.09	1.96	8.5
Foam fulvic acid	4.45	1.47	10.4
Foam humic acid	4.01	1.96	11.4
Foam-extract fulvic acid	5.19	1.76	8.7
Foam-extract humic acid	3.42	1.96	13.2
Typical Values (25)			
Stream fulvic acid	5.5 - 6.2		
Stream humic acid	4.0		
Bog and wetland humic substances	5 - 5.5	1-2 *(for humic substances for most environments)*	

solubility comparable to the foam filtrate. Rather, it is simply too large in size to pass through the filter. It is hypothesized that the foam extract may represent a colloidal aggregate of polar and nonpolar compounds of significant compositional complexity (see biomarker analyses) (a colloid is a particle which is less than 0.00024 mm in size and is in suspension with a charged surface (29)). It is further hypothesized that the extract was able to pass through the 0.45-μm filter in 0.1N sodium hydroxide solution because this may have broken some interactions within the colloidal aggregate, decreasing its size as well as improving solubility by dissociation of all carboxylic acid moieties.

Further evidence for the existence of colloidal aggregates is shown from the XAD isolation procedure of humic substances. Chemical characterization of humic acid from the foam extract revealed a low carboxyl content (one carboxyl group for every 19 carbon atoms) compared to stream humic acid (one carboxyl group for every 12 carbon atoms). Thurman et al. (30) and Thurman and Malcolm (31) showed that a ratio of one carboxyl group per 12 carbon atoms was an important ratio in XAD isolation procedures. They showed that compounds containing less than one carboxyl group per 12 carbon atoms are too insoluble for effective elution from the XAD column and remain adsorbed to the XAD column. The successful XAD elution of the foam-extract humic acid, however, indicates that this hydrophobic fraction may have an increased solubility due to the presence of other charged groups aggregated together in a colloidal conformation. Previous studies have suggested that colloidal carbon can contribute as much as 30% of the DOC (32) and 50 to 70% of the DOC in stagnant water (33). Colloidal carbon becomes a significant contribution to the DOC in wetlands as the low ionic strength of the water is insufficient to cause its flocculation and removal. Additional evidence for the existence of aggregates includes the adsorption and elution from XAD of extremely nonpolar compounds present in humic substances (e.g. phospholipids, steroids), which is discussed in more detail later (see Biomarker Analyses).

Fulvic and humic acids from Como Creek and Suwannee River stream samples contained phenolic hydroxyl contents within the typical range shown in Table V. However, the phenolic hydroxyl content of Como Creek samples was highly variable from stream, to foam, to foam-extract fractions and between fulvic and humic acids, and did not show any clear trends. Suwannee River foam samples were much more uniform in concentration between stream, foam, and foam-extract fractions, with humic acids consistently higher than fulvic acids in all samples. The variable phenolic hydroxyl content of Como Creek samples and lack of clear trends compared to the more consistent content of Suwannee River samples indicates that in contrast to carboxyls, the phenolic hydroxyl moiety may not play a very important role in controlling surface activity or in affecting foam formation.

^{13}C NMR. Figures 1a, b, and c show the ^{13}C NMR spectra of fulvic and humic acids from Como Creek foam compared to the ^{13}C NMR spectra for humic substances isolated from Como Creek stream water by Shamet (10). A striking difference is the increased aliphatic I region and decreased aromatic and carbonyl regions for foam fulvic and humic acid compared with stream samples. This is consistent with the increased H:C ratio and decreased carboxyl content for foam humic substances compared to stream humic substances. Furthermore, the carbonyl peak in foam fulvic acid is split into two regions, with the 180 peak slightly higher than the 176 peak. This indicates that there are more aliphatic than aromatic carboxyl moieties (34). In comparison, the foam humic acid shows a greater 176 peak and only a slight 180 peak, indicating that humic acid contains mainly aromatic carboxyl moieties.

Quantification of each spectral region was carried out by cut and weigh methods (35) and are shown in Table VI for Como Creek. The aliphatic region I increased from 24.5% in stream humic substances to 43 and 40% in foam fulvic and humic acids, respectively. In contrast, the aromatic peak decreased from 34% in stream humic substances to 21 and 27% in foam fulvic and humic acids, respectively, while the

Table VI. 13C NMR integration results (reported as a percentage of peak area) for fulvic and humic acids from stream and foam samples from Como Creek and the Suwannee River.

Sample	Aliphatic region I	Aliphatic region II	Aromatic Content	Carbonyl Content	Ketone Content
			Percentage of peak area		
Como Creek					
Stream Humic Substances (*10*)	23.0	17.0	36.0	20.0	5.0
Foam fulvic acid	43.2	13.1	21.1	15.6	7.1
Foam humic acid	40.0	12.8	27.2	12.0	8.0
Suwannee River					
Stream fulvic acid	29.7	11.0	29.7	18.9	10.7
Stream humic acid	24.5	9.0	40.0	17.7	8.8
Foam fulvic acid	45.6	11.7	19.5	14.1	9.1
Foam humic acid	38.5	10.3	28.2	14.8	7.7

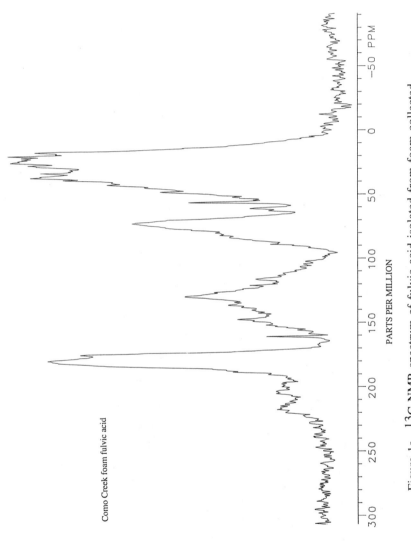

Figure 1a. 13C NMR spectrum of fulvic acid isolated from foam collected from Como Creek.

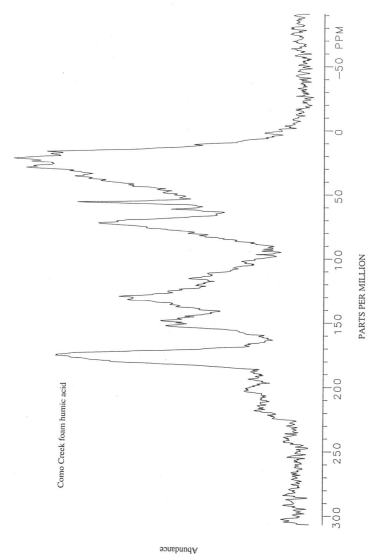

Figure 1b. ^{13}C NMR spectrum of humic acid isolated from foam collected from Como Creek.

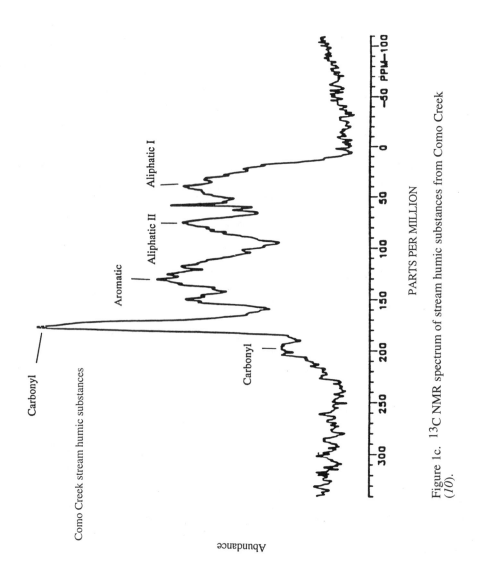

Figure 1c. 13C NMR spectrum of stream humic substances from Como Creek (*10*).

carbonyl peak decreased from 19% in stream humic substances to 15% and 12% in foam fulvic and humic acids, respectively.

The ^{13}C NMR spectra for Suwannee River stream and foam fulvic and humic acids are shown in Figures 2a, b, c, and d. The stream humic- and fulvic-acid ^{13}C NMR spectra are almost identical to published spectra (*34*). Quantification of the spectral regions are shown in Table VI. Similar trends are seen between the spectra of stream versus foam for Suwannee River as for Como Creek samples (Figures 1a, b, and c). Foam spectra show increased aliphaticity in region I and decreased aromaticity and carbonyl content. Differences for fulvic versus humic acids for stream and foam samples are similar to Como Creek, including increased aliphatic I and II regions and decreased aromaticity for the fulvic acid compared to humic acid but with comparable carbonyl contributions.

Although visually the ^{13}C NMR spectra of stream and foam humic substances from Como Creek and Suwannee River samples appear quite different from one another (e.g., broader, smoother peaks for all Suwannee samples), the quantitative contributions of each spectral region are comparable between the sampling sites. In particular, the fulvic acids from Como Creek and Suwannee River foam samples show great similarities, as do foam humic acids. This would indicate that, in terms of general aliphaticity and aromaticity between the two sites, the foam humic substances are comparable.

Quantification of Oxygen-Containing Functional Groups. Quantitative estimates of the contributions of the different carbonyl carbons present in the 160 to 180 signals for Como Creek and Suwannee River samples were made using a combination of ^{13}C NMR data, elemental analysis, and potentiometric titration data. For example, titration data indicated that foam humic acid from Como Creek contained 2.45 meq/g of carboxyl carbons. Elemental data indicate that 53.2% by weight of the sample is carbon, which translates to 44.3 mmol/g of carbon of which 12% is shown to be of carbonyl nature by ^{13}C NMR, or 5.432 mmol/g (meq/g). Thus, the difference between carboxyl content from titration (2.45 meq/g) and the size of the carbonyl peak (5.32 meq/g) can be assigned to other carbonyl bonds such as ester, lactone, and amide bonds. Table VII shows the relative contributions of different carbonyl functionalities for Como Creek and Suwannee River stream and foam samples. The data indicate that in all samples, more than 50% of the carbonyl peak is made up of carboxylic acid carbonyls, although Como Creek samples contain a smaller contribution of carboxylic acid moieties over ester/lactone/amide linkages compared to Suwannee River samples.

Surface Activity Measurements. The surface activity displayed by solutions of humic substances and raw foam samples from Como Creek and Suwannee River stream and foam samples was compared to the surface activity of an impure standard of commercial surfactant sodium dodecyl sulfate (SDS) and surface-tension measurements for both sites are shown in Figures 3a and 3b. Como Creek raw foam and foam-extract humic acid showed the greatest surface activity, with foam humic acid contributing to a lesser extent (Figure 3a). In contrast, Como Creek foam and foam-extract fulvic acid and stream humic substances showed little surface activity. Fulvic and humic acids from Suwannee River foam and foam extract showed comparable surface activity to the raw foam, and all samples were less surface active than the SDS (Figure 3b). Stream humic substances showed little surface activity and were comparable to Como Creek stream humic substances.

The differences between the surface activity of Como Creek and Suwannee River samples may be explained in terms of carboxyl content and carboxyl:carbon ratios.

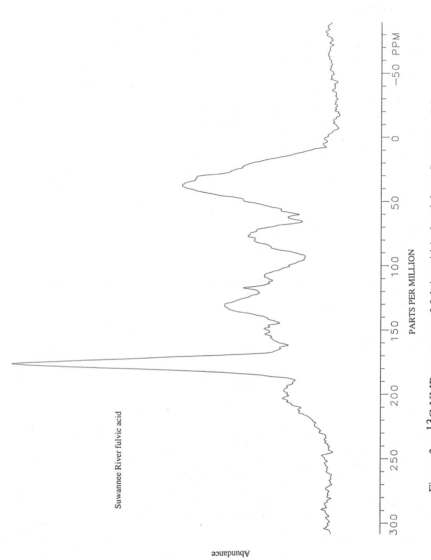

Figure 2a. 13C NMR spectrum of fulvic acid isolated from Suwannee River stream water.

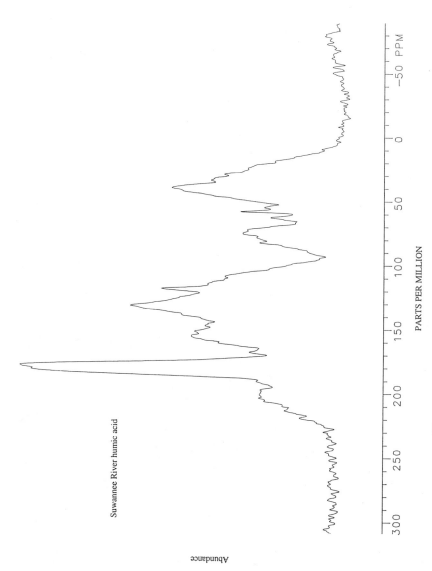

Figure 2b. ^{13}C NMR spectrum of humic acid isolated from Suwannee River stream water.

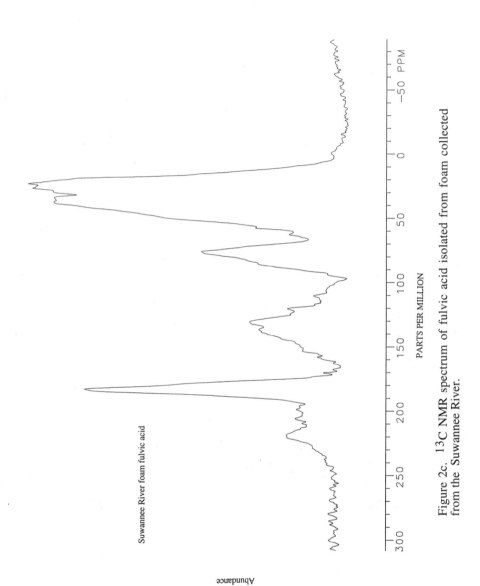

Figure 2c. ^{13}C NMR spectrum of fulvic acid isolated from foam collected from the Suwannee River.

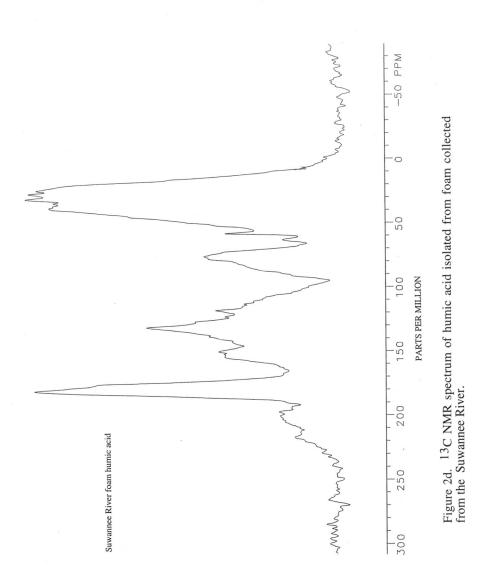

Figure 2d. 13C NMR spectrum of humic acid isolated from foam collected from the Suwannee River.

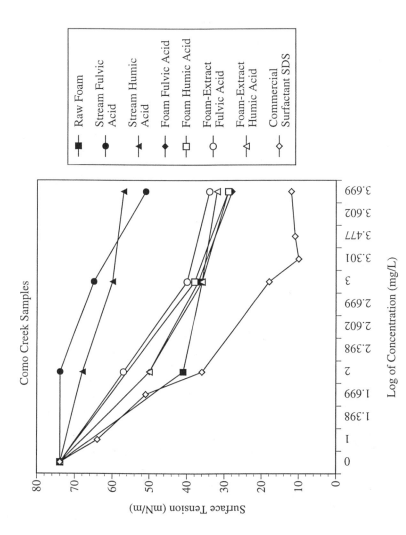

Figure 3a.   Surface-tension measurements (mN/m) of raw foam and humic substances from stream, foam, and foam-extract samples from Como Creek and commercial surfactant SDS as a function of concentration (mg/L).

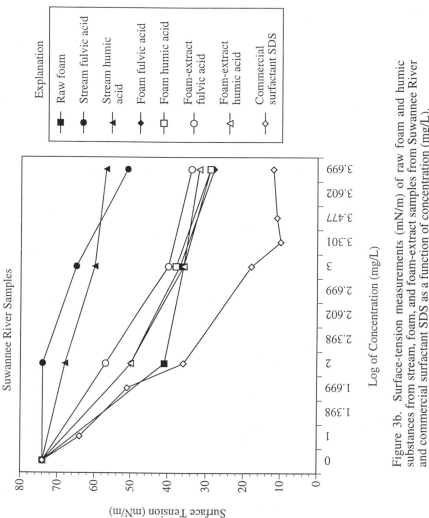

Figure 3b. Surface-tension measurements (mN/m) of raw foam and humic substances from stream, foam, and foam-extract samples from Suwannee River and commercial surfactant SDS as a function of concentration (mg/L).

Table VII. Concentrations (in meq/g) of oxygen-containing functional groups in fulvic and humic acids from stream and foam samples from Como Creek and the Suwannee River, using a combination of ^{13}C NMR data, elemental analysis, and potentiometric titration data.

Sample	Carboxylic acid (meq/g)	Ester, amide, lactone (meq/g)	Percentage contribution of ester, amide, lactone to carbonyl peak (%)	Ketone (meq/g)	Phenolic hydroxyl (meq/g)
Como Creek					
Como humic substances	4.50	3.62	44	2.03	1.08
Foam fulvic acid	4.16	2.77	40	3.15	1.27
Foam humic acid	2.45	2.87	54	3.55	1.56
Suwannee River					
Stream fulvic acid	5.87	1.93	25	4.44	1.47
Stream fulvic acid (32)	*6.10*	*2.00*	*25*	*2.60*	*1.50*
Stream humic acid	5.09	2.43	32	3.74	1.96
Foam fulvic acid	4.45	1.76	28	4.01	1.47
Foam humic acid	4.01	2.51	39	3.39	1.96

Como Creek samples show an overall trend of increasing surface activity with decreasing carboxyl content (Figure 4a). The foam-extract humic acid from Como Creek which has the lowest carboxyl content of 2.44 meq/g or a COOH:C ratio of 1:19 has the greatest amount of surface activity, and is comparable to the surface activity of raw foam and commercial surfactant, SDS (Figure 3a). Suwannee River samples also follow the trend of increasing surface activity with decreasing carboxyl content and increasing COOH:C ratios (Figure 4b). The smaller difference in surface activity between fulvic and humic acids in the foam and foam-extract fractions from Suwannee River may be due to the smaller differences in carboxyl content between these samples compared to Como Creek samples (Table V). Furthermore, the largest COOH:C ratio of 1:13 for the foam-extract humic acid from Suwannee samples may be the reason this fraction shows decreased surface activity compared to Como Creek foam-extract humic acid (COOH:C ratio of 1:19) and sodium dodecyl sulfate.

The surface activity of the different stream and foam humic-substance fractions and its relation to carboxyl content are strong evidence for the importance of the carboxyl groups in controlling the surface activity of the humic substances. Furthermore, the comparable surface-active behavior of the foam-extract fractions to raw foam is indicative that foam-extract humic substances are the main foaming agent in raw foam, and a chemical characterization of these fractions may reveal the chemical nature of the foaming constituents.

Foaming Ability and Foam Stability.   Initial and final foam heights of stream, foam, and foam-extract humic substances were compared to the commercial surfactant SDS as a function of pH (Figure 5). SDS showed the greatest foaming ability and foam stability compared to all other samples and was not affected by pH. The sulfonic acid moiety has a pKa below 3.0 and remains in dissociated form throughout the pH range tested. In contrast, both the foaming ability and foam stability of raw foam samples and fulvic and humic acids from both sites were pH dependent and were maximized at low pH values (3.0) and minimized at high pH values (10.0). Furthermore, the trend was one of increasing foaming ability and foam stability from stream, to foam, to foam extract and from fulvic to humic acids within each of these divisions.

The pH dependence of foaming indicates that protonation of carboxyl functionalities is required to increase hydrophobicity. In dissociated form, the humic- or fulvic-acid aggregate may have numerous polar or dissociated functional groups that hinder adsorption of hydrophobic sections of the aggregate at the air-water interface, thus hindering the aggregates ability to lower surface tension. Protonation of the carboxyl moieties serves to decrease the overall solubility of the substance and increase the hydrophobicity of the aggregate, resulting in increased surface adsorption, a lowering of surface tension, and the production of foam. Fulvic acids show little foaming ability or foam stability above pH 3.0, indicating that deprotonation of carboxylic moieties makes them too soluble for surface activity. In contrast, the humic acids retain sufficient hydrophobicity at higher pH's to remain slightly surface active. This is a direct reflection of the number of carboxyl moieties present within each fraction.

The pH dependence of foaming ability and stability has important ramifications for foaming events in the environment. For example, foaming at the headwaters of the Suwannee River is a sizeable event (three days was needed to collect a 40 liter sample) not only because more surfactant is available due to high DOC concentrations, but also because the acidic pH of the swamp water causes protonation of carboxyl moieties which enhances foaming ability and stability. In comparison, foaming is much less prolific on Como Creek (3 weeks were needed to collect a 40 liter sample) due to the

Como Creek Samples

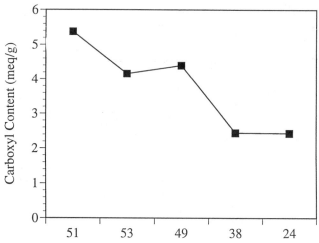

Surface Tension (mN/m) at 1000mg/L Concentration

Figure 4a.   Surface-tension measurements (mN/m) versus carboxyl content (meq/g) for Como Creek stream and foam humic substances.

Suwannee River Samples

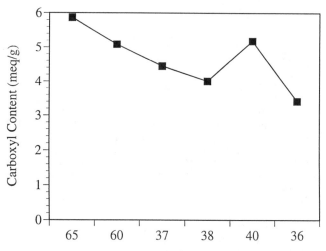

Surface Tension (mN/m) at 1000mg/L Concentration

Figure 4b.   Surface tension measurements (mN/m) versus carboxyl content (meq/g) for Suwannee River stream and foam humic substances.

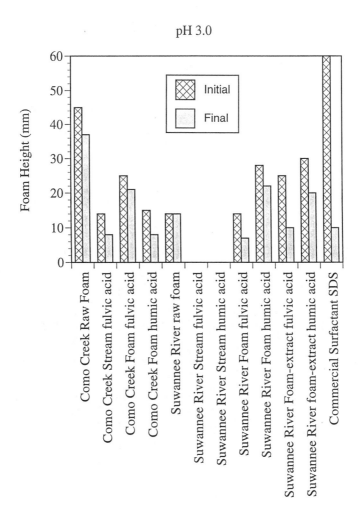

Figure 5. Measurements of foaming ability and foam stability for samples of raw foam, and humic substances from stream, foam, and foam extract samples from Como Creek and Suwannee River. Final foam heights were recorded after 5 minutes.

Continued on next page

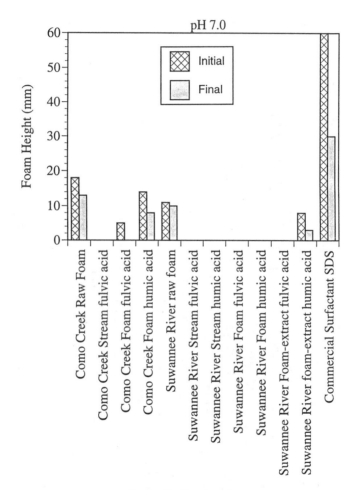

Figure 5. *Continued*

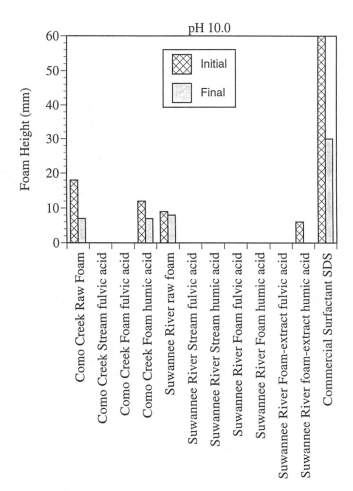

Figure 5. *Continued*

combined effects of lower DOC and a higher pH of the host stream (6.6) causing carboxyl groups to remain dissociated. Furthermore, foaming may be decreased further in natural waters that contain a high dissolved solids concentration, as the carboxyl moiety may form insoluble salts with magnesium and calcium ions forming scum. Both host streams in this study however, had unusually low total dissolved solids contents.

Summary of Chemical Characterization Data

In summary, the chemical characterization data show that humic substances in foam are an aliphatic, carboxyl-poor fraction, that are enriched in humic acid compared to stream humic substances. The carboxyl moiety appears to be the dominant ionic species affecting the surface activity of the humic substances, and foaming ability and foam stability are strongly dependent on pH as the average pKa of the organic acids is 4.2 (25).

The similarities in surface activity between raw foam and the foam-extract humic acid from Como Creek and Suwannee River indicate that the foam-extract humic acid is the main foaming agent. The low carboxyl content of the humic-acid extract, the increased aliphaticity, and the increased contribution of ester/lactone bonds to the carbonyl content indicate that this fraction of the DOC may contain lipids and other hydrophobic substances, which have undergone minimal oxidative degradation. This fraction, therefore, may represent fresh plant material that has encountered limited degradation.

Biomarker Analyses

Phospholipid Fatty Acid Analysis of Selected Suwannee River Samples. Phospholipid fatty acid concentrations and compositional make-up were determined for selected Suwannee River samples. Table VIII shows the general compositional makeup of the phospholipid fatty acid grouped into classes (e.g., branched, monounsaturated, polyunsaturated, and saturated). Individual fatty acids identified and their percentage contribution by weight to the overall phospholipid fatty acid profile are shown in Table IX. The largest phospholipid fatty acid content was in the raw foam, contributing 30,286 ng/g (30 ppm) (Table VIII) of fatty acids to the organic carbon pool (<0.003%, assuming 40% carbon content), with monounsaturated fatty acids (C16:1 and C18:1) making the largest contribution (63.2%, Table VIII) followed by saturated fatty acids (20.7%, Table VIII) (mainly C14:0, C16:0, and C18:0). The trend between samples was one of increasing phospholipid fatty acid content and compositional complexity from fulvic acid to humic acid, and from stream, to foam, to raw foam. After the raw foam, the foam-extract humic acid contained the largest and most diverse contribution of phospholipid fatty acids (1,481 ng/g of organic carbon) followed by foam humic acid (817 ng/g of organic carbon). As in the raw foam, monounsaturated fatty acids made the greatest contribution to the fatty acid profile in foam and foam-extract humic acid (50.1% and 51.3% by weight of total fatty acids identified, respectively) followed by saturated fatty acids (42.1% and 35.1% by weight of total fatty acids identified, respectively), with lesser amounts of polyunsaturated and branched fatty acids (Table VIII). Fulvic acids from stream and foam samples were compositionally the most simplified, containing only monounsaturated and saturated fatty acids.

Table VIII. Percentage contributions by weight of different classes of fatty acids bound to phospholipids in Suwannee River samples of raw foam and stream, foam, and foam-extract humic substances.

Sample	Percentage Contribution by Weight (%)						Total fatty acid (ng/g organic C)
	Terminally branched saturates	Mono-unsaturates	Poly-unsaturates Sat.	Branched mono-unsaturates	Mid-chain branched saturates	Satur-ates	
Stream fulvic acid	0	43.7	0	0	0	56.3	291
Foam fulvic acid	0	23.0	0	0	0	77.3	61
Foam humic acid	3.3	50.1	2.5	0	2.0	42.1	817
Foam-extract humic acid	8.3	51.3	4.7	0	.7	35.1	1,481
Raw foam	8.0	63.2	6.0	1.3	.8	20.7	30,286

Table IX. Individual fatty acid compositions of phospholipids isolated from raw foam and stream, foam, and foam-extract humic substances from the Suwannee River, in percentage of total fatty acids recovered.

Fatty acid	% of Total fatty acids				
	Stream fulvic acid	Foam fulvic acid	Foam humic acid	Foam-extract humic acid	raw foam

Terminally branched saturates

i 13:0	--	--	--	--	0.1
i14:0	--	--	--	--	.3
i15:0	--	--	3.3	3.7	3.4
a15:0	--	--	--	1.6	1.5
i16:0	--	--	--	1.1	.9
a16:0	--	--	--	--	.3
i17:0	--	--	--	1.5	.8
a17:0	--	--	--	.5	.7

Monoenoics

15:1 w6c	--	--	--	--	.3
16:1w9c	5.0	--	--	2.5	.2
16:1w7c	11.9	--	10.9	12.0	15.1
16:1w7t	--	--	--	--	.5
16:1w5c	--	--	--	--	.9
17:1w8c	--	--	--	--	.3
cy17:0	--	--	4.9	5.0	6.0
18:1w9c	10.1	23.0	4.8	7.2	5.2
18:1w7c	11.9	--	24.1	20.1	26.9
18:1w7t	--	--	--	--	.3
18:1w5c	--	--	--	--	.3
cy19:0	4.8	--	5.4	4.5	6.3
20:1w9c	--	--	--	--	.4
20:1w7c	--	--	--	--	.3
22:1w9c	--	--	--	--	.3

Table IX -- Continued.

Fatty acid	% of Total fatty acids				
	Stream fulvic acid	Foam fulvic acid	Foam humic acid	Foam-extract humic acid	raw foam
Polyenoics					
18:3w6	--	--	--	1.1	0.5
18:2	--	--	--	--	.3
18:2w6	--	--	2.5	3.7	4.2
20:4w6	--	--	--	--	.4
20:5w3	--	--	--	--	.3
20:3w3	--	--	--	--	.3
Branched monoenoics					
i17:1	--	--	--	--	.7
br18:1	--	--	--	--	.2
br19:1	--	--	--	--	.3
Mid-chain branched saturates					
10me16:0	--	--	--	--	0.5
10me17:0	--	--	2.0	0.7	.2
10me18:0	--	--	--	--	.2
Normal saturates					
12:0	--	--	--	--	0.2
13:0	--	--	--	--	--
14:0	7.8	--	3.6	3.5	1.6
15:0	3.5	--	13.0	2.6	.9
16:0	33.0	46.4	19.7	21.9	14.4
17:0	--	--	--	1.1	.8
18:0	11.9	31.0	5.8	6.1	2.2
20:0	--	--	--	--	.2
22:0	--	--	--	--	.2
23:0	--	--	--	--	.1
24:0	--	--	--	--	.2

Table X. Steroid content of Suwannee River samples of raw foam and humic substances from stream, foam, and foam extract.

Sample	Steroid content (ng steroid/g of organic carbon)						
	Choles-terol	Brassicas-terol	Campes-terol	Stigmas-terol	Sitomas-terol	Unks-terol	Total
Stream fulvic acid	426	---	---	---	---	---	426
Foam fulvic acid	---	---	---	---	---	---	---
Foam humic acid	1,347	---	---	---	---	---	1,347
Foam extract	1,242	---	---	338	623	---	2,204
Humic acid							
Raw foam	12,487	2,980	4,892	8,016	12,497	3,869	44,745

The results show that fatty acids derived from phospholipids are a small contribution to the organic carbon pool (picogram to nanogram/gram of organic carbon; <0.001% of the organic carbon). Furthermore, the complexity and large concentration of phospholipid fatty acids in raw foam versus humic substances from stream and foam indicate that most of these phospholipids are not structurally associated with humic substances and that the foam may simply be acting as a concentration medium for nonpolar compounds in solution.

Using the fatty acids as biomarkers reveals important information on the origin of the lipids and possible sources of the foam constituents (*36*). The phospholipid fatty acids identify bacteria, algae, fungi, diatoms, and higher plants as contributing sources of phospholipids. Bacterial input is indicated by the presence of monoenoic fatty acids (such as C18:1w7c and C16:1w7c), terminally branched saturates, branched monoenoic fatty acids, and mid-chain branched saturates, all of which are present in the raw foam. The raw foam also contains polyenoic fatty acids such as 20:4w6, which is indicative of a eukaryote source such as protozoa; 20:5w3, which is indicative of diatoms; and C18:2w6, which is indicative of higher plants (*36*). The fatty acid makeup of foam and foam-extract humic acid is very similar to raw foam, but at lower concentrations.

Steroid Analysis of Selected Suwannee River Samples. The same selected Suwannee River samples used for phospholipid analysis were also analyzed for the presence of six different steroids. The steroid compositional makeup and concentrations in stream, foam, and foam-extract samples are shown in Table X. Raw foam again showed the most complex steroid profile, with total steroid contributing 0.0045% to the organic carbon pool (assuming 40% organic carbon content). All six steroids were detected, indicating input from plants and algae (cholesterol, campesterol, sitosterol, and stigmasterol) and diatoms (brassicasterol). There is a possibility of animal input also. Bacteria characteristically lack steroids and, therefore, cannot contribute (*36*). The foam-extract humic acid contained the next largest steroid component consisting only of three steroids of plant origin, which contributed <0.0004% to the organic carbon pool. Similarly, stream humic substances contained small concentrations of cholesterol.

Steroids are hydrophobic molecules and sparingly soluble with only one hydroxyl functional group. Normally steroids are found concentrated in the sediment phase via partitioning onto organic matter on sediments (*25*). Their concentration in the foam and humic-acid fraction is indicative of the hydrophobic nature of these fractions. The increasing compositional complexity from stream, to foam, to foam-extract samples is also consistent with other characterization trends of increasing compositional complexity and hydrophobicity.

Lignin Oxidation Analysis. The oxidation of lignin under CuO conditions (*24*) results in phenolic oxidation products which display different characteristic patterns according to their vascular plant origin -- either angiosperm (flowering) or gymnosperm (nonflowering). For example, syringyl phenols are common only to angiosperms, whereas cinnamyl phenols are common only to nonwoody parts of angiosperms (e.g., grasses, needles, leaves, etc.) (*37*). Table XI shows the percentage contribution by weight of each phenol class to the total sum of phenols detected in Como Creek and Suwannee River stream, foam, and foam-extract samples. The samples from Como Creek show distinctly different phenolic contents compared to Suwannee River samples, reflecting the different flora surrounding the streams. A trend seen at each site, however, is lignin concentrations increase from stream, to foam, to foam-extract, and from fulvic to humic acids, with stream fulvic acid containing the least and foam-extract humic acid containing the most.

Overall, trends showed that Como Creek samples have higher lignin content than Suwannee River samples (1.34 to 3.47% of the organic carbon) and are enriched in vanillyl phenols, whereas syringyl, β-hydroxyl, and cinammyl phenols make smaller contributions. Suwannee River lignin contributes 0.6 to 1.46% of the organic carbon and again is enriched in vanillyl phenols. The presence of vanillyl, syringyl, and cinnamyl phenols in humic substances from stream and foam samples indicates that a lignin-derived structure contributes to the DOC. Furthermore, the fulvic acids uniformly yield less lignin-derived phenols that are depleted in syringyl and cinnamyl phenols relative to humic acids, which is in agreement with the findings of Ertel (37). Ertel (37) notes that it is not clear whether the differences between fulvic and humic acids are due to diagenesis and degradation or are simply a function of the partitioning process, but concludes that the lignin components of humic acid could be transformed to fulvic acid but not vice versa. All stream and foam samples contain all building blocks, and individual plant sources cannot be recognized, although all apparently contribute.

The relative contributions of acidic versus aldehydic phenols from vanillyl and syringyl families have been used as an indication of the extent of oxidative degradation of the lignin source material (37). Higher acid to aldehyde ratios indicate greater oxidative degradation. Vanillyl acid:aldehyde ratios and syringyl acid:aldehyde ratios are shown in Table XII for Como Creek and Suwannee River samples. Acid:aldehyde ratios for Como Creek samples show a trend of increasing from humic acids to fulvic acids (i.e., increased oxidation in fulvic acids). Furthermore, ratios tend to increase in the fulvic acids from stream, to foam, and to foam-extract fulvic acid. However, the opposite is true for humic acids, which decrease in ratio size from stream, to foam, to foam extract. This may reflect a partitioning effect, with less-oxygenated species partitioning into the humic- acid fraction and especially in the foam-extract fraction. In Suwannee River samples the trend of higher acid:aldehyde ratios in fulvic acid compared to humic acid is not as evident as in Como Creek samples, with much smaller differences between the two fractions, similar to other characterizations. However, the trend remains of increasing ratios from stream through foam extract for fulvic acid and the opposite for humic acids.

Summary of Biomarker Analyses. The combination of phospholipid fatty acids, steroids, and lignin phenols indicates that the chemical makeup of raw foam includes input from bacterial, algal, diatoms, fungal, and higher plant sources. The total of these compounds account for less than 5% of the organic carbon present in raw foam, and it is not possible, therefore, to assess which is the largest source. These compounds however, do reveal interesting compositional trends between the raw foam and the stream and foam humic substances. The compositional complexity of humic substances increases from stream, to foam, to foam extract and from fulvic acids to humic acids.

Discussion of Characterization Results and Hypotheses on Foam Chemistry

The chemical and physical evidence presented here shows that aquatic foams collected from pristine environments may have an entirely natural origin. The carboxyl moiety is the dominant anion of the surfactant as foaming ability and stability change dramatically over the pKa range of the carboxyl group (pH 3-5) (25). Most commercial anionic surfactants contain sulfur-containing anions which are not

Table XI. Percentage contributions by weight of the different phenol classes to the overall lignin phenol content for humic substances from stream, foam, and foam-extract samples from Como Creek and the Suwannee River.

Sample	% Contribution to total phenol content				Sum of * phenols (µg/g organic carbon)	% phenol carbon ** total organic carbon
	Vanillyl phenols	Syringyl phenols	Cinnamyl phenols	β-Hydroxyl phenols		
Como Creek						
Stream fulvic acid	62	15	4	19	21260	1.34
Stream humic acid	74	13	4	9	54949	3.47
Foam fulvic acid	50	17	6	27	18739	3.25
Foam humic acid	58	24	6	11	26222	1.66
Foam-extract fulvic acid	57	17	8	18	33460	2.11
Foam-extract humic acid	53	30	6	11	38497	2.44
Suwannee River						
Stream fulvic acid	24	21	5	49	10012	0.62
Stream humic acid	39	21	4	36	16962	1.06
Foam fulvic acid	37	27	5	31	9606	0.6
Foam humic acid	40	31	3	26	18344	1.15
Foam-extract fulvic acid	41	25	6	28	21958	1.38
Foam-extract humic acid	47	30	3	20	23172	1.46

* B-hydroxyl phenols may have a non lignin source.
** Phenol carbon contributes approximately 60% to the total weight of phenol.

Table XII. The ratio of acid to aldehyde for vanillyl and syringyl phenols from stream, foam, and foam-extract humic substances from Como Creek and the Suwannee River.

	Vanillyl acid:aldehyde ratio	Syringyl acid:aldehyde ratio
Como Creek		
Stream fulvic acid	1.20	0.46
Stream humic acid	.89	.52
Foam fulvic acid	1.22	.56
Foam humic acid	.90	.42
Foam-extract fulvic acid	1.58	.74
Foam-extract humic acid	.73	.39
Suwannee River		
Stream fulvic acid	0.80	0.43
Stream humic acid	.96	.62
Foam fulvic acid	.82	.58
Foam humic acid	.86	.49
Foam-extract fulvic acid	1.28	.77
Foam-extract humic acid	.84	.46

protonated over the pH range of 3-10. Furthermore, surface activity increases linearly with decreasing carboxyl content indicating that the carboxyl moiety is the dominant anion controlling surfactancy. As a result, the occurrences of natural foaming events are largely dependent on the chemistry of the host stream. For example, prolific foaming events should be expected on host streams with a more acidic pH where the carboxyl moiety is in protonated form and surface activity is enhanced.

Biomarker analyses (phospholipids, steroids, lignin) of Suwannee River river and foam samples identified less than 5% of the DOC in humic substances and revealed both allochthonous and autochthonous sources for humic substances. The fingerprinting of plant input from the presence of lignin, steroids, and phospholipids and the fingerprinting of bacteria, fungi, algae, and diatoms from the presence of specific phospholipid fatty acids highlights the plethora of sources of DOC. From these biomarker analyses and the evidence of a low carboxyl content and minimal oxidative degradation of the foam humic substances fraction, it is hypothesized that foams may be most prevalent during wet periods due to the flush of compounds of limited oxidative decay and high surface activity from senesced plant material into surface water, which triggers increased fresh bacterial and algal excretions. These fresh leachates can then be manifest as foam due to increased host stream turbulence from the rain event.

The biomarker analyses also revealed that nonpolar, relatively insoluble compounds exist in the water column (e.g. steroids, phospholipids) indicating that interactive processes may be at work to enhance the ability of these compounds to remain in solution, such as the formation of an aggregate. The existence of a foam fraction that will not pass through a 0.45 micron filter yet maintains similar solubility characteristics to the foam fraction that did pass through the filter is evidence for the existence of aggregates which are simply too large in size to pass through the 0.45 micron filter.

The existence of aggregates in natural waters has been discussed previously. Evidence for weak bonding interactions and the formation of aggregates have been reported using x-ray scattering experiments, which showed that hydrogen bonding is an important mechanism of aggregation (*38*). An aggregate is similar to a protein, with exterior hydrophobic sections which are surface active interspersed with hydrophilic sections to maintain solubility. D'Arrigo et al. (*9*) have postulated that a water-soluble, surfactant-rich extract from a Hawaiian forest soil was a glycopeptide-lipopolysaccharide complex containing phenolic, lignin-derived materials as well as other aliphatic materials, held together via hydrogen-bonding interactions to form a surfactant complex. Similarly, hydrogen bonding between proteins and carbohydrates in monomolecular films at an air-water interface has also been described (*39,40*). Humic substances have been shown to associate with the negative surface charge of the aluminosilicate by cation-exchange interactions via positively charged nitrogen groups (*41*), or via a negatively charged carboxyl moiety interacting with the positively charged surface of a metal colloids such as hydrous aluminum or iron oxide (*42-45*). Humic substances can also form charge-transfer complexes with metal ions such as copper (II) and iron (III) via carboxyl groups and carbonyl groups (*42,43,46-50*).

In summary, the existence of stream humic substances, foam humic substances, and foam filtrate humic substances that all contain increasing amounts of nonpolar organic components yet still remain solubilized indicates that aggregates of different sizes and complexity exist. The foam-extract humic acid is the largest and most hydrophobic of the aggregates and has undergone the least oxidative degradation (low carboxyl content, increased ester/lactone/amide contributions, low acid:aldehyde ratio of lignin oxidation products). Stream humic and fulvic acids, on the other hand, represent the oxidized and degraded end-member skeletons of these initially more complex aggregates. Further work is currently underway to characterize the size, molecular weight, and composition of particles making up the foam.

Conclusions

Foam is a concentrated organic medium with 10 to 20 times the dissolved organic carbon content of the bulk water phase. Humic substances make up more than 90% of the dissolved organic carbon in the foam compared to 55% in stream water and 81% in wetlands. Foam humic substances are made up of two distinct fractions -- a fraction that will pass through a 0.45-μm silver filter and a fraction that is too large and collects on the filter (foam extract) but which is entirely soluble in 0.1N sodium hydroxide. In Suwannee River samples both foam fractions have comparable carboxyl contents (i.e. similar solubilities). Both foam and foam-extract fractions are enriched in humic acid (36% and 55% by weight, respectively for Como Creek and 61% and 83% by weight, respectively for Suwannee River) compared to stream humic substances (10% and 14% for Como Creek and Suwannee River samples, respectively). Characterization of humic substances reveals uniform trends from stream, to foam, to foam extract and from fulvic acid to humic acid of increasing carbon and hydrogen content, decreasing oxygen content, increased aliphaticity, increased surface activity, foaming ability, and foam stability, increased fatty acid, steroid, and lignin concentrations and compositional complexity, decreased carboxyl content, and a decrease in the number of carboxyl moieties per carbon atom. In general, the trend is of increasing hydrophobicity from stream, to foam, to foam-extract samples. Furthermore, biomarkers indicate both allochthonous and autochthonous sources contribue to the DOC, including plants, algae, bacteria, fungi, and diatoms.

Chemical and physical evidence indicates that foams may contain organic aggregates, with varying degrees of hydrophobicity, surface activity, and compositional complexity. The most hydrophobic and chemically complex aggregate appears to exist in the foam extract humic acid which displays the greatest surface activity. Stream humic and fulvic acids are the highly degraded end member skeletons of these initially more complex aggregates that possess limited surface activity. It is proposed that aquatic foams may be important media for the concentration and transport of polar and nonpolar organic substances in the aquatic environment.

Acknowledgments

The authors wish to thank the following group and individuals for their contributions to this research: The Surface and Ground Water Toxic Program, U.S. Geological Survey, and Steve Seibold and Bill Bowman, Colorado Mountain Research Station, Nederland, Colorado, and Sarah Brown, Okefenokee Wildlife Refuge, Folkston, Georgia, for allowing sample collection. The use of brand, trade, or firm names in this paper are for identification purposes only and do not constitute endorsement by the U.S. Geological Survey.

Literature Cited

1   Walstra, P. In *Foams: Physics, chemistry and structure*; Wilson, A.J., Ed.; Springer Verlag; 1989.

2   Tschapek, M; Wasowski, C. *Geochim. Cosmochim. Acta* **1976**, *40*, 1343-1345.

3   Chen, Y; Schnitzer, M. *Soil Science* **1978**, *125*, 7-15.
4   Hullett, D.A. *The characterization of biologically-important organic acids in the Upper Mississippi River*; University of Minnesota: Minnesota, MN, 1979; Thesis, Master of Science
5   Van Beneden, G., *Compt. Rend. Journee Hydraulique, Soc., Hydrotech. France, Chemical Abstracts* **1962**, *59*, 6127.
6   Goldacre, R.J. *J. Anim. Ecol.* **1949**, *18*, 36-39.
7   Duce, R.A.; Quinn, J.G.; Oiney, C.E.; Piotrowicz, S.R.; Ray, B.J.; Wade, T.L. *Science* **1972**, *176*, 161-163.
8   D'Arrigo, J.S. *J. Chem. Phys.* **1981**, *75*, 962-968.
9   D'Arrigo, J.S.; Saiz-Jimenez, C.; Reimer, N.S. *J. Colloid and Interface Sci.* **1983**, *100*, 96-105.
10   Shamet, K.A. *Surfactants in a natural aquatic foam. Characterization and role in enhanced degradation of herbicide;* University of Kansas: Lawrence, KS, 1992; Masters Thesis.
11   Garrett, W.D. *Deep Sea Res.* **1967a**, *14*, 221.
12   Garrett, W.D. *Deep Sea Res.* **1967b**, *14*, 661-672.
13   MacIntyre, F. *Sci. American* **1974**, *230*, 62-77.
14   Seiburth, J. McN.; Willis, P.J.; Johnson, K.M.; Burney, C.M.; Lavoie, D.M.; Hinga, K.R.; Caron, D.A.; French, F.W. III; Johnson, P.W.; Davis, P.G. *Science* **1976**, *19*, 1415-1418.
15   Carlson, D.J. *Limnol. Oceanogr.* **1983**, *28*, 415-431.
16   Marty, J.C.; Zutic, V.; Precali, R.; Saliot, A.; Cosovic, B.; Smodlaka, N.; Cauwet, G. *Mar. Chem.* **1988**, *25*, 243-263.
17   Thurman, E.M.; Malcolm, R.L; *Environ. Sci. Technol.* **1981**, *15*, 463-466.
18   Defay, R.; Prigogine, I. *Surface Tension and Adsorption;* J. Wiley and Sons, Inc.: New York, NY, 1966, pp 113.
19   Rosen, M.J. *Surfactants and interfacial phenomena;* John Wiley and Sons, Inc.: New York, NY, 1989.
20   White, D.L.; Davis, W.M.; Nickels, J.S.; King, J.D.; Bobble, R.J. *Oecologia (Berl)* **1979**, *40*, 51-62.
21   Gehron, M.J.; Davis, J.D.; Smith, G.A.; White, D.C. *J. Microbial. Meth.* **1984**, *2*, 165-176.
22   Guckert, J.B.; Antworth, C.P.; Nichols, P.D.; White, D.C. *FEMS Microbiol. Ecol.* **1985**, *31*, 147-158.
23   Nichols, P.D.; Volkman, J.K.; Johns, R.B. *Phytochem.* **1983**, *22*, 1447-1452.
24   Hedges, J.I.; Ertel, J.R. *Anal. Chem.* **1982**, *54*, 174-178.
25   Thurman, E.M. *Organic geochemistry of natural waters;* Martinus Nijhoff/Dr. W. Junk Publishers: Dordrecht, The Netherlands, 1985.
26   Malcolm, R.L.; McKnight, D.M.; Averrett, R.C. In *Humic substances in the Suwannee River, Georgia: Interactions, properties, and proposed structures;* Averett, R.C.; Leenheer, J.A.; McKnight, D.M.; Thorn, K.A., Eds.; U.S. Geological Survey Open-File Report # 87-557, Denver, CO,1989.
27   Malcolm, R.L. In *Humic substances in soil, sediment, and water;* Aiken, G.R.; McKnight, D.M.; Wershaw, R.L.; MacCarthy, P., Eds.; 1985, Ch. 7, 181-209.
28   Bates, R.C.; Jackson, J.A., Eds. *Dictionary of Geological Terms;* 3rd Edition, American Geological Institute: Doubleday: New York, NY, 1984.
29   Wade, L.G.Jr. *Organic Chemistry*, Prentice Hall, Inc.: New Jersey, USA, 1995, pp 24.
30   Thurman, E.M.; Malcolm, R.L.; Aiken, G.R. *Anal. Chem.* **1978**, *50*, 775- 779.

31   Thurman, E.M.; Malcolm, R.L. *Concentration and fractionation of*
        hydrophobic   organic acid constituents from natural waters by liquid
        chromatography; U.S. Geological Survey Water Supply Paper 1817-G:
        Reston, VA, 1979.
32   Koenings, J.P.; Hooper, F.F. *Limnol. Oceanogr.* **1976**, *21*, 684-696.
33   Mullholland, P.J. *Limnol. Oceanogr.* **1981**, *26*, 790-795.
34   Thorn, K.A.; Folan, D.W.; MacCarthy, P. *Characterization of the*
        international   humic substances society standard and reference fulvic and
        humic acids by solution state carbon-13 (^{13}C) and hydrogen-1 (^{1}H) nuclear
        magnetic resonance spectrometry; U.S. Geological Survey Water-Resources
        Investigations Report 89-4196: Denver, CO, 1989.
35   Thorn, K.A. In *Humic substances in the Suwannee River, Georgia:*
        Interactions,   properties, and proposed structures; Averrett, R.C.;
        Leenheer, J.A.; McKnight, D.M.; Thorn, K.A., Eds.; U.S. Geological Survey
        Open-File Report 87-557: Denver, CO, 1989; pp 377.
36   Harwood, J.L.; Russell, N.J. *Lipids in plants and microbes*; George Allen
        and Unwin Ltd: London, 1984.
37   Ertel, J.R. *The lignin geochemistry of sedimentary and aquatic humic*
        substances; University of Washington: Seattle, WA, 1985; Ph.D. Thesis.
38   Wershaw, R.L. *J. Contam. Hydrol.* **1986**, *1*, 29-45.
39   MacRitchie, F.; Alexander, A.E. *J. Colloid Sci.* **1961**, *16*, 57-61.
40   MacRitchie, F; Alexander, A.E. *J. Colloid Sci.* **1961**, *16*, 61-67.
41   Wershaw, R.L.; Pinckney, D.J. In *Contaminants and sediments*; Baker, R.A.,
        Ed.; Ann Arbor Science Publishers: Ann Arbor, MI, 1980, Vol. 2; pp 207-219.
42   Wershaw, R.L.; McKnight, D.M.; Pinckney, D.J. In *Proceedings of the second*
        international symposium on peat in agriculture and horticulture; Schallinger,
        K.M., Ed.; Hebrew University: Jerusalem, 1983, 205-222.
43   Gamble, D.S.; Langford, G.H.; Underdown, A.W. In *Complexation of trace*
        metals in natural waters; Kramer, C.J.M.; Duinker, J.C., Eds.; Martinus
        Nijhoff/Dr. W. Junk: The Hague, The Netherlands, 1984.
44   Davis, J.A. *Geochim. Cosmochim. Acta* **1982**, *46*, 2381-2393.
45   Tipping, E; Cooke, D. *Geochim. Cosmochim. Acta* **1982**, *46*, 75-80.
46   Boyd, S.A.; Sommers, L.E.; Nelson, D.W.; West, D.X. *Soil Sci. Soc. Am. J.*
        1981, *45*, 945-949.
47   Kallianon, C.S.; Yassoglon, N.J. *Geoderma* **1985**, *35*, 209-221.
48   Senesi, N.; Sposito, G.; Martin, J.P. *Sci. Total Environ.* **1986**, *55*, 351-362.
49   Ghosh, K; Schnitzer, M.; *Soil Sci. Soc. Am. J.* **1981**, *45*, 25-29.
50   Giesy, J.P.; Briese, L.A. *Chem. Geol.* **1977**, *20*, 109-120.

METAL BINDING

Chapter 12

Complexing of Metal Ions by Humic Substances

Ying-Jie Zhang, Nicholas D. Bryan, Francis R. Livens,
and Malcolm N. Jones

Department of Chemistry, University of Manchester,
Manchester M13 9PL, England

The interaction of metal ions with humic substances is being studied
using two different techniques. UV-scanning ultracentrifugation is being
used to determine molecular weights and to investigate changes in
aggregation brought about by metal ion complexation. The relationship
between cation charge and conformation of the humic ligands is being
investigated.
    The complexation of U by humic substances from soils contami-
nated by natural processes is also being studied. Gel permeation
chromatography has been used to show that different fractions of humic
substances vary greatly in their effectiveness as ligands. These studies
have also shown that uranium desorption is redistributed slowly between
different fractions of humic substances following its initial adsorption.

It is well known that humic substances (humin, humic and fulvic acids) have a
substantial capacity to complex dissolved species such as metal ions and cationic organic
molecules and to interact with mineral surfaces [e.g. 1-5]. The complexation process may
affect the solubility of both the humic ligand and the species bound. For example, a
complex geochemical interrelationship has been identified between fulvic acids, poorly
crystalline Fe oxides and Fe(II) and Fe(III) ions [6]. In addition, scavenging by organic
coatings has been found to remove a range of metal ions efficiently from solution [e.g.
3,7]. Similarly, the humic ligands themselves are a complex mixture of colloidal
macromolecular species, whose state of aggregation and solubility may be affected by
concentration, by ionic strength and by metal binding to specific coordination sites [8-
15].
    These processes make the study of metal ion-humic interactions particularly
complex since the humic ligands cannot be treated as isolated, simple molecules and
changes in metal ion binding can significantly affect their state of aggregation and thus
their behaviour. The humic macromolecule may be viewed as presenting a range of
binding sites to a dissolved species. These may be either "specific" binding sites; that is,
particular functional groups or combinations of functional groups may co-ordinate metal

0097-6156/96/0651-0194$15.00/0

ions, or alternatively, metal binding may be "non-specific" [16-18]. This latter binding mode arises from the electrostatic interaction between the negatively charged humic macromolecule and positively charged cations. A widely used and relatively successful approach to the modelling of metal-humic complexation is to define a number of specific metal binding site groups (commonly 2, but on occasions more) with characteristic "stability constants" and then to make a correction to these "constants" to take account of progressive neutralisation of charge on the humic macromolecule (Figure 1) [e.g. 2,16]. Increasing metal loading of humic macromolecules causes increased aggregation of the individual macromolecules both through compression of the electrical double layers and also through the ability of metal ions to bridge between different parts of a macromolecule or even between different molecules. Thus, non-specific binding, for example of a Group I or II cation should have a lesser effect than binding (for example of a d-transition ion) in a specific site. In this paper we describe two techniques currently being used for the study both of metal ion binding by humics and of the physical state of humic-metal ion complexes.

Methods

Analytical Ultracentrifugation. This technique provides a direct means of measuring the molecular weights (usually, the weight-averaged molecular weights) of humic substances and of humic-metal complexes and has been described in detail elsewhere [20-22]. In principle, it is relatively straightforward to determine molecular weights, since a dynamic equilibrium (Figure 2) can be achieved between the centrifugal force exerted by the centrifuge (proportional to the square of angular velocity) and diffusion in the opposite direction of the humic macromolecules (proportional to the concentration gradient established in the cell). The distribution of humic material in the centrifuge cell is monitored during the experiment by measuring UV absorbance at 280 nm. Since the system is at equilibrium and all the variables except molecular weight are known or can be found (although a buoyancy correction is required, for which the partial specific volume of the humic substances must be known), a value for the molecular weight can be calculated (Equation 1).

$$\overline{M}_W = \frac{2RT\dfrac{d\ln c}{dx^2}}{(1-\overline{v}_2\rho)\,\omega^2} \tag{1}$$

where R = Gas constant, T = temperature (K), c = solute concentration, x = radial distance from the axis of rotation, $\overline{v}_2$ = partial specific volume of solute; ρ = solution density and ω = angular velocity. An example of a plot of ln(absorbance) against x^2 for a fulvic acid is illustrated in Figure 3a. The plot is approximately linear for low values of x^2 but develops curvature at higher x^2 values due to the polydispersity of the sample. The main drawbacks with sedimentation equilibrium are the time required to attain equilibrium (several days) and the need to try and keep the majority of the sample in solution. With very polydisperse mixtures, or samples of very high molecular weight, there may be significant sedimentation of the sample out of the solution, thus biasing the results towards the lower molecular weight species.

   An alternative method for molecular weight measurement is the Archibald or approach to equilibrium method [22]. In this, rather than establishing an equilibrium

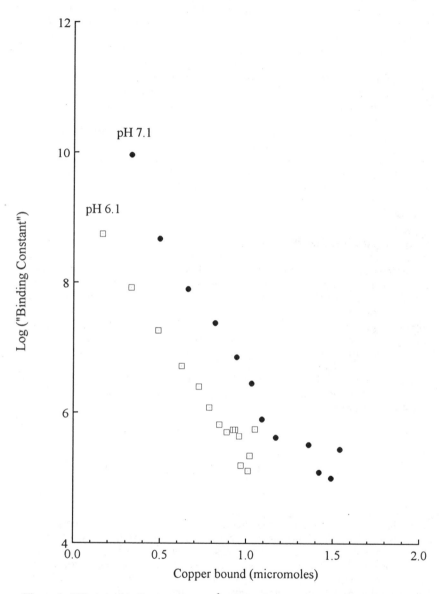

Figure 1: Effect on binding constant (K^* [19]) of neutralisation of humic charge (from [35]). Humic concentration 35 mg l^{-1}; $I = 0.05$. Note that an approximate 10-fold increase in bound Cu^{2+} concentration leads to about 6 orders of magnitude decrease in K^*.

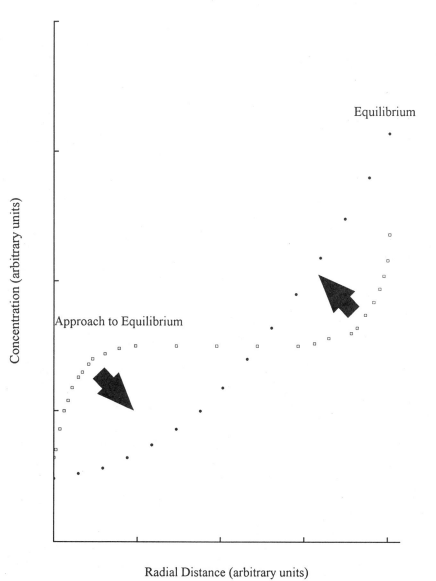

Radial Distance (arbitrary units)

Figure 2: Concentration distributions within sample cell during an ultracentrifuge experiment. i) approach to equilibrium and ii) at equilibrium.

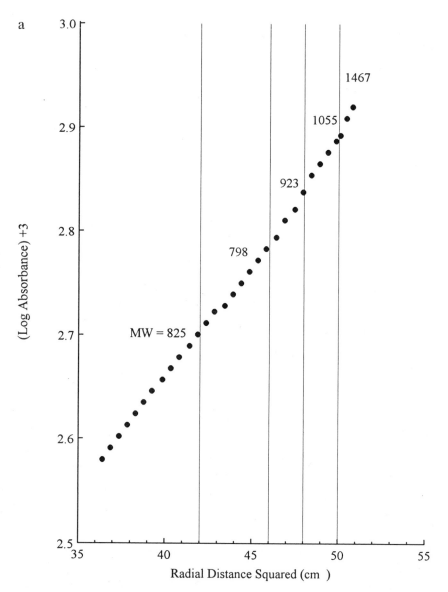

Figure 3: a) Molecular weight determination by equilibrium ultracentrifugation; b) Comparison of molecular weights determined by the Archibald method and by sedimentation equilibrium.

b

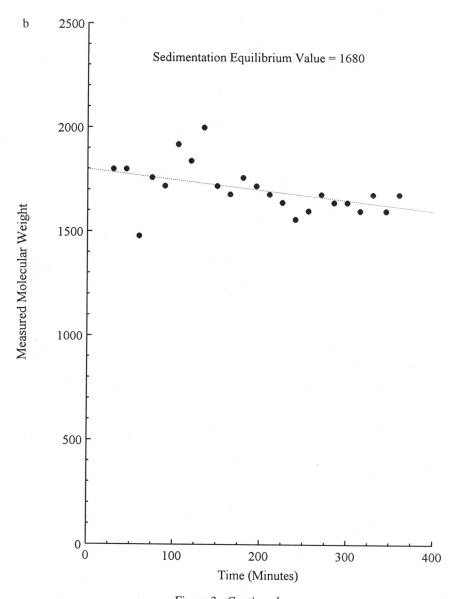

Figure 3. *Continued*

between settling and resuspending forces, advantage is taken of the fact that there can be no mass flux across the ends of the sample cell. Changes in the humic concentration at one end of the solution column are then monitored and a molecular weight value derived from the rate of change of concentration. In practice, since the UV detector is effective only over a limited absorbance range, it is usual to follow the decrease in concentration at the meniscus, rather than the increase in concentration at the base of the column. Data are then derived from Equation 2, a specific case of Equation 1.

$$\overline{M}_W = \frac{RT \left(\dfrac{dc}{dx}\right)_m}{(1-\overline{v}_2\rho)\, \omega^2 c_m x_m} \tag{2}$$

where the subscript m denotes a value at the meniscus. The concentration gradient $(dc/dx)_m$ is measured during approach to equilibrium and used to obtain $\overline{M}_w$ as a function of time. Extrapolation of $\overline{M}_w$ to zero time gives the weight average molecular weight for the whole sample. In many cases, both sedimentation equilibrium and approach to equilibrium give comparable results (Figure 3b), although for high molecular weight humics, this is not the case if, during sedimentation equilibrium, material is lost from the concentration profile. Clearly, both equilibrium ultracentrifugation and the Archibald method can be applied to metal-humic complexes as readily as to isolated humics.

Gel Permeation Chromatography. This is again a potentially valuable technique since it can, in principle, separate complex mixtures of macromolecules directly on the basis of molecular size. Indeed, in cases such as proteins, with molecules of very different sizes, and where appropriate calibration standards are available, it is possible to obtain well resolved, chemically pure fractions and to determine molecular weights directly. There is, however, the possibility of artefacts arising due to interaction with the gel [23,24]. This may be irreversible; that is sample material may simply be retained on the top of the gel column, or it may be more subtle, in which case the retarded fractions are eluted, but later than would be expected solely on the basis of molecular size.

There are a number of disadvantages in the application of gel permeation chromatography to studies of humics. There are no usable molecular weight standards, so it is difficult to make credible molecular weight measurements directly [25-28]. The polydispersity of humics can often lead to broad, overlapping bands, so that individual fractions may well not be cleanly separated [e.g. 23; see later]. There is also often extensive interaction with the gel, and whilst sample losses may be minimised by careful control of experimental conditions, it is impossible to prevent interaction with the gel entirely. Thus, the separation achieved is probably best viewed as arising from both size effects and interaction effects. In spite of these difficulties, gel chromatography of humic substances has been widely used, both to fractionate and characterise humic substances and also to study the reactions of humic substances with species such as metal ions [e.g. 29-32]. Both ultracentrifugation and gel permeation chromatography are presently being applied to the characterisation of metal ion-humic interactions.

Results and Discussion

Ultracentrifugation. Recently, ultracentrifugation has been used in conjunction with metal binding measurements, microcalorimetry and modelling studies. A range of well-characterised humic and fulvic acid samples are being used in these experiments and the effects on measured molecular weight of metal binding are presently being investigated. In particular, whilst it is well known that, at high concentrations of metal ions, humic substances precipitate [33,34], very little is known about the processes of metal-humate colloid formation and precipitation [35]. Initial investigations into metal-humic aggregation have been carried out using the Archibald method to follow changes in molecular weight in response to metal binding. In general, a significant increase in weight averaged molecular weight ($\overline{M}_w$) occurs as metal ion concentration increases. An example is given in Figure 4, which shows changes in $\overline{M}_w$ with total copper and lanthanum concentrations, $[Cu]_T$ and $[La]_T$.

At copper concentrations greater than 2×10^{-4} M, complete precipitation occurs. In these experiments, the aggregation processes have been found to occur over a relatively narrow range of metal concentrations, about an order of magnitude in the case of copper. Moreover, the humics can tolerate addition of a certain amount of copper ($[Cu]_T < 1.0 \times 10^{-5}$ M) before the onset of aggregation in spite of the fact that there is very significant metal binding in this concentration range since the strongest metal binding sites are being occupied [36].

Since the humic macromolecules have large negative charges in solution [37], it is to be expected that cation charge will have a substantial effect on the interactions between metal ions and humics [38]. This can be illustrated by comparing the binding of lanthanum and that of copper. Complete humic precipitation occurs at $[La]_T = 2.2 \times 10^{-5}$ M, almost an order of magnitude lower than the copper concentration required for precipitation. Moreover, whereas aggregate sizes increase only relatively slowly on addition of copper, they increase much more rapidly with lanthanum. Should complete flocculation leading to precipitation follow the Schulze-Hardy rule for a lyophobic colloid, the ratio of flocculation concentrations for Cu^{2+} and La^{3+} ions would be $(3/2)^6 = 11.4$, which compares with the experimental value of approximately 9.

Gel permeation. For several years, gel permeation has been used to study the association of a number of actinide elements with humic substances [39,40]. This work has concentrated on a number of sites in the vicinity of the Sellafield nuclear fuel reprocessing plant in north west England. More recently, other work has shown that humic substances may play a dominant role in determining actinide migration. This is exemplified by some studies carried out at the Needle's Eye natural analogue site in south-west Scotland. At this site, uranium migration from a mineralised vein has been extensively studied over a number of years [*e.g.* 41-43]. In a small, highly organic area at the foot of the cliff in which the vein outcrops, uranium has been found to be strongly complexed by humic substances; indeed over 90% of the uranium in the soil is organically bound [42]. In addition, recent work [44] has shown that the desorption of uranium from this soil is kinetically hindered, thus substantially reducing uranium mobility [43]. This suggests that the kinetics of sorption-desorption reactions are a fundamental control on element migration. Although many studies exist in which metal-humic complexation has been treated as an equilibrium process (some examples include [16-18,45-48]), there are fewer in which reaction rates are considered (though see [12,49-54]).

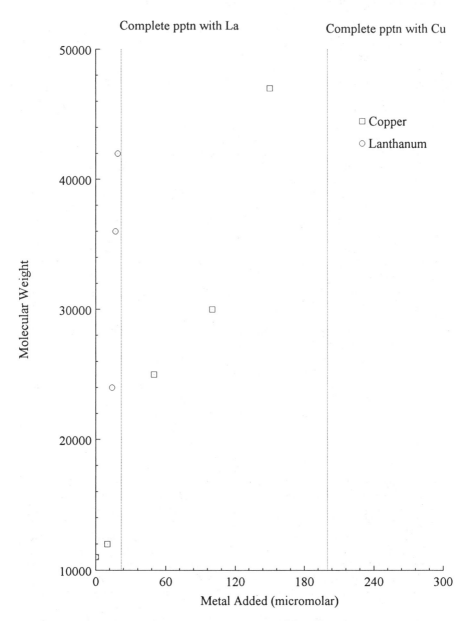

Figure 4: Molecular weight change in response to addition of copper and lanthanum. Humic concentration 21 mg l⁻¹; pH 6-7; I = 0.05.

The desorption of uranium from these soils can be described by a number of rate constants of the order of 10^{-4} to 10^{-6} hr^{-1} [44]. Gel permeation fractionation experiments using the humic substances from the Needle's Eye site suggest that they comprise three poorly-resolved fractions (Figure 5). This behaviour is quite different to that observed with humic substances from a range of other sites, where a larger number of better-resolved fractions are usually found. In the Needle's Eye humic substances, the majority of the uranium is held in the late-eluting fraction, with relatively little in the earlier ones. This U-binding fraction has a molecular weight (measured by ultracentrifugation) of 10000 or below, compared with about 70000 for the more rapidly eluted fraction. The dominance of U binding by this relatively low molecular weight material suggests that it is the chemistry of this fraction which, to a large extent, controls U behaviour.

Radiotracer experiments, using ^{235}U and Needle's Eye humic substances have also shed light upon changes in U binding which take place following an initial, fairly rapid complexing step. The ^{238}U pattern represents the end result of a long equilibration time (hundreds of years) whilst the ^{235}U patterns represent much shorter equilibration times (Figure 5). It appears that the ^{235}U is proportionately less strongly associated with the high molecular weight fraction than ^{238}U. However, as equilibration time increases, the distributions of the two isotopes resemble one another more closely. This suggests that a slow change in U speciation occurs following sorption. This may represent a change in the state of aggregation of the humic macromolecules, brought about by the presence of the U cations, or it may arise from the gradual transfer of uranium from a non-specific "surface" association to specific metal binding sites [54,55]. In any event, the observed combination of complex sorption-desorption kinetics and slow, post-sorption speciation changes will complicate predictive modelling of such systems.

Conclusions

Whilst it is well known that complexation by humic substances is an important control on the environmental behaviour of metal ions in many contexts, the precise nature of the metal-humic reaction remains poorly understood. Analytical ultracentrifugation and gel permeation chromatography both have the potential to provide information on these reactions and such studies illustrate fully the complexity of metal-humic systems. Ultracentrifugation allows identification of changes in the state of aggregation of humic substances in response to metal binding, and the data illustrate the ready formation of high molecular weight aggregates. Gel permeation allows the fractionation of metal-humic complexes and shows clearly that metal ion-humic interaction is selective. This suggests that both the chemistry of the metal ions and of the humic ligands will play a part. Moreover, the rates of sorption and desorption reactions have an important effect on metal ion migration. Slow, post-adsorption speciation changes are also identified which may also have an influence. As a result of such complex processes and our relatively poor understanding of the underlying chemistry, the incorporation of metal-humic interactions into speciation and migration models remains a major challenge.

Acknowledgments

The authors are very grateful to British Nuclear Fuels plc, the British Council and the UK Natural Environment, and Science and Engineering Research Councils for financial support. We would also like to thank the Scottish Wildlife Trust for permission to sample at the Needle's Eye Nature Reserve.

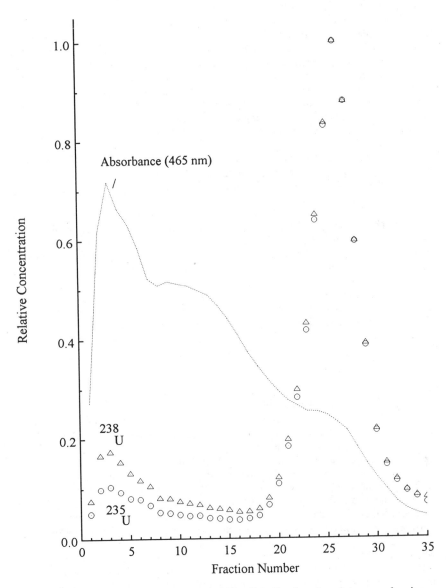

Figure 5: Gel permeation separation of Needle's Eye humic substances, showing distribution of ^{235}U *radiotracer (circles) and* ^{238}U *(triangles) between fractions.* ^{238}U *concentration 1500 mg kg^{-1};* ^{235}U *concentration 500 mg kg^{-1}; pH 8-9; I = 0.1. UV absorbance at 465 nm is used as an indicator of organic matter concentration.*

Literature Cited

1 Piccolo, A.; Celano, G. & Desimone, C. *Sci. Tot. Environ.* **1992**, *118*, 403-412.
2 Langford, C. H. *ACS Symp. Ser.* **1994**, *565*, 404-417.
3 Hering, J. G. *Adv. Chem. Ser.* **1995**, *244*, 95-110.
4 Burba, P. *Fres. J. Anal. Chem.* **1994**, *348*, 301-311.
5 Gamble, D.S.; Langford, C.H. & Webster, G.R.B. *Rev. Environ. Contam. Toxicol.* **1994**, *135*, 63-91.
6 Goodman, B.A.; Cheshire, M.V. & Chadwick, J. *J. Soil Sci.* **1991**, *42*, 25-38.
7 Young, L.B. & Harvey, H.H. *Geochim. Cosmochim. Acta* **1992**, *56*, 1175-1186.
8 Filella, M. & Buffle, J. *Colloids Surf. A* **1993**, *73*, 255-273.
9 Kipton, H.; Powell, J. & Town, R.M. *Anal. Chim. Acta* **1992**, *267*, 47-54.
10 Shinozuka, N. & Lee, C. *Mar. Chem.* **1991**, *33*, 229-241.
11 Engebretson, R.R. & Vonwandruszha, R. *Environ. Sci. Technol.* **1994**, *28*, 1934-1941.
12 Ephraim, J.H. *Anal. Chim. Acta* **1992**, *267*, 39-45.
13 Reid, P.M.; Wilkinson, A.E.; Tipping, E. & Jones, M.N. *J. Soil Sci.* **1991**, *42*, 259-270.
14 Tombacz, E. & Meleg, E. *Org. Geochem.* **1990**, *15*, 375-381.
15 Teasdale, R.D. *J. Soil Sci.* **1987**, *38*, 433-442.
16 Gessa, C.; Deiana, S.; Maddau, V.; Manunza, B. & Rausa, R. *Agrochimica* **1994**, *38*, 85-90.
17 Ephraim, J.H. & Allard, B. *Environ. Intl.* **1994**, *20*, 89-95.
18 Tipping, E. *Radiochim. Acta* **1993**, *62*, 141-152.
19 Buffle, J.; Altmann, R.S.; Filella, M. & Tessier, A. *Geochim. Cosmochim. Acta* **1990**, *54*, 1535-1553.
20 Rietz, E. Land. *Volkenrode* **1987**, *37*, 235-244.
21 Reid, P.M.; Wilkinson, A.E.; Tipping, E. & Jones, M.N. *Geochim. Cosmochim. Acta* **1990**, *54*, 131-138.
22 Wilkinson, A.E.; Hesketh, N.; Higgo, J.J.W.; Tipping, E. & Jones, M.N. *Colloids Surf. A* **1993**, *73*, 19-28.
23 Town, R.M. & Powell, H.K.J. *Anal. Chim. Acta* **1992**, *256*, 81-86.
24 Shaw, P.J.; Dehaan, H. & Jones, R.I. *Environ. Technol.* **1994**, *15*, 753-764.
25 Berden, M. & Berggren, D. *J. Soil Sci.* **1990**, *41*, 61-72.
26 Tao, S. *Environ. Technol.* **1994**, *15*, 1083-1088,
27 Shaw, P.J.; Jones, R.I. & Dehaan, H. *Environ. Technol.* **1994**, *15*, 765-774.
28 Finger, W.; Post, B. & Klamberg, H. *Z. Pflanz. Bodenk.* **1991**, *154*, 287-291.
29 Krajnc, M.; Stupar, J. & Milicev, S. *Sci. Tot. Environ.* **1995**, *159*, 23-31.
30 Edwards, M. & Benjamin, M.M. *J. Am. Water Works Ass.* **1992**, *84*, 56-66.
31 Cameron, D.F. & Sohn, M.L. *Sci. Tot. Environ.* **1992**, *113*, 121-132.
32 Conzonno, V.; Dirisio, C.; Tudino, M.; Ballsells, R.E. & Cirelli, A.F. *Fresenius Anal. Bull.* **1994**, *3*, 1-5.
33 Rashid, M.A. *Soil Sci.* **1971**, *111*, 298-306.
34 Bloomfield, C.; Kelso, W.I. & Pruden, G. *J. Soil Sci.* **1976**, *27*, 435-444.
35 Livens, F.R. *Environ. Poll.* **1991**, *70*, 182-208; *Environ. Poll.* **1991**, *74*, 86-87.
36 Bryan, N.D. *Ph.D. Thesis*, University of Manchester **1994**.
37 de Wit, J.C.M.; van Riemsdijk, W.H. & Nederlof, N.M. *Anal. Chim. Acta* **1990**, *232*, 189-207.
38 Tipping, E. & Hurley, M.A. *Geochim. Cosmochim. Acta* **1992**, *56*, 3627-3641.
39 Livens, F.R.; Baxter, M.S. & Allen, S.E. *Soil Sci.* **1987**, *144*, 24-28.
40 Livens, F.R. & Singleton, D.L. *J. Environ. Radioactivity* **1991**, *13*, 323-340.

41 Hooker, P.J. *British Geological Survey Report WE/90/5*; British Geological Survey: Keyworth, England, **1990**.

42 Scott, R.D.; MacKenzie, A.B.; Ben-Shaban, Y.A.; Hooker, P.J. & Houston, C. M. *Radiochim. Acta* **1991**, *52/53*, 357-365.

43 Jamet, P.; Hooker, P.J.; Schmitt, J.M.; Ledoux, E. & Escalier des Orres, P. *Mineral Dep.* **1993**, *28*, 66-76.

44 Braithwaite, A. *Ph.D. Thesis*, University of Manchester **1994**.

45 Lead, J.R.; Hamilton-Taylor, J.; Hesketh, N.; Jones, M.N.; Wilkinson, A.E. & Tipping, E. *Geochim. Cosmochim. Acta* **1994**, *294*, 319-327.

46 Finger, W. & Klamberg, H. *Z. Pflanz. Bodenk.* **1993**, *156*, 19-24.

47 Maes, A.; DeBrabandere, J. & Cremers, A. *Radiochim. Acta* **1991**, *52/53*, 41-47.

48 Czerwinski, K.R.; Buckau, G.; Scherbaum, F. & Kim, J.I. *Radiochim. Acta* **1994**, *65*, 111-119.

49 Gerke, J. *Geoderma* **1994**, *63*, 165-175.

50 Burpa, P. *Fresenius J. Anal. Chem.* **1994**, *348*, 301-311.

51 Rate, A.W.; McLaren, R.G. & Swift, R.S. *Environ. Sci. Technol.* **1992**, *26*, 2477-2483.

52 Choppin, G.R. & Clark, S.B. *Mar. Chem.* **1991**, *36*, 27-38.

53 Langford, C.H. & Gutzman, D.W. *Anal. Chim. Acta* **1992**, *256*, 183-201.

54 Clark, S.B. & Choppin, G.R. ACS Symp. Ser. **1990**, *416*, 519-525.

55 Rao, L.F., Choppin, G.R. & Clark, S.B. *Radiochim. Acta*, **1994**, *66/67*, 155-161.

Chapter 13

A Comparison of the Dissociation Kinetics of Rare Earth Element Complexes with Synthetic Polyelectrolytes and Humic Acid

Sue B. Clark[1,3] and Gregory R. Choppin[2]

[1]Savannah River Ecology Laboratory, University of Georgia,
P.O. Drawer E, Aiken, SC 29802
[2]Department of Chemistry, Florida State University,
Tallahassee, FL 32306–3006

The kinetics of Sm(III) and Eu(III) interactions with size fractionated samples of a humic acid (HA) and polyacrylic acid (PAA) have been studied by two techniques - passage through an ion exchange resin column and ligand exchange. The kinetics with the PAA samples were significantly faster, presumably reflecting the homogeneity and monofunctional nature of PAA relative to HA. The HA kinetics involved pathways on a fast time scale similar to those for PAA as well as much slower pathways. For both PAA and HA, the kinetics were faster with the lower molecular weight fractions, reflecting the role of decreased conformational change as the size decreases.

Due to the strength of their complexing capacities, natural humic substances can dominate metal speciation in aquatic systems; however, describing metal binding to these heterogeneous, ill-defined ligands is an unresolved challenge. The polyelectrolytic nature of humic materials results in two general types of cation interaction. In "site-binding", the cations may be completely or partially dehydrated due to interactions with specific ionized groups on the polyion (1). Conversely, "territorial binding" arises from non-specific, hydrated metal interactions with the high negative charge density on the polyelectrolyte from the net concentration of ionized binding groups (2).

Humic substances are composed of heterogeneous macromolecules consisting of multiple functional groups randomly arranged, which leads to uncertainty about the model to use for metal ion binding. The configuration of the polyion is influenced by processes such as internal repulsion between ionized groups and/or hydrogen bonding and aggregation of smaller polyions. These processes are dynamic and controlled, in part, by solution conditions such as pH and ionic strength. Binding of rare earth cations by humics can be assigned to ionic bonding, and is dominated by the

[3]Current address: Department of Chemistry, Washington State University, P.O. Box 644630, Pullman, WA 99164–4630

propensity of these cations for complexation with donor oxygen groups. Further, in the pH range below ca. 7-8, the phenolic groups on the humic substance remain protonated, leaving the carboxylate groups as the primary binding sites for the lanthanides, unless coordination to the -OH group involved chelation via a carboxylate group.

Well-characterized synthetic polyelectrolytes are sometimes used as models for humic and fulvic acids to gain insight into a description of metal binding (3-5). We have chosen polyacrylic acid (PAA) as a model for studies of rare earth element binding to polyelectrolytes, and to compare to studies of lanthanide binding to humic acids. PAA is a homogeneous, linear polyelectrolyte with regularly spaced carboxylate groups, and has been used in studies of transition metal binding to macromolecules. NMR and ESR studies of Mn(II) binding to PAA suggest three types of metal interactions: one in which the metal mobility is reduced and its hydration shell disrupted, a second environment involving a hydrated metal with reduced mobility, and a third type of metal that retains its mobility and hydration shell (6). By comparison, Lis, et al. (5) has shown via luminescence that lanthanide binding to PAA is rapid (within minutes), and is accompanied by partial dehydration of the metal.

Previously, we have described the binding mechanism for f-elements to humic acid (HA) as a cooperative process in which metal complexation results in polyion charge neutralization and subsequent configurational changes in the polyion's three dimensional structure (7-10). Studies of dissociation kinetics have been used to show that, although complexation is rapid, initial interactions are labile (i.e. "weak"). As the metal-polyion complex ages over a period of 48 hours, however, lability for a fraction of the metal is reduced significantly, and additional "strong" interactions occur. These strong binding sites arise presumably from folding of the HA about the metal, and the migration of metal to binding sites within the tertiary structure of the HA. Such conformational changes in the bulky polyelectrolyte are slow; hydrogen bonds are disrupted and the internal electrostatic repulsion is reduced. This aging time required to achieve equilibrium in the dissociation kinetics is referred to as "binding time" in this text.

Here, we report on dissociation kinetic studies of the rare earth elements Sm(III) and Eu(III) from PAA and HA. The solution chemistry of both metals is essentially identical, and they are frequently used as analogs for trivalent actinide elements. We have studied the rare earth dissociation kinetics from different size fractions of PAA and HA using ion exchange and ligand exchange methods. The differences in metal labilities for complexes of the two polyelectrolytes and the effect of polyelectrolyte molecular weight on binding times are discussed in the context of possible polyion-cation interactions in these systems.

Materials and Methods

All chemicals were reagent grade, unless noted otherwise. Stock solutions of the rare earth elements were prepared by dissolving the appropriate oxide (from Cereac, Inc.) in warm concentrated $HClO_4$, diluting, and filtering. The solution was standardized by titration with EDTA using xylenol orange (11). A perchlorate solution of [152]Eu was obtained from Oak Ridge National Laboratory, and standardized using gamma spectrometry. Stock solutions were prepared by dilution with 0.10 M $NaClO_4$. Arsenazo III (3,6-bis[2-arsonophenyl]-azo-4,5-dihydroxy-2,7-napthalene-disulfonic acid) was obtained from Aldrich (99.9%).

Polyacrylic acid was purchased from Polysciences, Inc., and used with no further purification. HA was obtained from lake sediments (Lake Bradford, FL), and isolated and purified as described earlier *(12)*. Two different size fractions of HA were obtained by ultrafiltration, as described below. All kinetic studies were performed at 0.10 M ionic strength (adjusted by the addition of $NaClO_4$), and were buffered by the addition of acetate (acetic acid and sodium acetate, pH 4.2). Buffer concentrations were 0.01 M. TES (N-tris[hydroxymethyl]methyl-2-aminoethane sulfonic acid, pH 7.0) buffer was used to maintain hydrogen ion concentrations during humate fractionation.

Equipment. Measurements of the gamma radiation of ^{152}Eu were performed using a 2" well-type NaI(Tl) crystal with a single channel analyzer. Solutions were counted in 13 x 100 borosilicate tubes to a precision of $\leq 1\%$. Hydrogen ion activity was measured using a Radiometer PHM 84 research pH meter in conjunction with a Corning Semimicro Ag/AgCl combination pH electrode. The electrode was calibrated daily using Fisher standard buffer solutions (pH 4.00 and 7.00 at $25^{\circ}C$). Visible spectra were obtained with a rebuilt (On-Line Instruments Systems) Cary-14 spectrophotometer interfaced to a personal computer. Stopped-flow measurements were completed at Argonne National Laboratory using a Hi-Tech SF-51 stopped-flow spectrophotometer interfaced to an SLI-1123 computer linked to a VAX 780 system. All kinetic data were analyzed using personal computers.

Humic Acid Fractionation. The two different size fractions of humic acid used in this study were obtained by ultrafiltration of previously characterized HA from Lake Bradford, FL. An Amicon ultrafiltration cell (model 8050) was used for fractionation with a filtration membrane of the desired pore size. Molecules exceeding that pore size were retained above the membrane. The nominal molecular weight cut-off (MWCO), or pore sizes, were reported by the manufacturer (Amicon), and are based on globular protein standards. Membranes used in this study were XM300 (MWCO 300,000 daltons, lot number AI), YM100 (MWCO 100,000 daltons, lot number AT), and XM50 (MWCO 50,000 daltons, lot number AG). A 5 liter fiberglass reservoir (Amicon model RG5) was filled with the solution to be fractionated. The solution was ultrafiltered under pressure (50 psi) with constant stirring.

Following the recommendations of Buffle, et al. *(13)*, solutions of 150 mg/liter HA, pH 7.0 (TES), 0.10 M $NaClO_4$ were ultrafiltered through the membranes sequentially, starting with the largest MWCO. The concentrated fraction of HA retained above the membrane was rinsed twice with a solution of TES buffer, 0.10 M $NaClO_4$. The remaining solution of fractionated HA was dialyzed and lyophilized to isolate the solid material which was characterized by acid-base titration. Characteristics of the polyelectrolytes used in this study are given in Table I.

Table I: Polyelectrolyte Characteristics

	Size Fraction (daltons)	Carboxylate Capacity (meq/g)
Polyacrylic Acid	$18,100 \pm 500$	11.7 ± 0.8
	$450,000 \pm 50,000$	11.8 ± 0.4
Humic Acid	$50,000\text{-}100,000^{(a)}$	6.89 ± 0.6
	$\geq 300,000^{(b)}$	3.52 ± 0.2
	unfractionated[c]	3.86 ± 0.03

[a] Comprised ca. 50 weight % of unfractionated HA.

[b] Comprised ca. 5 weight % of unfractionated HA.

[c] From reference (*12*).

Dissociation Experiments. Solid polyelectrolyte material was dissolved in a minimum volume of 0.1 M NaOH and diluted in distilled, deionized water; these solutions were stored in darkness at 4°C until ready to use. All PAA and HA concentrations used in this work are based on the milliequivalents of carboxylate binding groups per gram of polyion in Table I. All dissociation studies employed metal concentrations that corresponded to 5% of the PAA or HA carboxylate capacity. Consequently, the following experimental matrix was used unless noted otherwise: [polyelectrolyte] = 1 x 10^{-4} eq/L, [metal] = 5 x 10^{-6} M, [buffer] = 0.01 M, I = 0.1 M (NaClO$_4$). All experiments were conducted at room temperature.

Ion Exchange Experiments. As described by Rao, et al. (*10*), ion exchange experiments were performed using columns of 1.2 cm inner diameter, with 5.7 cm of the column filled with Dowex 50 x 4, 100-200 mesh cationic exchange resin. This technique is designed to provide rapid separation of metal bound strongly to the polyelectrolyte from other metals (i.e., uncomplexed or labile metal complexes) in the system. The resin was used in the sodium form, and buffered to the desired pH. To start an experiment, a solution of polyelectrolyte was mixed with a solution of Eu(III) spiked with [152]Eu. Immediately, a small aliquot was removed and forced through a column in les than 30 s., followed by passage of a minimum of ten free column volumes of the buffer solution adjusted to 0.1 M ionic strength (NaClO$_4$). The fraction of Eu(III) eluting as the complex was determined by counting the [152]Eu in the eluant. Rapid separation inhibited adsorption of the polyelectrolyte to the resin. Complete recovery of the HA was verified using absorption spectroscopy (*10*).

The remaining solution of Eu(III)-polyelectrolyte was allowed to "age" at room temperature (i.e. the binding time was allowed to increase). At various time intervals, an aliquot of the complex solution was removed and run through other columns as described above. Changes in the fraction of [152]Eu eluting as the complex as a function of binding time were determined.

Ligand Exchange Experiments. The ligand exchange technique has been described previously (*4,7-9,14*). A solution of Sm(III) complexed by either PAA or HA was allowed to equilibrate for a desired amount of binding time. The rates of dissociation of this complex was followed by introducing a colorimetric exchanging ligand, arsenazo III. The Sm(III)-arsenazo III complex has a higher formation constant than for complexes with PAA or HA, and consequently, the polyelectrolyte-Sm(III) complex equilibrium is shifted to dissociation as the arsenazo III binds the free Sm(III). The rate limiting step for ligand exchange is the dissociation of the polyelectrolyte complex, as the complexation with arsenazo III does not influence the dissociation kinetics (*11*). The reaction was monitored as a function of time by measuring the formation of the Sm(III)-arsenazo complex at 640 nm. Stopped-flow was used to monitor fast processes ($t_{1/2} \leq 30$ s.), while slower processes were monitored in separate experiments with conventional visible spectrophotometry.

Sm(III) dissociation from the polyelectrolytes occurred by multiple first-order processes. To determine the number of first-order processes and estimate rate constants and the concentration of metal dissociating by each process in the reaction, a kinetic spectrum method was used (*15*). The kinetic spectrum H(k,t) is defined as the following distribution function:

$$H(k,t) = \frac{\delta^2(\Delta Absorbance)}{\delta(\ln t)^2} - \frac{\delta(\Delta Absorbance)}{\delta(\ln t)} \tag{1}$$

Using absorbance data collected as a function of time, the distribution function H(k,t) was calculated, and plotted versus ln(time) to yield a curve providing a maximum for each first-order process in the reaction. The position of each maximum yields an estimate of the rate constant (t = 2/k), and the area under the maximum provides an estimate of the amount of metal dissociating by that process. Due to concerns of using higher-order derivatives to calculate rate constants for distributions of pathways in multiple first-order mechanisms as described in the literature (*16-20*), the kinetic spectrum method was used only to obtain initial estimates for the appropriate rate equations. The actual rate parameters reported herein were obtained from a simplex non-linear regression (*21*) of the original experimental data. When dissociation occurred in both fast ($t_{1/2} \geq 30$ s) and slow ($t_{1/2} < 30$ s) time domains, each data set was treated independently.

Results and Discussion

The polyelectrolyte characteristics reported in Table I suggest the simplicity of the PAA systems relative to HA. For PAA, the two fractions used were composed of significantly different sizes. The large size fraction contained a wide range of molecular weights, whereas the smaller fraction represents a narrow range and is a primary molecular weight standard. However, the carboxylate capacities of the two fractions are the same. For HA, both the size fractions and carboxylate capacities were very different. The carboxylate capacity for the unfractionated material is also given for reference. The quantity of each size fraction obtained from the bulk material suggest that the original HA is composed of macromolecules predominantly in the range of 50,000-100,000 daltons. However, the carboxylate capacities of the two size fractions suggest that the larger size fraction ($\geq$ 300,000 daltons) is similar to the unfractionated material. While extreme care was exercised to obtain a discreet size fraction of HA, they are operationally defined. For example, the large size fraction may indeed represent aggregates of smaller humate molecules, in spite of attempts to prevent this.

Ion Exchange Studies. Ion exchange has been used previously to probe the different types of Eu(III) and UO_2^{2+} interactions with unfractionated HA as a function of metal binding time ($8,10$). The sulfonic acid group on the ion exchange resin can compete for metal bound in labile sites of the polyelectrolyte, causing retention of the metal by the resin. A significant observation was that changes in the amount of metal retained by the resin as a function of binding time were observed. As the binding time was increased, more of the metal-humate complex eluted intact from the resin, suggesting a reduction in the lability of the humate complex for an increased percentage of the metal with time. Similar results were observed in this study for the size fractionated HA and PAA.

Figure 1 shows the amount of metal-polyelectrolyte complex that elutes intact from the column as a function of time. For PAA, little difference is observed between the two size fractions. Approximately 24 hours of binding time is required to achieve an equilibrium in which 10% of the Eu(III)-PAA elutes. In other words, after equilibration about 90% of the Eu(III) is bound to PAA in labile sites that dissociate on the ion exchange column. Interestingly, more binding time is required to achieve an equilibrium with the two HA fractions. For the small size fraction, 48 hours of binding time is required, similar to results for unfractionated HA reported previously (8). Binding time for the large fraction is even longer, taking three days to equilibrate. At equilibrium, significantly higher percentages of complexed Eu(III) were observed, indicating less labile interactions with HA. At equilibrium, 28% of Eu(III) elutes as a complex with the small fraction of HA, whereas 38% elutes with the large HA fraction.

The observed similarity for the two size fractions of PAA is likely due to the homogeneity of this polyelectrolyte, regardless of the size of the macromolecule. HA, however, represents a much more complicated polyelectrolyte compared to PAA which results in increased binding time for the HA systems.

Ligand Exchange Studies. Previous studies of lanthanide and actinide dissociation from synthetic polyelectrolytes and unfractionated HA revealed a complex mechanism characterized by multiple dissociation pathways ($4,7$-9). As shown in the kinetic spectrum in Figure 2, Sm(III) dissociation from the small size fraction of PAA occurred by three first order processes with half-lives of 0.02, 0.8, and 32 s. The amount of metal dissociating by each pathway was a function of the binding time for the polyelectrolyte complex prior to dissociation. While the most labile process (log t = -1.8, $t_{1/2}$ = 0.02 s) was the dominant mode for dissociation regardless of the binding time, the percentage of Sm(III) dissociating by that process decreased when the binding time prior to dissociation was increased from 15 minutes to 24 hours (Fig. 2).

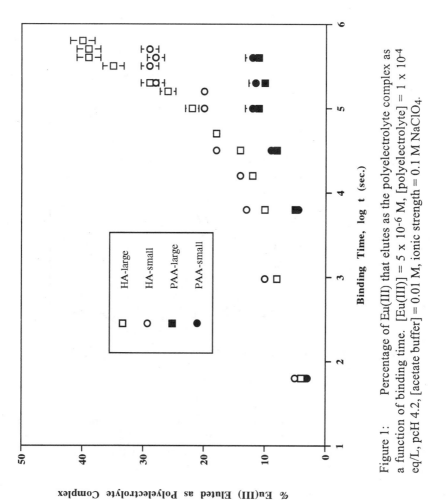

Figure 1: Percentage of Eu(III) that elutes as the polyelectrolyte complex as a function of binding time. $[Eu(III)] = 5 \times 10^{-6}$ M, [polyelectrolyte] $= 1 \times 10^{-4}$ eq/L, [acetate buffer] $= 0.01$ M, pcH 4.2, ionic strength $= 0.1$ M NaClO$_4$.

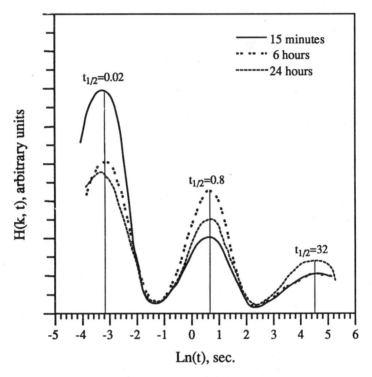

Figure 2:   Kinetic spectrum for Sm(III) dissociation from small size fraction of PAA as a function of Sm(III)-PAA binding time. Conditions were [Sm(III)] = 5 x 10^{-6} M, [PAA] = 1 x 10^{-4} eq/L, pcH 4.2, [acetate buffer] = 0.01 M, ionic strength = 0.1 M $NaClO_4$.

After 24 hours of binding time, the amount of Sm(III) dissociating by the slowest pathway (log t = 1.5, $t_{1/2}$ = 32 s) had increased noticeably, resulting in an overall decrease in the rate of dissociation with increased binding time. Increasing the binding time beyond 24 hours resulted in no observable change in the dissociation kinetics.

Twenty-four hours of binding time was also required for constant dissociation kinetics for Sm(III) bound to the large molecular weight fraction of PAA. After a binding equilibrium was achieved, dissociation from the large PAA was slower and more complicated than the kinetics observed for the small PAA fraction. The rate equation providing the best description of the experimental data included four first-order terms with half-lives of 0.2, 0.5, 1.8, and 47 s. Similar to observations with the small size fraction, as the binding time increased up to 24 hours more dissociation to the longer-lived components of the rate equation occurs, although the most labile term was always the primary dissociation process.

Values of the parameters for the rate equations describing Sm(III) dissociation from the two fractions of PAA are given in Table II. Interestingly, the observed overall rate of dissociation is much faster than the time required to establish a binding equilibrium, indicating hysteresis in the complexation/dissociation reactions. Also, the amount of Sm(III) dissociating by the slowest pathway at equilibrium for both PAA fractions agrees with the equilibrium amount of strongly-bound Eu(III) observed by ion exchange (i.e., amount of Eu(III) eluting as a PAA complex, Fig. 1).

The rate of Sm(III) dissociation from size fractionated HA was also a function of metal-HA binding time, although the overall rate of dissociation from HA was much slower than from PAA. Similar to PAA, increasing the binding time resulted in higher percentages of Sm(III) dissociating by the slower processes. Figure 3 shows the kinetic spectra obtained for the two time domains for Sm(III) dissociation from the small HA after 48 hours of binding time. Three first-order terms were observed in each time domain, yielding six dissociation pathways. No change in the kinetics of dissociation from the small size HA fraction was observed after 48 hours of binding time, similar to the ion exchange results for this HA fraction. The equilibrium kinetic parameters calculated for the rate equations are given in Table II. The half-life of the longest-lived pathway for dissociation from the small HA is 48 minutes. The most labile interaction is also the primary pathway for dissociation, accounting for approximately half of the Sm(III)-small HA binding. The small HA complex is significantly less labile than Sm(III) bound to either fraction of PAA, but like the PAA complexes, hysteresis in the complexation/dissociation reactions are evident as dissociation is much faster than the time required to reach binding equilibrium.

For the large size HA fraction, three days of binding time was required to obtain consistent dissociation kinetics, as observed in ion exchange experiments. Like the small size fraction of HA, six different dissociation terms were required to model the data observed in two time domains. From the rate constants reported in Table II, the estimated half-life of the slowest dissociation process from the large HA is almost 3 hours. Unlike the other systems studied, the most labile interaction is significant, but does not clearly dominate the Sm(III)-large HA interactions. The first (k = 2.1 s^{-1}, $t_{1/2}$ = 0.3 s) and third processes (k = 0.047 s^{-1}, $t_{1/2}$ = 15 s) together represent 60% of the Sm(III) binding.

Table II: Calculated parameters for rate equations describing the dissociation
             of Sm(III) from size fractionated PAA and HA using ligand exchange.
             Conditions were metal conc., poly conc., pH 4.2, I = 0.1 M NaClO$_4$.

	Rate Constants (s^{-1})	% Sm(III)
Small PAA	41 ± 2	51 ± 5.0
	$(8.5 \pm 0.1) \times 10^{-1}$	32 ± 3.4
	$(2.2 \pm 0.3) \times 10^{-2}$	17 ± 2.6
Large PAA	38 ± 2	45 ± 4.8
	1.5 ± 0.1	30 ± 2.7
	$(3.8 \pm 0.3) \times 10^{-1}$	15 ± 1.5
	$(1.5 \pm 0.2) \times 10^{-2}$	10 ± 2.0
Small HA	2.3 ± 0.2	44 ± 4.8
	$(3.7 \pm 0.3) \times 10^{-1}$	16 ± 1.7
	$(4.7 \pm 0.4) \times 10^{-2}$	11 ± 1.0
	$(1.7 \pm 0.3) \times 10^{-2}$	6.5 ± 0.7
	$(2.1 \pm 0.3) \times 10^{-3}$	8.5 ± 0.8
	$(2.4 \pm 0.2) \times 10^{-4}$	14 ± 3.7
Large HA	2.1 ± 0.3	28 ± 3.0
	$(6.6 \pm 0.5) \times 10^{-1}$	6.2 ± 0.8
	$(6.0 \pm 0.4) \times 10^{-2}$	30 ± 4.9
	$(6.4 \pm 1.0) \times 10^{-3}$	7.8 ± 0.7
	$(5.5 \pm 1.9) \times 10^{-4}$	10 ± 0.9
	$(6.5 \pm 2.3) \times 10^{-5}$	18 ± 2.8

    The increased binding time necessary to reach an equilibrium for dissociation
from the two fractions of HA compared to PAA is likely related to the heterogeneity
of the HA.   The HA fractions probably experience significant intermolecular
interactions such as hydrogen bonding and aggregation that are not as important in
the homogenous PAA systems.   Complexation of the rare earth cations alters these
interactions, which slowly change the HA configuration as well as the lability of a
fraction of the metal interactions. Interestingly, the equilibrium amounts of Sm(III)
dissociating from either fraction of HA in the slow time domain appear to be similar
to the strong interactions for Eu(III) observed at equilibrium by ion exchange. At
equilibrium, $29 \pm 5\%$ of the Sm(III) dissociates from the small HA by the three
slowest processes ($t_{1/2} > 30$ s.) compared to $32 \pm 2\%$ Eu(III) that elutes from the ion

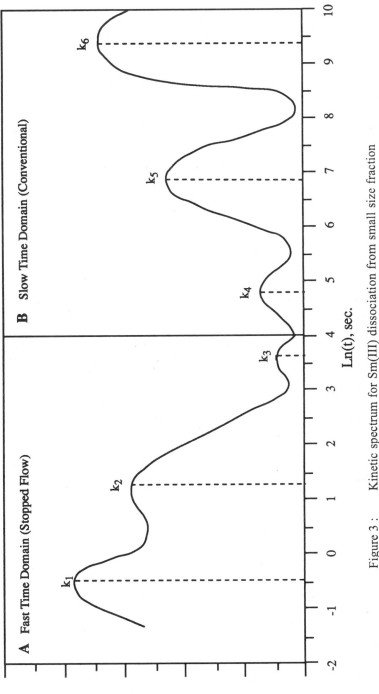

Figure 3 : Kinetic spectrum for Sm(III) dissociation from small size fraction of HA. Conditions were $[Sm(III)] = 5 \times 10^{-6}$ M, $[HA] = 1 \times 10^{-4}$ eq/L, pcH 4.2, [acetate buffer] = 0.01 M, ionic strength = 0.1 M $NaClO_4$.

exchange resin as a complex with the small fraction of HA. Ligand exchange results for dissociation from the large HA also agree well with ion exchange studies; 36 ± 4% Sm(III) dissociates by the three least labile processes, whereas 38 ± 8% Eu(III) eluted bound to the large HA in ion exchange.

Conclusions

Rare earth element complexation by PAA and HA is fast (5). However, most of the initial interactions are labile or "weak"; stronger interactions take over as the metal reside longer on the polyelectrolyte. Weak and strong interactions are indicated by both ion exchange and ligand exchange dissociation experiments. It is difficult to infer site binding versus counterion condensation from these methods, and such a distinction must await methods that probe more directly the molecular character of the different interactions.

Sm(III) and Eu(III) interactions with the simple polyelectrolyte PAA are significantly more labile at equilibrium than interactions with HA. These observations are consistent with a comparison of Th(IV) interactions with unfractionated HA and the synthetic polyelectrolytes polymaleic acid and polymethylvinylethermaleic acid (4). Greater than 95% of Th(IV) is observed as having strong interactions with HA compared to 70% of Th with either of the synthetic polyions. PAA seems to be a good model for lanthanide binding to a homogeneous polyelectrolyte in that the simplicity of the PAA ligand is reflected in the similarity of dissociation kinetics regardless of the size of the PAA fraction, and in the increased rates of equilibration and dissociation for PAA relative to HA systems.

For both PAA and HA, the lability of rare earth element interactions is greater for the smaller molecular weight fractions of each polyelectrolyte. Similar observations have been reported for Cu(II) dissociation from size fractionated HA (22). Interestingly, the smaller size fractions of HA have been shown to the most effective in transporting Am(III) and Cm(III) through sandy aquifers (23). Consequently, the influence and affect of polyelectrolyte size must be considered when predicting the mobility of these complexes in natural systems.

Acknowledgments

We wish to thank Dr. J. C. Sullivan, Chemistry Division, Argonne National Laboratory for the use of stopped-flow instrumentation. This research was supported at Florida State University by a grant from the U. S. Department of Energy, Office of Basic Energy Sciences. Manuscript preparation was supported at the Savannah River Ecology Laboratory under contract number DE-AC09-76SR00819 with the U. S. Department of Energy.

Literature Cited

(1)   Marcus, R. A. *J. Chem. Phys.* **1955**, *23*, 1057.
(2)   Manning, G. S. *Acct. Chem. Res.* **1979**, *12*, 443-449.

(3) Koppold, F. X.; Choppin, G. R. *Radiochim. Acta.* **1987**, *42*, 29-33.
(4) Choppin, G. R.; Cacheris, W. *Inorg. Chem.* **1990**, *29*, 1370-1374.
(5) Lis, S.; Wang, Z.; Choppin, G. R. *Inorg. Chim. Acta.* **1995**, *239*, 139-143.
(6) Meurer, B.; Spegt, P.; Weill, G. *Biophys. Chem.* **1982**, *16*, 89-97.
(7) Cacheris, W. P.; Choppin, G. R. *Radiochim. Acta.* **1987**, *42*, 185-190.
(8) Clark, S. B.; Choppin, G. R. In *Chemical Modeling of Aqueous Systems II*; D. C. Melchoir and R. L. Bassett, Ed.; ACS Symposium Series; American Chemical Society: 1990; Vol. 2; pp. 519-525.
(9) Choppin, G. R.; Clark, S. B. *Marine Chem.* **1991**, *36*, 27.
(10) Rao, L.; Choppin, G. R.; Clark, S. B. *Radiochim. Acta.* **1994**, *66/67*, 141-147.
(11) Clark, S. B. Ph.D. Thesis, Florida State University, 1989.
(12) Bertha, E. L.; Choppin, G. R. *J. Inorg. Nuc. Chem.* **1978**, *40*, 655-658.
(13) Buffle, J.; Deladoey, P.; Haerdi, W. *Analytica Chim Acta.* **1978**, *101*, 339-357.
(14) Choppin, G. R.; Cacheris, W. P. *J. Less-Common Metals.* **1986**, *122*, 551-554.
(15) Olson, D. L.; Shuman, M. S. *Anal. Chem.* **1983**, *55*, 1103-1107.
(16) Fish, W.; Morel, F. M. M. *Can. J. Chem.* **1985**, *63*, 1185-1193.
(17) Fish, W.; Dzombak, D. A.; Morel, F. M. M. *Environ. Sci. Tech.* **1986**, *20*, 676-683.
(18) Turner, D. R.; Varney, M. S.; Whitfield, M.; Mantoura, R. F. C.; Riley, J. D. *Geochim. Cosmochim. Acta.* **1986**, *50*, 289-297.
(19) Nederlof, M. M.; vanRiemsdijk, W. H.; Koopal, L. K *Environ. Sci. Tech.* **1994**, *28*, 1037-1047.
(20) Nederlof, M. M.; vanRiemsdijk, W. H.; Koopal, L. K *Environ. Sci. Tech.* **1994**, *28*, 1048-1053.
(21) Caceci, M. S.; Cacheris, W. P. *BYTE.* **1984**, *9*, 340-362.
(22) Olson, D. L.; Shuman, M. S. *Geochim. Cosmochim. Acta.* **1985**, *49*, 1371-1375.
(23) Marley, N. A.; Gaffney, J. S.; Orlandini, K. A.; Cunningham, M. A. *Environ. Sci. Tech.* **1993**, *27*, 2456-2461.

Chapter 14

Isotopic Characterization of Humic Colloids and Other Organic and Inorganic Dissolved Species in Selected Groundwaters from Sand Aquifers at Gorleben, Germany

M. Ivanovich[1,4], M. Wolf[2], S. Geyer[2], and P. Fritz[3]

[1]Atomic Energy Authority Technology, B551 Harwell,
Oxfordshire OX11 0RA, United Kingdom
[2]Gesellschaft für Strahlen und Umweltforschung, Institut für
Hydrologie, Ingolstadter Landstrasse 1, D–85764 Neuherberg, Germany
[3]Umweltforschungzentrum Leipzig Halle GmbH, Permoser
Strasse 15, D–07050 Leipzig, Germany

The aim of this work was to determine the significance of groundwater colloids in far-field radionuclide migration. To this purpose, isotopic data presented in this paper were obtained from several selected Gorleben groundwaters as part of the colloid characterisation programme. The contents of major and minor ions, light isotopes (^{2}H, ^{3}H, ^{13}C, ^{14}C, ^{18}O and ^{34}S), and the U/Th isotopes were measured. Radiocarbon and ^{13}C were measured in dissolved inorganic carbon (DIC), ion the humic acid (HA-colloids) and fulvic acid (FA-solution) fractions of dissolved organic carbon (DOC). The ^{18}O and ^{34}S were also determined in dissolved sulphate phase. The U/Th isotope measurements were carried out on total and surface solid phases, colloid fraction (1-1000 nm particle size, HA) and solution (<1.5 nm, FA).

The ^{3}H data using a piston-flow model have yielded groundwater ages of less than 40a before present (BP) for a near surface groundwaters and much older ages for all other deeper organically-rich Gorleben groundwaters. The ^{14}C content of DIC is low and probably influenced by strong dilution with mineralised fossil carbon. The resulting high ^{14}C-model ages (up to 27 ka BP) can only be interpreted as upper limits of the true groundwater ages.

[4]Current address: Enterpris Limited, Philip Lyle Building, University of Reading, Whiteknights, Berkshire, RG6 6BX, United Kingdom

0097–6156/96/0651–0220$16.00/0

The ^{14}C content of the HA fractions is partly very low or partly below detection limit yielding very high ^{14}C-model groundwater ages (up to >30 ka BP) whereas, some of the FA fractions contain higher ^{14}C yielding model groundwater ages more consistent with post-glacial recharge (<12 ka BP). These groundwater ages are consistent with Holocene recharge inferred from the ^{2}H and [18]O isotope data. These differences between the HA and FA fractions are also confirmed by the U/Th isotope ratios suggesting different source/histories of the two organic carbon fractions. Thus, it is possible that the FA fraction has originated mainly from the near-surface (soil) source, while the HA fraction derives mainly from organic matter from older sediments such as lignite lenses or from a mixture of an enhanced input of organically-rich water younger than some 12 ka BP and much older lignite source.

Mobility of radionuclides in groundwaters is governed by complex physico-chemical interactions which largely depend on the geochemical characteristics of the radionuclides, fluids and the solid surfaces of the geosphere. The presence of colloids (particles whose operational size range is 1 to 1000 nm), and natural organic substances such as humic and fulvic acids, may affect the behaviour of radionuclides by diverse processes such as sorption, complexation, dissolution/precipitation, steric exclusion, etc. The importance of colloids and humic substances to the migration of radionuclides in the natural system has been recognised [1]. Furthermore, the knowledge of the migration behaviour of actinides in the geosphere is of potential importance with respect to safe storage of radioactive waste [1,2]. The higher valency actinide ions ($Z \geq 3^+$) generate actinide 'pseudo colloids' by sorption onto groundwater colloids which, in turn, may be highly mobile in the geosphere. The importance of the pseudo colloid generation and their subsequent migration has been demonstrated for the Gorleben system of aquifers by Kim and his co-workers [3].

The main aim of this work was to determine the significance of groundwater colloids in the transport of radionuclides in the far field of a nuclear waste repository. Thus, as part of the colloid characterisation programme, groundwaters and sediment samples were collected from six boreholes which penetrate the geologically and chemically very heterogeneous Gorleben aquifers system [4-6]. These samples were then characterised both chemically and isotopically in order to establish their mobility.

Earlier work on Gorleben aquifers [1-3] has identified two organic phases in the sampled groundwaters; the distinction between the two phases was dictated by the ultrafiltration methodology deployed to sample natural aquatic colloids. Thus, colloids (operational size range 1 to 1000 nm) were identified as made up mostly of humics (HA) and the organic macromolecules associated with solution (passed through the 10,000 MW cut-off ultrafilter) were identified as fulvics (FA). Throughout this paper colloids/HA and solution/FA are synonymous terms used for the two organic groundwater phases studied.

The Gorleben site is located in Lower Saxony (Northern Germany) in the district of Lüchow-Dannenberg near the River Elbe (Fig 1). Groundwater samples were collected from boreholes GoHy 201, 572, 573, 611, 612 and 2227. These boreholes intersect at least three different aquifers as can be seen in the geological cross-section through the overlying sediments of the Gorleben salt dome (Fig 2). The sampled aquifers are situated within sand, silt, marl and clay layers. GoHy 201, 572 and 611 correspond to shallow Pleistocene sand and silt aquifers with low salt content, whereas GoHy 573, 612 and 2227 intersect a relatively deep and saline aquifer with intercalations of reworked tertiary lignites. A total of six solid samples originating from three different horizons in drill cores from GoHy 573 and 2227 boreholes has been studied. In each case, two horizons represented quartz-sand type material and the third that of lignite suspected to be a local source of humic materials found to be the main constituent of the colloids separated from the groundwater samples.

Sampling and Measurement Methods

Groundwater and Colloid Sampling. During the field campaign of May/June 1991, six Gorleben boreholes were sampled for groundwaters and four also for groundwater colloids. Groundwaters from boreholes GoHy 201, 611, 612, 572, 573 and 2227 were sampled for the following light isotopes: ^{2}H, ^{3}H, ^{13}C, ^{14}C, ^{18}O and ^{34}S. Groundwaters from boreholes GoHy 611, 572, 573 and 2227 were also sampled for U series radionuclides as well as for colloid in-situ and off-line separation and characterisation.

For ^{2}H and ^{18}O analyses two 30 ml aliquots were collected, for ^{3}H analyses 1 ℓ or two 1 ℓ samples were collected in glass bottles and, for ^{34}S and ^{18}O determinations on dissolved sulphate, 5 ℓ were collected in polyethylene (PE) bottles. The sulphate species were precipitated in the field by adding Cd acetate. For ^{14}C and ^{13}C analyses on dissolved inorganic carbon (DIC), one or two 60 ℓ samples were collected in PE containers. Further samples were also collected for ^{14}C and ^{13}C measurements on dissolved organic carbon (DOC). In the case of high DOC content (GoHy 612, 572, 573 and 2227) two or three 50 ℓ samples were collected in stainless steel (or Al) containers. To avoid ^{14}C contamination by atmospheric CO_2, the sampling vessels were flushed with pure N_2 or Ar. Samples with low DOC content (GoHy 201 and 611) were processed in the field *(8)*. In this case, the DOC content was enriched by pumping about 300 ℓ (GoHy 611) and 400 ℓ(GoHy 201), after filtration and acidification, through Teflon-capped glass columns filled with XAD-8 resin *(8,9)* and silicalite *(10,11)*.

Before sampling for U series radionuclides and groundwater colloids, boreholes GoHy 611, 572, 573 and 2227 were flushed by pumping off three times their volume. Possible contamination of the groundwater samples which

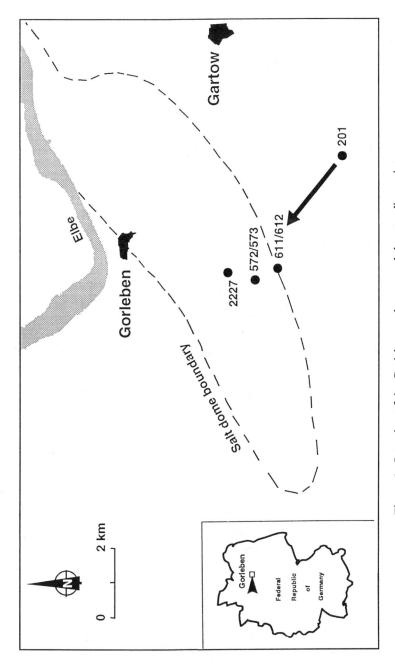

Figure 1. Location of the Gorleben study area and the sampling points together with the salt dome boundary and the general groundwater flow direction.

may be caused during the pumping procedure was monitored by checking the physical and chemical properties (pH, Eh, Temp, O_2, HCO_3^-, and electrical conductivity). The sampling was carried out under anaerobic conditions (N_2 + 1% CO_2 atmosphere) in order to prevent significant changes in the physical and chemical characteristics of the colloids through interaction with air. Sample processing in the field comprised the concentration of 500 ℓ of 1 µm prefiltered groundwaters using an Amicon DC10-LA ultrafiltration system fitted with a 10,000 MW cut-off hollow fibre tangential flow cartridge filter (~1.5 nm cut-off pore size) having an effective surface area of 5 m^2 *(12)*. This system was used to produce 20 ℓ of colloid concentrate giving a concentration factor of about 25 and colloidal size distribution between 1000 and ~1 nm. Twenty litre samples of ultrafiltrate (the fraction representing 'solution' that passed through the 10,000 MW ultrafilter) were also collected. Two 50 ℓ unfiltered groundwaters samples were also collected and stored anaerobically in Al-containers with a special coating (Sigma Coating 9255-0300) to prevent contamination during transportation and storage. Further samples included 25 ℓ of 1 µm prefiltered and acidified water for U/Th isotopic analysis and 1 µm prefilters preserved for the U/Th analysis of particulates.

Measurement Techniques. Light isotopes Concentrations of ^{3}H were determined radiometrically after synthesis of propane *(13)* in a proportional counter after about twenty-fold electrolytical enrichment of [3]H *(14)*. The ^{14}C content of large samples (≥ 0.5 g C) were determined by liquid scintillation counting after synthesis of benzene *(15)* from CO_2 which was extracted from the water sample after acidification. The ^{14}C content of small samples (<0.5 g C) were measured by accelerator mass spectrometry (AMS) by the Isotrace Laboratory of the University of Toronto. The ^{2}H, ^{13}C, [18]O and [34]S isotopic analyses were carried out by standard mass spectrometry.

Uranium series radionuclides The partition of the activities of the longer-lived isotopes of U and Th between the colloid and solution phases, as well as the isotopic activity ratios and relative concentrations were measured using isotope dilution/alpha spectrometry *(16)*. The U/Th isotope concentrations in the colloid fraction were derived by subtracting the U/Th isotope concentrations in the ultrafiltrate from that in the colloid concentrate corrected by the concentration factor of the colloid concentrate. In addition to total sample analysis, each solid sample was subjected to a sequential leaching procedure designed to allow the extraction of various solid phases: (i) organics and ion-exchangeable (ORG + IE) (4% NaOH shaken for 24 h); (ii) sorbed (AD) (Na_4 EDTA/1M Na_2CO_3 shaken for 24 h); (iii) amorphous Fe/Mn oxides (AM Fe/Mn) (Tamms acid oxalate shaken in complete darkness for 24 h); (iv) crystalline Fe/Mn oxides (CR Fe/Mn) (Tamms acid oxalate sat in light for 1 to 2 days); and (v) resistates (RES) (HF/$HClO_4$ complete dissolution). Each phase was analysed for U and Th isotopes to yield information on U and Th

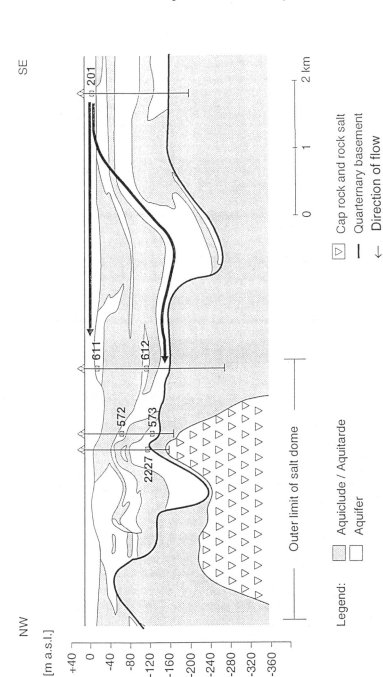

Figure 2. Geological cross-section of the sediments above the Gorleben salt dome (after ref (7)) together with the borehole locations and the probable groundwater flow lines.

distribution and the corresponding state of radioactive equilibrium in the solid phases. This approach should allow an approximate characterisation of the solid surface (made up of organic/ion exchangeable phase, sorbed phase and amorphous Fe/Mn oxyhydroxides) in terms of U/Th isotopic contents and subsequent comparison with the state of U/Th series disequilibrium in the colloid and solution phases.

Results

Light Isotopes. The results of the light isotope (^{2}H, ^{3}H, ^{13}C, ^{18}O and ^{34}S) measurements are presented in Tables I and II.

The ^{3}H data are expressed as TR (Tritium Ratio; 1 TR corresponds to a ^{3}H/H ratio of 10^{-18} or to an activity concentration of 0.118 Bq/kg water). The ^{14}C data are expressed as percent modern carbon (pmc; 100 pmc corresponds to 95% of the NBS oxalic acid standard or to an activity concentration of 0.226 Bq/g C). The ^{2}H and ^{18}O data are expressed as $^0/_{00}$ deviations(δ-values) relative to SMOW (Standard Mean Ocean Water). The ^{13}C data are expressed as $^0/_{00}$ deviations relative to PDB (Pee Dee Belemnite) standard. The ^{34}S data are expressed as $^0/_{00}$ deviations relative to CDT (Canion Diablo Troilite) standard.

Uranium Series Radionuclides. Table III presents radiometric data obtained from the four Gorleben groundwaters GoHy 611, 572, 573 and 2227. Approximately 60% of the total U in the fluid phase after 1 μm prefiltering remains in solution for samples GoHy 611, 572 and 573 and, about 30% of total U remains in solution in GoHy 2227. In contrast, relatively young groundwater from GoHy 611 has a reduced (but still high) Th content of which only 16% is associated with the organic colloid phase whereas, the other three groundwaters have enhanced Th content relative to published data for most groundwaters *(17)* and over 99% of total Th is associated with organic colloids. Relevant radiometric data obtained for the six solid samples are given in Table IV (for more detail see ref *(18)*).

Discussion

Light Isotopes. Tritium With one exception (GoHy 611), tritium concentrations of all other samples are either below detection limit or are very low. From a piston flow model assumption, the corresponding tritium model-age of GoHy 611 is below 40a BP, whereas the ages of all other samples are well above this technique's age limit (40a). The low but detectable ^{3}H content of samples GoHy 572 and 612 could be interpreted either as representing relatively young waters or, more probably, as representing admixtures of small

Table I Light isotope contents of groundwaters from selected Gorleben aquifers

GoHy number	Sampling date	^{3}H (TR)	$^{14}C_{DIC}$ (pmc)	$d^{13}C_{DIC}$* (‰ PDB)	d^2H** (‰ SMOW)	$d^{18}O$+ (‰ SMOW)	$d^{34}S$++ (‰ CDT)	$\delta^{18}O_{SO_4}$ # (‰ SMOW)
201	14.05.91	<1.1	26.5 ±2.1	-13.6	-62.3	-9.1	2.7	12.8
611	16.05.91	8.0 ±0.7	26.5 ±1.7	-13.3	-62.0	-8.8	2.7	9.4
572	14.05.91	0.58 ±0.19	5.2 ±1.1	-4.1	-59.8	-8.8	-	-
612	16.05.91	0.52 ±0.21	1.9 ±0.8	-11.9	-60.4	-9.0	-	-
573	15.05.91	<0.40	3.7 ±0.8	-9.2	-61.5	-8.9	-	-
2227	13.05.91	<0.30	5.0 ±0.8	-12.8	-62.4	-9.0	34.0	15.2

All quoted errors are 2σ uncertainties including counting statistics and other experimental errors.
* All uncertainties are ± 0.4‰   ** All uncertainties are ± 1‰   + All uncertainties are ± 0.15‰
++ All uncertainties are ± 0.3‰   # All uncertainties are ± 0.4‰

Table II   Light isotope contents of DOC fractions, DOC concentrations and DOC fractionation in groundwaters from selected Gorleben aquifers

GoHy number	Sampling date	$^{14}C_{HA}$ (pmc)	$d^{13}C_{HA}$* (‰ PDB)	$^{14}C_{FA}$ (pmc)	$d^{13}C_{FA}$* (‰ PDB)	DOC (mg/ℓ)	HA (%)	FA (%)
201	14.05.91	28.8 ±0.6	-26.9	58.6 ±2.4	-26.0	0.9	6	41.7
611	16.05.91	14.5 ±0.4	-26.3	38.2 ±5.4	-24.8	1.4	8	32.5
572	14.05.91	<3.8	-26.2	23.6 ±1.8	-26.2	14.4	66	30.3
612	16.05.91	1.7 ±1.1	-26.1	28.3 ±1.6	-25.1	184	92.7	5.3
573	15.05.91	<1.1	-27.1	7.0 ±3.2	-25.9	95.2	61.5	6.7
2227	13.05.91	1.7 ±1.1	-27.1	10.6 ±3.3	-25.9	73.4	78.5	8
2227#	13.05.91	-	-	11.7 ±1.0	-	-	-	-

All quoted errors are ±2σ uncertainties including counting statistics and other experimental errors.
*All uncertainties are ±0.4‰   #Measurement carried out by AEA Technology for comparison purposes.

Table III U and Th isotope distribution between colloids and solution

Phase (Harwell code)	[U] (mBq/l)	[U] (% total)	[Th] (mBq/l)	[Th] (% total)	$^{234}U/^{238}U$ (Activity ratios)	$^{230}Th/^{234}U$ (Activity ratios)
GoHy 611						
Colloid (3094 + 3096)	0.23 ±0.02*	37.7	0.003 ±0.002	15.8	3.84 ±0.73	0.73 ±0.27
Solution (3095)	0.38 ±0.03	62.3	0.016 ±0.001	84.2	1.70 ±0.20	0.84 ±0.08
GoHy 572						
Colloid (3097 + 3099)	1.51 ±0.05	39.6	3.69 ±0.12	99.1	1.00 ±0.05	1.88 ±0.08
Solution (3098)	2.30 ±0.08	60.4	0.033 ±0.001	0.9	2.01 ±0.09	0.10 ±0.01
GoHy 573						
Colloid (4000 + 4002)	1.18 ±0.07	36.3	11.75 ±0.31	99.6	2.56 ±0.18	5.22 ±0.23
Solution (4001)	2.07 ±0.07	63.7	0.05 ±0.02	0.4	2.51 ±0.10	0.07 ±0.01
GoHy 2227						
Colloid (3056 + 3059)	5.85 ±0.33	71.6	4.38 ±0.18	98.9	1.28 ±0.10	1.02 ±0.07
Solution (3057)	2.32 ±0.32	28.4	0.05 ±0.02	1.1	2.26 ±0.40	0.10 ±0.05

* All quoted errors are 1σ uncertainties due to counting statistics only.

Table IV A   Selected radiometric data for GoHy 573 solid phases

Sample phase* (Harwell code)	[U] (ppm)	[U] (% total)	[Th] (ppm)	[Th] (% total)	$^{234}U/^{238}U$ (activity ratio)	$^{230}Th/^{234}U$ (activity ratio)
(135.5-136.0 m) - Quartz sand 1						
ORG + IE** (1306)	0.07 ±0.01+	12.5	0.06 ±0.02	4.7	1.63 ±0.34	0.67 ±0.13
AD (1307)	0.13 ±0.01	23.8	0.27 ±0.03	23.2	1.07 ±0.13	1.53 ±0.18
AM Fe/Mn (1308)	0.02 ±0.004	3.0	0.12 ±0.02	9.9	1.83 ±0.56	1.73 ±0.40
TOTAL (3282)	0.34 ±0.02	100	0.97 ±0.09	100	1.03 ±0.09	1.59 ±0.15
(136.0-136.5 m) - Quartz sand 2						
ORG + IE (1311)	0.15 ±0.02	31.8	0.07 ±0.02	6.9	1.04 ±0.17	0.29 ±0.07
AD (1312)	0.05 ±0.01	10.1	0.29 ±0.03	28.4	1.60 ±0.36	2.03 ±0.32
AM Fe/Mn (1313)	0.02 ±0.01	3.9	0.11 ±0.01	10.4	1.35 ±0.43	3.40 ±0.69
TOTAL (3283)	0.41 ±0.03	100	0.97 ±0.11	100	1.06 ±0.09	1.27 ±0.13
(136.5-137.0 m) - Lignite						
ORG + IE (1316)	1.11 ±0.05	42.3	1.12 ±0.07	19.0	1.40 ±0.03	0.45 ±0.03
AD (1317)	0.25 ±0.02	9.5	1.77 ±0.08	29.9	1.18 ±0.10	2.86 ±0.21
AM Fe/Mn (1318)	0.05 ±0.01	2.0	0.55 ±0.04	9.3	1.78 ±0.27	2.80 ±0.31
TOTAL (3284)	2.75 ±0.08	100	6.61 ±0.18	100	1.15 ±0.03	0.76 ±0.03

*   Primary phases' data have been omitted as they are not expected to take part in rock/water/colloid interactions.
**   See text for the definition of phase abbreviations.
+   All quoted errors are 1σ uncertainties due to counting statistics only.

Table IV B   Selected radiometric data for GoHy 2227 solid phases

Sample phase* (Harwell code)	[U] (ppm)	[U] (% total)	[Th] (ppm)	[Th] (% total)	$^{234}U/^{238}U$ (activity ratio)	$^{230}Th/^{234}U$ (activity ratio)
(129.6-130.0 m) - Sand/marl 1						
ORG + IE** (1321)	0.10 ±0.01[+]	6.3	0.06 ±0.02	1.3	1.35 ±0.20	0.63 ±0.10
AD (1322)	0.41 ±0.03	27.3	1.01 ±0.07	22.6	1.21 ±0.10	0.83 ±0.07
AM Fe/Mn (1323)	0.14 ±0.02	9.5	1.18 ±0.08	26.5	1.21 ±0.19	2.42 ±0.31
TOTAL (3285)	1.46 ±0.05	100	4.18 ±0.13	100	1.07 ±0.04	1.09 ±0.05
(130.0-130.5 m) - Sand/marl 2						
ORG + IE (1326)	0.10 ±0.01	5.6	0.03 ±0.01	0.8	1.51 ±0.24	0.39 ±0.07
AD (1327)	0.48 ±0.03	28.3	0.97 ±0.08	22.9	1.05 ±0.09	0.80 ±0.08
AM Fe/Mn (1328)	0.16 ±0.02	9.5	1.12 ±0.10	26.2	0.81 ±0.12	2.34 ±0.35
TOTAL (3286)	1.44 ±0.06	100	3.73 ±0.18	100	0.95 ±0.05	0.97 ±0.06
(130.5-131.2 m) - Lignite						
ORG + IE (1331)	0.83 ±0.07	48.2	0.89 ±0.11	40.7	1.61 ±0.17	2.47 ±0.19
AD (1332)	0.59 ±0.07	34.0	0.83 ±0.14	37.8	0.96 ±0.15	1.19 ±0.19
AM Fe/Mn (1333)	0.04 ±0.02	2.2	0.12 ±0.05	5.4	4.0 ±2.0	0.87 ±0.28
TOTAL (3287)	0.71 ±0.05	100	1.43 ±0.09	100	1.57 ±0.12	0.72 ±0.05

*   Primary phases' data have been omitted as they are not expected to take part in rock/water/colloid interactions.
**   See text for the definition of phase abbreviations.
[+]   All quoted errors are 1σ uncertainties due to counting statistics only.

amounts of young and old waters. The observed ^{3}H presence has also been ascribed to contamination caused by drilling (19).

Carbon-14 and ^{13}C in DIC and DOC   The DIC ^{14}C and ^{13}C data (see Table I) show relatively low ^{14}C contents (26.5 pmc) and the typical δ^{13}C range of -13.6 to -13.3$^0/_{00}$ for the near-surface groundwaters such as GoHy 201 and 611 representing relatively young waters in the Gorleben system. With respect to ^{14}C, these values also show that mineralisation of old organic matter plays an important role in the genesis of these groundwaters. Therefore, ^{14}C model ages calculated by the usual geochemical correction models (20-26) can only be interpreted as upper limits on the true groundwater ages.

All other groundwaters from deeper horizons, GoHy 572, 612, 573 and 2227, have yielded low ^{14}C content (1.9 to 5.2 pmc) and higher δ^{13}C values (-12.8 to - 4.1 $^0/_{00}$) indicative of either high groundwater ages or, more likely, high dilution of the DIC by mineralised fossil organic matter and subsequent dissolution of carbonate minerals. One possible source of this ^{14}C-free carbon is lignite present at various horizons in the sediments of the Gorleben aquifers. The corresponding δ^{13}C values in these deeper groundwaters are in the range of -12.8 to -9.2$^0/_{00}$ (one exception being GoHy 572 with δ^{13}C of -4.1$^0/_{00}$) typical of groundwaters which evolved in carbonate-rich aquifers. The high δ^{13}C value measured for GoHy 572 is a possible result of methanogenesis and/or carbon isotope exchange processes taking place in the aquifer.

The ^{14}C content in DOC fractions (see Table II) tends to decrease from shallower to deeper aquifers with the humic acid (HA) fraction (>10,000 MW) ^{14}C content range of <1.1 to 28.8 pmc and the corresponding range for fulvic acid (FA) fraction (<10,000 MW) of 7.0 to 58.6 pmc. The significant difference in ^{14}C content of the two DOC fractions may be indicative of their different origins, the FA fraction originating mainly from soils (near surface horizons) and the HA fraction deriving mainly from sedimentary organic matter. The ^{13}C content of all DOC fractions of the sampled Gorleben groundwaters fall in the range of Calvin cycle plant matter (27) inferring, as expected, a terrestrial origin.

Only minor amounts of HA were found in the near surface groundwaters such as GoHy 201 and 611. On the other hand, the FA fraction representing between 30 and 40% of DOC in the shallower Gorleben groundwaters is typical of the waters from other sandy aquifers in this region (28). The FA fraction drops to below 10% of DOC in deeper Gorleben groundwaters (GoHy 612, 573 and 2227).

Deuterium and [18]O   The δ^{2}H and δ^{18}O values given in Table I are typical for recent groundwaters in this region and are not influenced by processes such as rock/water interactions or evaporation. These data show no influence of colder climate (29), and hence the Gorleben groundwaters are all likely to be of Holocene origin (younger than 10 to 12 ka).

Sulphur-34 and ^{18}O in dissolved sulphate The $\delta^{34}S$ values from GoHy 201, 611 and 2227 have been plotted against the $\delta^{18}O$ values of the dissolved sulphate together with similar data measured in rock samples of different geological formations from the Asse salt mine, north Germany *(30)* (see Fig. 3). The ^{34}S and ^{18}O contents of the dissolved sulphate in GoHy 201 and 611 are characteristic of sulphates from atmospheric fallout. In contrast, the high ^{34}S content of GoHy 2227 is a possible result of strong isotopic enrichment due to sulphate reduction. Similar measurements in groundwaters GoHy 572 and 573 were not possible due to their very low sulphate content.

Radiocarbon-model Ages of Gorleben Waters. Based on different assumptions, several sets of ^{14}C-model ages (t in years before present (BP)) have been calculated from the ^{14}C content (A_t) of carbonate components (DIC) and DOC (FA and HA) using the following expression:

$$t = 8267 \ln (A_O/A_t) \tag{1}$$

The initial ^{14}C content (A_O) for DIC was calculated in two different ways:

Method 1 In the first method, the initial ^{14}C content (A_O) for DIC was calculated using the measured hydrochemical data (see Table V) by the geochemical code PHREEQE *(31,32)* and assuming the closed system conditions represented by the following equation:

$$A_o = A_g \, [N_{CO_2(aq.)} + N_{H_2CO_3} + 0.5\,(N_{HCO_3^-} + N_{NaHCO_3} + N_{KHCO_3}$$
$$+ N_{MgHCO_3^+} + N_{CaHCO_3^+} + N_{MnHCO_3^+})] \tag{2}$$

where A_g is ^{14}C content of the soil gas (~100 pmc or estimated from the measured ^3H concentrations assuming the validity of the exponential model), and N_i is the mole fraction of different carbon species i (the original thermodynamic data base in the PHREEQE code was modified after ref *(33)*). For both FA and HA components of DOC, the initial ^{14}C content, A_O was assumed to be 100 pmc for method 1.

Method 2 In the second method, the initial ^{14}C content (A_O) for DIC was assumed to be equal to the ^{14}C content of GoHy 201 representing a typical young groundwater in the system (A_O = 26.5 pmc). Similarly, for FA and HA components, the corresponding A_O was given the measured values for GoHy 201 of 58.6 pmc and 28.8 pmc, respectively.

Two sets of ^{14}C-model ages are shown in Fig 4. The comparison between the calculated ages derived for different carbonate components in each set shows

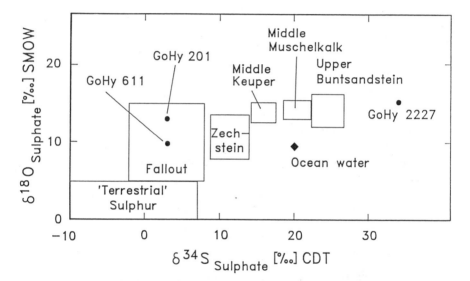

Figure 3. The d^{34}S and d^{18}O values of dissolved sulphate of Gorleben groundwaters GoHy 201, 611 and 2227 together with regions of fallout sulphate, terrestrial sulphate and sulphate of different geological formations at the Asse salt mine.

Table V Hydrochemical data for Gorleben groundwaters carried out in the field (Temp., pH, Eh, O_2, CO_2, HCO_3^-) and in the laboratory (Ca^{2+}, Mg^{2+}, Na^+, K^+, Cl^-, SO_4^{2-}, NO_3^-)

GoHy No.	Sampling date	Depth (m)	Temp. (°C)	pH	Eh (mV)	O_2	CO_2	Ca^{2+}	Mg^{2+}	Na^+	K^+	HCO_3^-	Cl^-	SO_4^{2-}	NO_3^-
						<----------------------------- (mg/ℓ) ----------------------------->									
201	14.05.91	30–35	9.6	7.75	121	<0.1	1.8	19.7	2.25	7.57	0.70	45.8	10.2	22.8	<0.02
611	16.05.91	21–24	8.9	8.4	127	<0.1	-	33.6	2.16	10.7	0.71	64.1	13.6	31.3	0.04
612	16.05.91	121–125	10.9	8.36	35	<0.1	-	22.6	3.64	456	2.93	802.8	234	29.0	1.78
572	14.05.91	70–73	12.3	8.9	17	<0.1	-	3.41	0.27	135	1.19	285.5	28.8	0.5	0.02
573	15.05.91	134–137	14.4	8.2	-3	<0.1	-	9.21	2.04	431	3.38	490	434	1.7	<0.02
2227	13.05.91	128–130	14.1	7.8	-58	<0.1	11.8	21.4	4.83	962	14.8	481.9	1303	31.1	<0.02

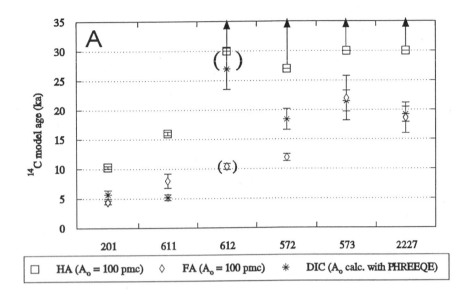

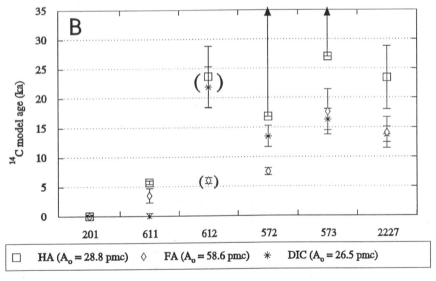

Figure 4. The ^{14}C-model ages in years together with their two-fold standard deviations (bars) for Gorleben groundwaters. DIC-^{14}C ages were calculated with A_0 derived from PHREEQE code, whereas for FA and HA an initial value A_0 of 100 pmc was assumed (Figure 4A). In the lower diagram the model ages of the same samples were calculated with the assumption that the measured values of the young sample GoHy 201 represent the initial conditions in the aquifer system. Therefore, values of A_0 of 26.5 pmc for DIC, 58.6 pmc for FA and 28.8 pmc for HA were used for the ^{14}C-age calculations (Figure 4B). The samples were arranged in the assumed flow direction.

that the ages calculated from DIC and FA data are systematically lower than those calculated for the HA component. This is not surprising considering the likelihood that HA in the groundwaters has derived mainly from fossil organic matter in the sediments whereas the FA origin is mainly in the soil horizon. Thus, a tentative conclusion based on the above observations is that the groundwater ^{14}C dating using DOC should be based on FA component and not HA.

For GoHy 201, 573 and 2227 groundwaters, the ^{14}C-model ages for DIC and FA are consistently similar. On the other hand, for GoHy 572 and 612 groundwaters the DIC-derived ages are higher than the FA-derived ages. In the case of GoHy 572, this apparent discrepancy may be explained by suggesting methanogenesis and/or isotope exchange processes as the possible causes of the 'open-system' conditions in this groundwater. In contrast, for GoHy 612, the large difference between the DIC- and FA-derived ^{14}C ages may be due to the mixing of younger and older waters at that horizon in the aquifer. Furthermore, in the case of GoHy 611, the lower DIC-derived groundwater age seems more realistic on hydrological grounds than the corresponding higher FA-derived age. The observed difference may be explained by addition of older (fossil) FA from the sediments (as well as mixing with ^{3}H-containing younger water) along the assumed flow path from borehole GoHy 201 to GoHy 611 (see Fig.2).

The second set of ^{14}C-model ages derived by method 2 above are generally lower than the corresponding ages in the first set. In the case of GoHy 572 (DIC), 573 (DIC and FA), 612 (DIC) and 2227 (DIC and FA) all the calculated groundwater ages are in the <25 ka) range. However, the corresponding stable isotope data discussed in Section 4.1 above has all the characteristics of Holocene groundwaters leading to an inevitable conclusion that the ^{14}C- model ages of these Gorleben groundwaters can only be interpreted as maximum ages in this hydrogeologically very complex region.

Uranium Series Radionuclides. The ^{234}U/^{238}U activity ratios in the colloid phases of the four Gorleben groundwaters vary between 1.0 and 3.8, the younger groundwater, GoHy 611 having the highest activity ratio (see Table III). Only GoHy 573 has the same ^{234}U/^{238}U activity ratio for the organic colloid (HA) and solution (FA) phases. For the other three groundwaters, the colloid and solution phases do not appear to be in isotopic equilibrium with respect to U. With the exception of GoHy 611 (0.83 ± 0.08), all ^{230}Th/^{234}U activity ratios in solution (FA) are low (<0.1) consistent with much lower Th solubility relative to U. In contrast, ^{230}Th/^{234}U activity ratios in the colloid phases of all four groundwaters are high (0.73 to 5.22) reflecting increased Th presence due to organic colloids (mostly HA) in these groundwaters.

In the case of GoHy 573 total solid samples (see Table IVa), the lignite sample is much richer in both U and Th relative to the quartz sand samples; this is consistent with rare-earth elements (REE) and other trace metal data *(6)*.

Whereas, for the quartz sand $^{234}U/^{238}U$ activity ratios are unity within the quoted 1σ uncertainty, the lignite total sample has a slight excess of ^{234}U relative to ^{238}U. The two sand samples from this core have yielded $^{230}Th/^{234}U$ activity ratios greater than unity indicating a net loss of U to groundwater by non-preferential leaching, whereas the total lignite sample has a $^{230}Th/^{234}U$ activity ratio much less than unity indicative of U uptake from groundwater.

The total solid samples of sand/marl from GoHy 2227 (see Table IVb) have yielded both activity ratios of unity within 2σ quoted uncertainties indicating little recent exchange with the corresponding groundwater. The total lignite sample from this core, however, has a $^{234}U/^{238}U$ activity ratio considerably higher than unity and the corresponding $^{230}Th/^{234}U$ activity ratio indicates 'recent' U uptake by the solid.

The apparent differences between the solids in the two Gorleben cores may be due to different sorption characteristics of the overlying sediments such as variable clay content. Thus, the lanthanide and other metal concentrations in the GoHy 573 groundwater are much higher than in the GoHy 2227 groundwaters (see ref *(6)*) possibly due to the lower sorption capacity of the relatively poorer quartz sand overlying the lignite in the GoHy 573 aquifer; similarly the lower concentrations in GoHy 2227 groundwater may be accounted for by higher sorption capacity of the overlying sand/marl layers.

The solid phase distributions of U and Th in both cores present contrasting patterns between the lignite samples, on one hand, and the quartz sand/marl, on the other (see Table IV). Thus, in the case of lignite samples large proportions of U (more than 40%) and Th (19 and 41%) are found associated with the organic and ion-exchangeable (ORG + IE) phase whereas, in the case of the quartz sand/marl, this phase contains minor fractions of the total solid content. The solid surface phases state of U/Th series disequilibrium is indicative of active radionuclide exchange taking place.

The question of whether this exchange is between the solid surfaces and solution or with organic colloids (HA) and whether the colloids derive their isotopic signatures from interaction with solution or solid surfaces has been considered. Figure 5 presents $^{234}U/^{238}U$ activity ratios for various solid surface, colloid and solution phases in GoHy 573 and 2227. In the case of GoHy 573 the colloid and solution phases are in isotopic equilibrium with respect to U. All of the $^{234}U/^{238}U$ activity ratios obtained from surface phases of the three solid samples are much lower than 2.5 obtained for the fluid phase leading to the conclusion that the solid phases are not necessarily or exclusively the local source of U to the fluid phases. In contrast, organic colloids in GoHy 2227 are not in isotopic equilibrium with the corresponding solution but, instead, their $^{234}U/^{238}U$ activity ratio is within less than 2σ of the (ORG + IE) phases of the three solid samples, the implication being that they may have acquired their isotopic signatures from that source. Although this evidence of colloid

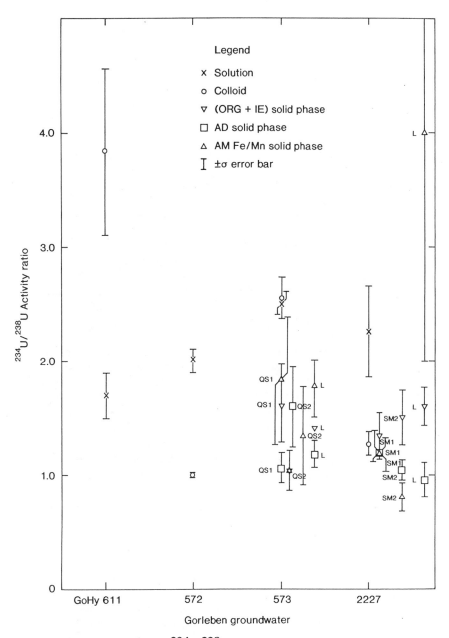

Figure 5. Comparison of $^{234}U/^{238}U$ activity ratios from liquid and solid phases of the four Gorleben groundwaters GoHy 601, 572, 573 and 2227.

provenance is not entirely conclusive, it is possible that the humic colloids in GoHy 2227 and GoHy 573 have different isotopic histories and possibly different sources. This conclusion is supported to a certain extent by the ^{14}C-derived groundwater ages discussed in the preceding section in which the HA component of DOC ^{14}C signatures have yielded consistently older groundwater ages possibly due to mixing to a variable degree with dead carbon from the lignite (or other fossil organics) surfaces.

Conclusions

Tritium measurements in a number of selected Gorleben groundwaters have indicated that a near surface water, GoHy 611 has a ^{3}H-model age of less than 40 a, while all the other groundwaters investigated are older than 40 a. From the stable isotope data of ^{2}H and [18]O it has been concluded that all the sampled Gorleben groundwaters are of Holocene origin (<12 ka BP).

The low ^{14}C concentrations in DIC component are probably influenced by strong dilution with mineralised fossil component carbon resulting in apparent high ^{14}C-model ages (up to 27 ka BP). The high δ^{34}S and δ^{18}O values of dissolved sulphate in the deep groundwaters GoHy 2227 indicate sulphate reduction processes as one source of ^{14}C free DIC in deeper horizons of the aquifer. In consequence, all ^{14}C-model ages of these groundwaters can only be interpreted as upper limit of the true groundwater ages in this hydrogeologically very complex region.

Different ages/origin of the HA and FA fractions of DOC follow from their different ^{14}C contents. Whereas the ^{14}C contents of the HA fraction are partly very low or below detection limit resulting in very high ^{14}C-model ages (up to >30 ka BP) some of the FA fractions contain higher amounts of ^{14}C giving model ages which are more consistent with post-glacial recharge (<12 ka BP). These differences between HA and FA fractions are also confirmed by U/Th isotope ratios suggesting different source histories of the two fractions. Thus, it is possible that the FA component has mainly originated from the near-surface (soil) source while the HA fraction derives mainly from organic matter present in older sediments either as local lignite lenses or as a mixture of an enhanced input of organically-rich water younger than some 12 ka BP and much older lignite source.

Acknowledgements

The authors wish to thank Professor J I Kim and Dr Gunar Buckau of TUM Radiochemistry Department for their support of this work. We are particularly grateful to Dr Bernd Delakowitz of the same Department for his leadership of the sampling campaigns and for his subsequent support to the project. The field sampling campaign was supported by BfS (Salzgitter) and was carried out in collaboration with DBE (Gorleben). We thank Dr R Artinger, Dr W Graf, Mr

W Rauert and Mr P Trimborn for the permanent support of this work. For the measurement of the isotopes ^2H, ^3H, ^{13}C, ^{14}C, ^{18}O and ^{34}S we thank the staff of GSF-Institut für Hydrologie, especially Mrs H Halder, Mrs C Rohde, Miss A Schmitt, Miss P Seibel and H Rast. The ^{14}C contents on small samples were analysed by the Isotrace Laboratory of the University of Toronto (Dr P Beukens). We thank John Fairchild for the ^{14}C measurements carried out at AEA Technology. For the groundwaters and colloid sampling and U series radionuclide analyses the authors thank the staff of the Environmental and Geochemistry Laboratory of the Analytical Sciences Centre, Harwell Laboratory and especially Ms Sandra Hasler. We are grateful to Mrs Boardman, AEA Technology for producing the manuscript. The work was partly funded by CEC (Contract No. F12W/0084) within the 3rd CEC R&D Programme on Management and Storage of Radioactive Waste (1990-1994). In the case of AEA Technology this work was also part-funded by HMIP of UKDOE. The results of this work will be used in the formulation of UK Government Policy but views expressed in this paper do not necessarily represent Government policy.

Literature Cited

1   Kim, J. I. ; Buckau, G. ; Klenze, R. In *Natural Analogues in Radioactive Waste Disposal;* Come B. and N. Chapman Eds., Graham and Trotman, London, 1987, pp289-299.

2   Kim, J.I. Radiochim.Acta., 52/53, **1991** pp71-81.

3   Kim, J.I. In *CEC Project MIRAGE, Third Summary Progress Report*, Côme, B. Ed ; EUR 12858 EN, Brussels, 1990, pp1-105.

4   Kim, J.I.; Delakowitz, B.; Zeh, P.; Probst, T.; Lin, X.; Ivanovich, M.; Longworth, G.; Hasler, S. E.; Gardiner, M.; Fritz, P.; Klotz, D.; Lazik, D.; Wolf, M.; Geyer, S.; Read, D.; Thomas, J. Colloid migration in Groundwaterss: geochemical interactions of radionuclides with natural colloids, First Progress Report (RCM 00292), TUM-Institut für Radiochemie, Garching, Germany, 1992.

5   Kim, J.I.; Delakowitz, B.; Zeh, P.; Probst, T.; Lin, X.; Ehrlicher, U.; Schauer, C.; Ivanovich, M.; Longworth, G.; Hasler, S. E.; Gardiner, M.; Fritz, P.; Klotz, D.; Lazik, D.; Wolf, M.; Geyer, S.; Read, D.; Thomas, J. 2nd Progress Report (RCM 00692), TUM-Institut für Radiochemie, Garching, Germany, 1992.

6   Kim, J.I.; Delakowitz, B.; Zeh, P.; Probst, T.; Lin, X.; Ehrlicher, U.; Schauer, C.; Ivanovich, M.; Longworth, G.; Hasler, S. E.; Gardiner, M.; Fritz, P.; Klotz, D.; Lazik, D.; Wolf, M.; Geyer, S.; Read, D.; Thomas, J: 3rd Progress Report (RCM 00293), TUM- Institut für Radiochemie, Garching, Germany, 1993.

7   BGR (Bundesanstalt für Geowissenschaften und Rohstoffe), Hydrogeologisches Untersuchungsprogramm Gorleben: Beprobung von

Grundwässern im Raum Gorleben für die Untersuchung der gelösten Huminstoffe, AP 9G/312121/10, Hannover, 1990, pp1-13.

8　　Geyer, S.; Wolf, M.; Wassenaar, L. I.; Fritz, P.; Buckau, G.; Kim, J. I.; In *Proc. Int. Symp. on Application of Isotope Techniques in the Study of Past and Current Environmental Changes in the Hydrosphere and the Atmosphere, April 19-23*, 1993, Vienna, 1993.

9　　Thurman, E M: In *Humic Substances in Soil, Sediment and Water;* Adkin, G. R.; McCarthy, P.; McKnight D. D.;and Wershaw, R.Eds. ; John Wiley and Sons Inc., New York, 1985, pp87-103.

10　　Flanigen, E. M.; Bennett, J. M.; Grose, R. W.; Cohen, J. P.; Patton, R. L.; Smith, J. V.; *Nature* 271, **1978**, pp512-516.

11　　Schultz-Sibbel, G. M. W.; Gjerde, D. T.; Chriswell, C. D.; Fritz, J. S.; Coleman, W. E.; *Talanta*, 29, **1982**, pp447-452

12　　Ivanovich, M.; Longworth, G.; Wilkins, M. A.; Harwell Report AERE-G 4290/UKDOE/RW/87.084; 1987.

13　　Wolf, M.; Rauert, W.; Weigel, F. *Int. J. Appl. Rad. Isot* 32, **1981**, pp919-928.

14　　Eichinger, L.; Forster, M.; Rast, H.; Rauert, W.;Wolf, M.. In *Low-Level Tritium Measurement*, IAEA, Vienna, TECDOC 246, **1980**, pp43-64.

15　　Eichinger, L.; Rauert, W.; Salvamoser, J.; Wolf, M. *Radiocarbon* 22/2, **1980**, pp417-427.

16　　*Uranium series disequilibrium: applications to earth, marine and environmental sciences* (second edition), Ivanovich, M.; Harmon, R. S., Eds. ; Clarendon Press, Oxford, 1992, pp34-61.

17　　Gascoyne, M. In *Uranium series disequilibrium: applications to earth, marine and environmental sciences* (2nd edition) ; Ivanovich M. and Harmon R. S. Eds. ; Clarendon Press, Oxford, 1992, pp34-61.

18　　Ivanovich, M.; Longworth, G.; Hasler, S. E.; Gardiner, M.P. In: *Colloid migration in Groundwaterss: geochemical interactions of radionuclides with natural colloids ;* Kim, I.J. and Delakowitz B, Eds. ; Second Progress Report RCM 00692, 1992.

19　　Sonntag, C.; Suckow, A. In:*Paleohydrological Methods and their Applications for Radioactive Waste Disposal*, NEA Workshop 9-10 Nov 1992, OECD Paris, 1993, pp251-265.

20　　Ingerson, E.; Pearson, F.J. Jr. In *Recent Researches in the Fields of Atmosphere, Hydrosphere, and Nuclear Geochemistry*, (Sugawara Festival Volume), Maruzen Co., Tokyo, 1964, pp263-283.

21　　Vogel, J. C. In *Isotope Hydrology*, IAEA, Vienna, 1970, pp225-240.

22　　Tamers, M. A. *Geophysical Surveys* 2, **1975**, pp217-239.

23　　Mook, W. G. In *Interpretation of Environmental Isotope and Hydrochemical Data in Groundwaters Hydrology*, IAEA, Vienna, **1976**, pp213-225.

24   Reardon, E.J.; Fritz, P. *J. Hydrol.* 36, **1978**, pp201-224.
25   Fontes, J.C.; Garnier, J.M. *Water Resour. Res.* 15, **1979**, pp399-413.
26   Eichinger, L. Radiocarbon, 25, **1983**, pp347-356.
27   Deines, P. In: *Handbook of Environmental Isotope Geochemistry*, Fritz P.and Fontes J. C. Eds.; Elsevier Scientific Publishing Company, Amsterdam-Oxford-New York, **1980**, Vol. 1A, pp329-406.
28   Geyer, S. *Isotopengeochemische Untersuchungen an Fraktionen von gelöstem organischem Kohlenstoff (DOC) zur Bestimmung der Herkunft und Evolution des DOC im Hinblick auf die Datierung von Grundwasser.* Thesis LMU München, **1993** pp175.
29   Wolf, M.; Moser, H. *Zwischenbericht zur Berechnung chemisch korrigierter ^{14}C-Modellalter von Grundwasserproben aus dem Raum Gorleben.* Internal Report (unpublished), GSF-Institut für Hydrologie, Neuherberg, 1983.
30   Wolf, M.; Batsche, H; Graf, W.; Rauert, W.; Trimborn, P.; Klarr, K.; von Stempel, C. In *Paleohydrogeological Methods and their Applications for Radioactive Waste Disposal*, Proc. NEA workshop 9-10 Nov 1992, OECD, Paris, 1993, pp.207-218.
31   Parkhurst, D. L.; Thorstenson, D. C.; Plummer, L. N. PHREEQE - A computer program for geochemical calculations. US Geol. Survey, Water Resources Investigations, 1980,pp80-96, (revised and reprinted, May 1985).
32   Cheng, S. and Long, A. In *Hydrology and Water Resources in Arizona and the Southwest*, Proc. of the 1984 meetings of the Arizona section - American Water Resources Assn. and the Hydrology section - Arizona-Nevada Academy of Science, April 7, 1984, Tucson, Arizona, 1984, pp21-135.
33   Wolf, M.; Rohde, H. In *Water-Rock Interaction* ; Kharaka Y K and Maest S., Eds.; Balkema, Rotterdam, 1992, Vol. 1, pp. 195-198 .

Chapter 15

Solution and Solid State ^{13}C NMR Studies of Alginic Acid Binding with Alkaline Earth, Lanthanide, and Yttrium Metal Ions

A. E. Irwin, C. M. De Ramos, and B. E. Stout

Department of Chemistry, University of Cincinnati,
Cincinnati, OH 45220–0172

Alginic acid is a binary polyuronide made up of β-D-mannuronic acid and α-L-guluronic acid arranged in three types of blocks: homopolymeric sequences of mannuronate residues (MM blocks), guluronate residues (GG blocks) and a region where the two residues alternate (MG blocks). The preference of various divalent and trivalent metal ions for GG blocks over MM blocks and extent of binding in whole and depolymerized alginic acid samples can be determined using ^{13}C NMR spectroscopy. When polyvalent metal ions bind with alginic acid, a gelatinous complex forms which is in equilibrium with the free metal ions and alginate residues in solution. The NMR spectra reflect the solution phase only, i.e., the signals arising from the free alginic acid rather than the bound. The signal intensity is indicative of the amount of binding occurring. For all metal ions studied, GG blocks are preferred. For the divalent metal ions, extent of binding increases as ionic size increases. For the trivalent metal ions, the extent of binding is related to both ionic size and coordination number. CPMAS ^{13}C NMR allows study of the gel phase, and signal assignments are made.

Through our studies of alkaline earth, lanthanide, and yttrium ions, and their interaction with the polyuronide alginic acid, we would like to predict the environmental behavior of the actinides. We chose to begin our study by investigating the interaction of soil constituents with the divalent alkaline earth metal ions and the trivalent lanthanides, which serve as convenient trivalent actinide analogs. Soil is primarily composed of polyelectrolytes. By choosing a simple polyelectrolyte such as the polyuronide alginic acid, we can begin to establish possible pathways of metal ion transport.

Alginic acid was chosen for study because a majority of the signals in the ^{13}C NMR spectrum are resolved. The properties of alginic acid, such as uronic acid composition, vary depending on the location and maturity of the algae from which the alginic acid is isolated (1). Like many polyuronides, alginic acid is made up of two uronic acid residues (2). The two acids, guluronic acid and mannuronic acid, vary in their connectivity. The various linkages give rise to what are referred to as three block types (3-5) that can be described as homopolymeric or heteropolymeric. There are two homopolymeric block types, GG blocks, where two or more guluronic acids are linked diaxially, and MM blocks, where two or more mannuronic acids are linked diequatorally. Heteropolymeric blocks are regions of alternating mannuronic acid and guluronic acid. These alternating regions, the MG blocks, behave similarly to the GG blocks in our studies and will not be discussed specifically.

Before a metal ion is added, the long alginate chains exist in random coils, typical of polymers in dilute solutions. Upon addition of a polyvalent metal ion, two chains are cross-linked through association of GG blocks, forming a gel. According to the "egg-box" model by Grant et al. (*6*), these associated regions of metal ions are regularly packed and coordinated in the GG block cavities. This crosslinking allows for the remaining residues in the chain, i.e. those not directly complexed, to be in close enough proximity to each other to allow for ionic interactions to stabilize the polymer chain. Such a stabilization may in turn lead to easier complexation of the remaining residues.

By examining the effect of alkaline earth, lanthanide and yttrium metal ions on alginates, we can begin to understand how these metals interact with simple components of soil matter. After laying the groundwork with these studies, we can then predict possible mechanisms for ion transport in soils.

Experimental

Materials. Alginic acid, isolated from the kelp *Macrocystis pyrifera* (Sigma), was used without further purification. Alkaline earth metal chlorides were reagent grade and were obtained from Fisher ($MgCl_2$, $CaCl_2$, $BaCl_2$) and Baker ($SrCl_2$). Lanthanide oxides (99.9%) were obtained from Sigma (La_2O_3), Aldrich (Eu_2O_3, Pr_6O_{11}, Lu_2O_3, Y_2O_3), Matheson, Coleman & Bell (Nd_2O_3) and Alfa Products (Tb_4O_7, Yb_2O_3). All other reagents used were analytical grade.

Preparation of Metal Ion Stock Solutions.

Alkaline Earths. 1.0 M stock solutions of the alkaline earth metal chlorides were prepared by dissolving the chloride salt in D_2O (99.9% D). Standardization was either by direct EDTA titration (for $CaCl_2$) or back titration of the metal-EDTA complex with standard $Mg(ClO_4)_2$ using Eriochrome Black T indicator. The solutions were buffered to pH 10 with NH_3/NH_4Cl and the determinations were done in triplicate.

Lanthanides and Yttrium. 1.0 M stock solutions of lanthanide and yttrium perchlorates were prepared in D_2O following the method described elsewhere (*7*) for the preparation of aqueous lanthanide salts. Terbium, ytterbium, and lutetium oxides required reflux for 6 to 8 hours in concentrated $HClO_4$ (Tb_4O_7, Yb_2O_3) or concentrated HNO_3 (Lu_2O_3). All solutions were standardized in triplicate by EDTA titration using xylenol orange indicator and hexamethylene tetramine buffer (*8*).

Caution! Ionic perchlorate salts are not explosive in general. However, exposure of pure compounds to easily oxidized organic matter at high temperature greatly increases the risk of fire or explosion (*9*).

Depolymerization of Alginic Acid. Alginic acid depolymerization is a process that removes the MG blocks by cleaving the MG glycosidic bond which is more susceptible to acid hydrolysis than the glycosidic bonds in homopolymeric blocks. The resulting solid is a mixture of the two homopolymeric GG blocks and MM blocks. The method used was a modification of the procedure described by Penman and Sanderson (*10*).

Characterization of Alginic Acid.

Carboxylic Acid Capacity. Alginic acid carboxylic acid capacity was determined in triplicate by acid-base titration. A known excess of 0.050 M NaOH was added to a suspension of 75 mg of alginic acid in water. The mixture was stirred until all the alginate was dissolved, and the unreacted base was back-titrated with 0.020 M $HClO_4$ using phenolphthalein as indicator.

Average Molecular Weight. Viscosity measurements using a Cannon-Fenske capillary viscometer on a thermostated water bath kept at 25.0 ± 0.1 °C were used to determine the average molecular weight. A 200 mL stock solution containing 10.00 g alginate/L water was prepared by first adding 1.0 M NaOH to a suspension of alginic acid in water to neutralize 100% of the carboxylate groups, followed by addition of NaCl to achieve a total concentration of 0.1 M NaCl. Eight different concentrations of sodium alginate were prepared from the stock solution ranging from 2.00 to 9.00 g/L. All viscosity measurements were performed in triplicate and the reported viscosity-average molecular weight was the result of two determinations.

Uronic Acid Composition and Block Length. A sample containing 75 mg alginic acid was first suspended in a small volume of D_2O, then fully neutralized by titration with 1.0 M NaOD. D_2O was added to achieve a final concentration of 25 mg alginic acid/mL D_2O. Equilibration of at least one hour was allowed prior to NMR acquisition. The uronic acid composition was determined by ^{13}C NMR spectroscopy, with a 2 s pulse repetition time and 16,000 scans. The reported composition and block lengths were the average of two determinations.

Preparation of Metal Alginate Samples. To prepare metal-alginate samples, 75 mg alginic acid was suspended in D_2O then 75% of the carboxylic acid (calculated from the carboxylic acid capacity) was neutralized by the addition of 1.0 M NaOD. The resulting solution was stirred for 30 minutes and heated to 70 °C prior to the addition of metal ion. To this solution, a pre-determined amount of preheated 0.050 M alkaline earth metal or 0.015 M lanthanide or yttrium metal solution was slowly added to achieve the desired metal loading. Percent metal loading is defined as the number of milliequivalents of metal ion per milliequivalent of carboxylate times 100 and is used in place of concentration in order to compare ions with different valencies. A final volume of 3.00 mL was achieved by addition of D_2O. Six samples were prepared with 4, 8, 12, 16, 20, and 24% metal loadings for all metals except terbium. Due to extensive signal broadening at high metal loadings, Tb^{3+} metal loadings were 4, 6, 8, 10 and 12%. All samples were equilibrated for 19 to 24 hours in a sand bath at a temperature of 50 to 60 °C prior to NMR acquisitions. The final pH of the solutions measured at room temperature ranged from 3.8 to 4.2.

Preparation of metal-depolymerized alginate solutions was carried out as above, without the addition of NaOD. Eight samples of 1, 2, 4, 8, 12, 16, 20, and 24% metal loadings were prepared.

Preparation of Calcium Depolymerized Alginate CPMAS NMR Sample. Calcium depolymerized alginate was prepared by suspending 0.25 g alginic acid in 2.0 mL H_2O. The resulting solution was stirred for 30 minutes to allow equilibration, and heated to 70 °C prior to the addition of metal ion. To this solution, 0.300 mL of preheated 0.95 M calcium chloride solution was slowly added to achieve the desired metal loading of 50%. The solution was then diluted with H_2O to a final volume of 10.0 mL. After equilibration for 20 hours, the sample was removed from the sand bath and allowed to cool. The resulting gelatinous mixture was centrifuged to remove excess water. The gel was exposed to air until all excess water evaporated. The resulting solid was ground to a fine powder to be placed in a rotor.

NMR Spectroscopy. All proton-decoupled ^{13}C NMR spectra were acquired on a Bruker AMX 400 NMR spectrometer operating at 100.62 MHZ. [13]C spectra were acquired using 32K data points, a spectral width of 31250 Hz, a 90° pulse with a pulse repetition time of 10.5 s, and a total of 1600 scans. For proton-decoupled ^{13}C NMR spectroscopy of depolymerized alginate samples, the pulse repetition time was reduced to 8.5 s. A probe temperature of 70 °C was used in order to decrease signal broadening due to viscosity. All determinations were performed in duplicate.

CPMAS NMR spectra were acquired on the same Bruker AMX 400 NMR spectrometer with a standard proton decoupled pulse program with a repetition time of 0.53 s, 8K data points, and Hartmann-Hahn conditions being met. For the whole alginate sample, a spin rate of 4300 rps was achieved and 6 scans were taken. No improvement was found for increases in delay time from 0.5 to 5 s. A spin rate of 4600 rps was achieved and 16,000 scans were taken for both depolymerized samples. Spinning sidebands at approximately 115 ppm were identified by varying the sample spin rate.

Reported average signal intensities for G's and M's were averages of the most resolved C1 and C2 carbons of GG blocks and C1, C2 and C4 carbons of MM blocks, where C1 is the anomeric carbon and C6 is the carboxylate carbon. For depolymerized samples, average signal intensities are for all ring carbon signals, C1-C5. Reported errors were the standard deviation between at least two determinations. In order to best be able to analyze the data acquired from the NMR spectra, the measured signal intensities for the resolvable signals for each metal ion loading are normalized to the lowest metal loading. The normalized signal intensity was plotted as a function of percent metal loading.

Results and Discussion

Solution ^{13}C NMR Studies. Carboxylic acid capacity is the number of milliequivalents of titratable carboxylic acid per gram alginate. For the alginic acid used, the carboxylic acid capacity was determined to be 4.50 ± 0.03 meq/g by repeated acid-base titration. The viscosity-average molecular weight was determined to be 16,700 ± 900 Da. This weight was calculated using the Mark-Houwink-Sakurada equation (*11-13*) with the constants for *Macrocystis pyrifera* determined by Martisen et al. (*14*). From our viscosity-average molecular weight, an average chain length of 75 residues was calculated. Following the method of Grasdalen et al. (*15*), the uronic acid composition was found to be 45 ± 3% guluronic acid and 55 ± 3% mannuronic acid. Also, the average length of a homopolymeric block, either GG or MM, was found to be approximately 5 residues.

^{13}C NMR was used to follow any changes in the alginic acid as metal ions were added. From the data acquired, there are two aspects of binding that can be discussed. The first is *GG-preference*, the preference each metal has for one block type over another. The second is *extent of binding*, an indicator of the strength of the metal-alginate interaction, which is roughly analogous to stability constant.

Divalent Alkaline Earth Metal Ions. When a polyvalent metal ion is added to the alginic acid, it crosslinks two or more alginate chains forming a gel which cannot be observed by solution NMR. The signals present are due to free alginic acid residues, i.e. residues that are not bound to the metal, while the signals of the complexed alginate residues are broadened into the baseline. Therefore, a decrease in signal intensity indicates an increase in metal-alginate binding. A series of calcium alginate ^{13}C NMR spectra are shown in Figure 1. Following the signal corresponding to the C4 (*16*) carbon on a guluronate residue, the G4 signal (which is representative of the GG blocks), an increase in metal-alginate binding was observed as metal ion loading was increased from 4% to 24%. However, a decrease in intensity was not seen for every signal. Looking at M4 (which is representative of the MM blocks), there was little change in signal intensity as metal loading increased. Similar studies were repeated for magnesium, and a different result was observed. Unlike calcium alginate, no significant binding was observed for magnesium alginate in the range of

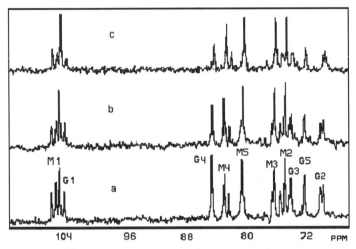

Figure 1. ^{13}C NMR spectra of calcium-alginate. Metal loadings of: a) 4% b) 12% and c) 24%.

4%-24% metal loading. For most of the metals studied, the maximum metal ion loading was limited by either a very small signal-to-noise ratio or high sample viscosity due to gelation. For the other divalent ions, this occurred at 24%, but for Mg^{2+}, precipitation, not gelation, began to occur at 24%. The high signal-to-noise, low sample viscosity, and absence of significant decrease in signal intensity indicate that magnesium ion does not effectively crosslink the alginate chain.

Figure 2 is a plot of percent calcium metal loading versus normalized peak intensity. Because a decrease in intensity indicates an increase in binding, a negative slope also indicates an increase in binding with increasing metal ion loading. The lower line illustrates an increase in metal ion binding to the GG blocks from 8% to 24%. The smaller slope of the upper line indicates that little binding occurred between calcium and the MM blocks. Calcium shows a positive GG-preference. This preference, as indicated by the gap between the two lines, means calcium ion prefers binding with GG blocks over MM blocks. Strontium and barium also show a positive GG-preference.

To explain the preference for one block type over another, it may be helpful to examine the structure of alginic acid, Figure 3. Two guluronic acids are linked diaxially, forming a deep cavity. Two mannuronic acids are linked diequatorially, also forming a cavity. However, the cavity is shallow and not as pronounced as the cavity produced by the GG blocks. This difference in cavity size may be a reason why GG blocks are preferred. To determine if cavity size was significant and to examine the difference in the two block types, molecular modeling studies were performed (*17*). From these studies it was found that the types of oxygens in close proximity, ca. 2.41 Å, to the metal ion varied between the two types of blocks. The oxygens predicted to be involved in binding with the GG blocks were found to be more basic than those involved in binding with the MM blocks. Therefore, a stronger interaction between the metal ion and the more basic oxygens of the GG blocks is expected.

In order to discuss extent of binding, the three divalent ions must be compared at the same metal loading. A plot of ionic radius versus normalized signal intensity at 8% metal loading for the GG and MM blocks is shown in Figure 4. The negative

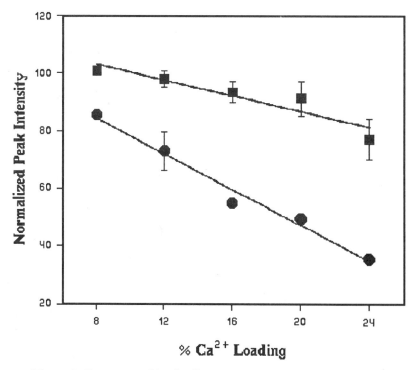

Figure 2. Percent metal ion loading versus normalized peak intensity for calcium-alginate. ● GG blocks ■ MM blocks

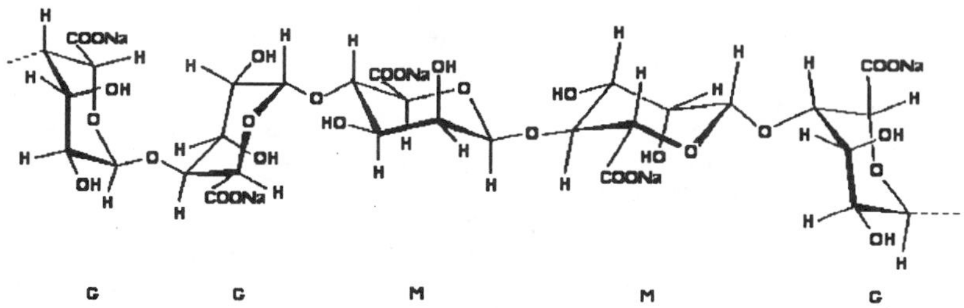

Figure 3. Structure of sodium alginate. G= guluronate and M= mannuronate

slope of the two lines indicates an increase in binding for both GG and MM blocks as ionic radius increases. This trend is expected if the cavities formed are size specific. The larger alkaline earth ions, Table I (*18*), will have stronger ionic bonds due to their ability to more effectively fill the cavity. Our previous lanthanide luminescence studies (*19*) have shown that crosslinking of alginate chains requires inner-sphere metal ion interactions which means that the metal ion is at least partially dehydrated. There are two possible explanations for the lack of binding between magnesium and alginic acid. Mg^{2+} has an enthalpy of dehydration that is 300 kJ/mol greater (*20a*) than that of Ca^{2+},

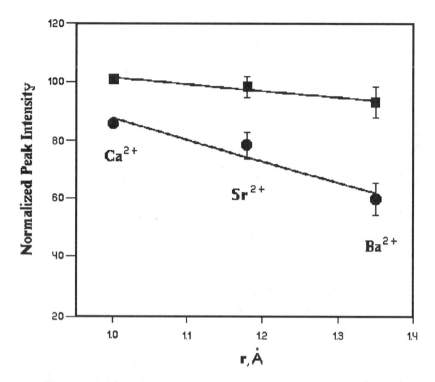

Figure 4. Ionic radius versus normalized peak intensity for divalent metal-alginates at 8% metal loading. ● GG blocks ■ MM blocks

suggesting that Mg^{2+} may not be dehydrated in these systems. The hydrated magnesium ion has a radius of 3.83 Å (*20b*) which is too large to fit in the MM or GG cavities. Alternatively, a completely dehydrated magnesium ion may be too small to effect crosslinking since this would bring two negatively charged alginate chains in close proximity to one another.

For higher divalent metal ion loadings, it is not apparent that one metal ion binds more effectively than another. This may indicate that a reduction in the significance of the ionic radius due to saturation of the binding sites occurs when a certain concentration of metal ion is reached. On the other hand, such a change in the extent of binding may be due to a change in the secondary structure of alginic acid. As the repulsive forces of the negatively charged carboxylate group are neutralized by the metal cation, the structure of the alginate becomes more condensed (*21*).

Trivalent Lanthanide and Yttrium Metal Ions. The trivalent lanthanide and yttrium ions were also studied by solution ¹³C NMR. A series of three spectra for europium alginate, representative of the trivalent ions, Figure 5, showed that an increase in metal ion binding to the GG blocks occurred as more metal was added. An increase in binding to the MM blocks, which was not seen for the divalent ions studied, was also observed. A similar plot of normalized signal intensity as a function of percent metal loading for europium alginate also showed a positive GG-preference. However, when evaluating the magnitude of this preference by comparison with the plot for calcium, it was seen that the trivalent ions showed only a moderate preference

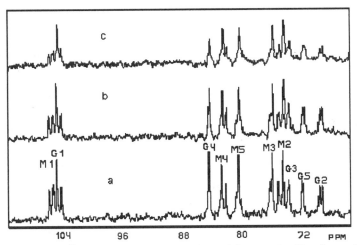

Figure 5. ¹³C NMR spectra of europium-alginate. Metal loadings of:
a) 4% b) 12% and c) 24%.

Table I. Metal Ion Coordination Number, Ionic Radius, and Charge Density.

Metal Ion	Coordination Number	r, Å	Z/r
Mg^{2+}	6	0.72	2.78
Ca^{2+}	6	1.00	2.00
Sr^{2+}	6	1.18	1.69
Ba^{2+}	6	1.35	1.48
La^{3+}	9	1.216	2.47
Pr^{3+}	9	1.179	2.54
Nd^{3+}	9	1.163	2.58
Eu^{3+}	9	1.120	2.68
Tb^{3+}	9	1.095	2.74
Y^{3+}	8	1.019	2.94
Yb^{3+}	8	0.985	3.05
Lu^{3+}	8	0.977	3.07

for GG blocks over MM blocks. This decrease in magnitude is due to an increase in binding to MM blocks rather than a decrease in binding to GG blocks.

Other trivalent ions were studied and are also listed in Table I. One notable change that occurs for these ions is the coordination number, changing from 9 to 8

(*22*) after Tb. This change in coordination number results in an interesting variation in binding that becomes evident when evaluating the extent of binding. A plot of charge density versus normalized signal intensity at low metal loadings, Figure 6, shows the same trend for both block types, an increase in binding as charge density increases.

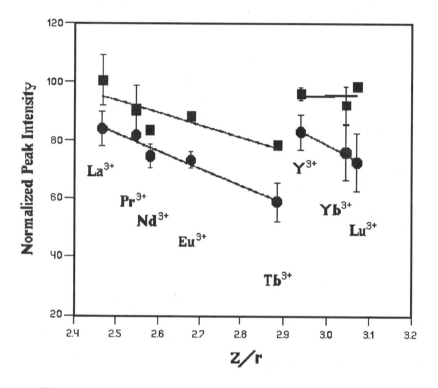

Figure 6. Charge density versus normalized peak intensity for trivalent metal-alginates at 8% metal loading. ● GG blocks ■ MM blocks

This in turn may be correlated with the stability of the metal-alginate complex. However, the ions appear to be divided into two groups, the nine coordinate and the eight coordinate. Since extent of binding is roughly analogous to stability constant, a comparison of a similar plot of the relationship of charge density and stability constant for similar chelating ligands is warranted. For example, a plot of charge density versus log stability constant for EDTA shows a smooth increase in stability constant from La to Lu, meaning charge density is probably the most important parameter in determining the strength of the metal-ligand interaction. In contrast, our work shows that coordination number as well as charge density plays a role in the metal-alginate interaction.

Depolymerized Alginic Acid Solution ^{13}C NMR Studies. Studies are being conducted with depolymerized alginic acid. In these samples, the MG blocks were removed by acid hydrolysis leaving shorter chains containing mostly GG or MM blocks. For whole alginic acid, complexation of one ion to a GG block leads to a zipper effect. If two or more metal ions bind to GG blocks that are separated by one or more MM or MG blocks, those blocks may be drawn into the gel phase and the signals arising from those residues will exhibit a decrease in intensity even though the M residues are not involved in binding. For the depolymerized alginic acid, there is a reduction in the significance of the zipper effect that leads to a decrease in the apparent binding to both GG and MM blocks because the MG blocks have been removed.

The ^{13}C NMR spectrum of depolymerized sodium alginate, Figure 7, is simplified when compared to the whole sodium alginate spectrum. The marked

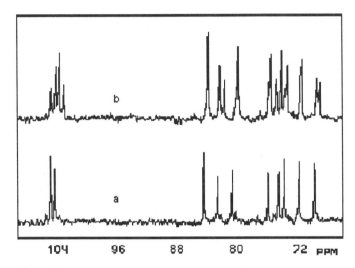

Figure 7. ^{13}C NMR spectra of: a) sodium-alginate and b) sodium-depolymerized alginate.

reduction in intensity of signals due to alternating residues indicates that most of these alternating regions have been removed. Continuing our previous studies, calcium was added to samples of depolymerized alginic acid. The resulting spectra showed the same increase in binding as metal loading increased. A plot similar to the previous calcium plot was obtained for calcium-depolymerized alginate, Figure 8. Again, calcium exhibited a positive GG-preference. However, at metal loadings of less than 12%, there appeared to be little differentiation between the two block types and the amount of binding was not significant. At high metal loadings, there were fewer bound MM blocks in the depolymerized sample in comparison to the whole alginic acid.

Cross-Polarization Magic-Angle Spinning NMR Studies. Another technique was required to follow any changes in complexed alginic acid, since observation of the uncomplexed alginic acid only can be made using solution NMR. Therefore, work has begun in solid state CPMAS ^{13}C NMR to evaluate any changes in the whole or depolymerized alginic acid when metal ions are added and gelation occurs.

In the sodium alginic acid spectrum, Figure 9, assignment of signals of complexed alginate was possible based on information from our early work on solid state studies of depolymerized and homopolymeric GG and MM samples. Also, assignments were consistent with solution spectra of sodium alginate. The most downfield signal at ca. 153 ppm is due to the carboxylate carbons, C6. Moving

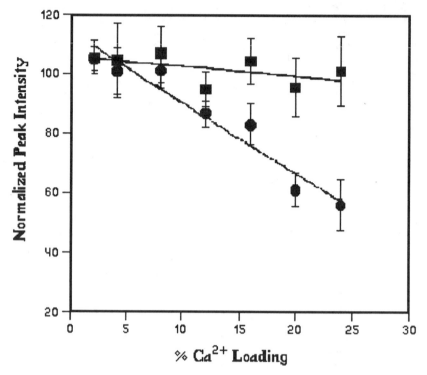

Figure 8. Percent metal ion loading versus normalize peak intensity for calcium-depolymerized alginate. ● GG blocks ■ MM blocks

upfield, the next distinct signal, centered at ca. 83 ppm, is due to the anomeric C1 carbons. The remaining signals, between 70 and 45 ppm, are due to the ring carbons.

In the sodium depolymerized alginic acid sample, Figure 10, some resolution of the signals is possible. Through studies of GG and MM homopolymeric samples, it is possible to attribute the signals occurring at ~67 ppm and ~53 ppm to G-residues. The signal at ~56 ppm results from M-residues. The signal occurring at ~61 ppm has not yet been assigned, but is believed to result from an M-residue. The small signal appearing at ~115 ppm is a spinning sideband.

When the whole alginic acid and the depolymerized alginic acid spectra are compared, an interesting change in the chemical shifts occur. In the whole alginic acidspectrum, the C6 and C1 signals have a signal separation of 70.1 ppm. In the depolymerized sample spectrum, the same separation between the two signals is 74.5 ppm. Studies of GG and MM homopolymeric samples show a similar widening of signal separation. One possible explanation is that the shifting of the signals is due to the shorter chain lengths of the depolymerized and homopolymeric samples. For the whole alginate sample, the chains are longer and the lengths of the chains vary. The NMR signals arise from the average of the signals in the polydisperse sample. However, in depolymerized alginate, shorter chains prevail and the average NMR signal arising from the chains is shifted.

Comparison of the calcium depolymerized spectrum in Figure 11 with the sodium depolymerized spectrum shows that the region between 70 and 45 ppm is simplified. Two of the four distinct signals appearing in the sodium depolymerized

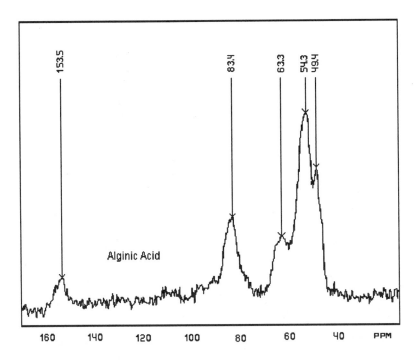

Figure 9. ^{13}C CPMAS NMR spectrum of sodium-alginate.

spectrum are not present in the calcium depolymerized spectrum. The fact that the signals at ~61 ppm and ~56 ppm are not present is most likely related to our sample preparation procedure. This can be explained by recalling that calcium showed a large, positive GG-preference at high metal ion loadings in Figure 8, indicating that most of the calcium is bound to G-residues for depolymerized samples. When the sample for solid state NMR was prepared, the excess water was removed by centrifugation. Since the M-residues were uncomplexed, they were removed in the supernatant. This accounts for the absence of signals arising from M-residues in the calcium depolymerized alginate spectrum. The shoulder appearing upfield of the last distinct signal is slightly more prominent in the calcium depolymerized spectrum than in the sodium depolymerized spectrum, indicating that it might also arise from a G-residue.

Conclusions

The divalent alkaline earth metal ions show a strong *GG-preference* because of the more basic oxygens involved in complexation. At low metal ion loadings, the ionic radius of the metal ion determines the *extent of binding*. For higher metal ion loadings, the binding that occurs does not indicate that one metal binds more effectively than another. This indicates, possibly, that when a certain concentration of metal ion is reached, there is a reduction in the significance of the ionic radius to the *extent of binding* due to a saturation of the binding sites. Such a change in the *extent of binding* may be due to a change in the secondary structure of alginic acid. Yttrium and the lanthanide ions all show a moderate, positive *GG-preference*. For both eight and nine coordinate ions, the *extent of binding* increases with increasing charge density. This trend is present at all metal ion loadings.

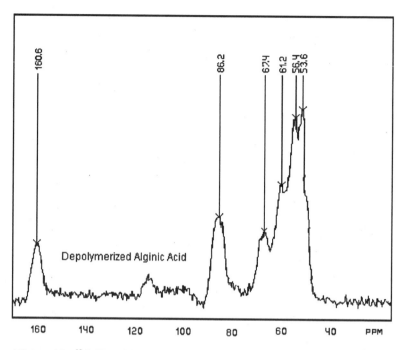

Figure 10. ^{13}C CPMAS NMR spectrum of sodium-depolymerized alginate.

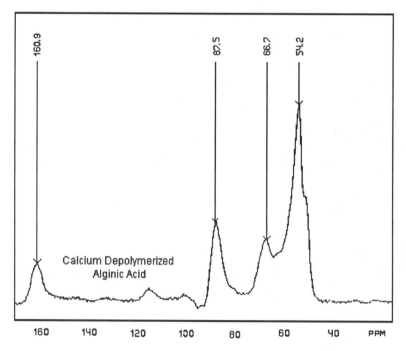

Figure 11. ^{13}C CPMAS NMR spectrum of calcium-depolymerized alginate.

Preliminary studies of depolymerized alginic acid indicate that the *GG-preference* of the divalent alkaline earths remains in mixtures of homopolymeric blocks. One conclusion from our data might be that there is a minimum chain length required for maximum metal ion binding. Until further studies are completed, the *extent of binding* for the divalent ions and depolymerized alginate cannot be determined.

Early work in solid state cross-polarization magic-angle spinning nuclear magnetic resonance spectroscopy has led to preliminary assignments of signals in sodium alginate and sodium depolymerized alginate spectra.

Acknowledgments

This project was funded in part by the National Science Foundation (#CHE-9109396) and the Petroleum Research Fund (#24372-G3). The authors wish to thank George Kreishman and Elwood Brooks for assistance with CPMAS NMR.

Literature Cited

(1) McKee, J. W. A.; Kavalieris, L.; Brasch, D. J.; Brown, M. T.; Melton, L. D., *J. Appl. Phycology* **1992**, *4*, 357.
(2) Fischer, F. G.; Dorfel, H., *Z. Phys. Chem.* **1955**, *301*, 186.
(3) Haug, A.; Larsen, B.; Smidsrød, O., *Acta Chem. Scand.* **1966**, *20*, 183.
(4) Larsen, B.; Smidsrød, O.; Painter, T.; Haug, A., *Acta Chem. Scand.* **1970**, *24*, 726.
(5) Haug, A.; Larsen, B.; Smidsrød, O., *Carbohydr. Res.* **1974**, *32*, 217.
(6) Grant, G. T.; Morris, E. R.; Rees, D. A.; Smith, P. J. C.; Thom, D., *FEBS Lett.*, **1973**, *32*, 195.
(7) Desreux, J. F., *Lanthanide Probes in Life, Chemical and Earth Sciences: Theory and Practice*; Bunzli, J.-C. G.; Choppin, G. R., Eds.; Elsevier: New York, **1989**.
(8) Lyle, S. J.; Rahman, M. M., *Talanta* **1963**, *10*, 1177.
(9) Schilt, A. A., *Perchloric Acid and Perchlorates*; The G. Frederick Smith Chemical Co.: Ohio, **1979**.
(10) Penman, A.; Sanderson, G. R., *Carbohydr. Res.* **1972**, *25*, 273.
(11) Mark, H., *Physical Chemistry of High Polymeric Systems*, Vol. 2; Wiley (Interscience): New York, **1940**.
(12) Houwink, R., *J. Prakt. Chem.* **1940**, *155*, 241.
(13) Sakudara, I., *Kasenkouenshyu* **1941**, *6*, 177.
(14) Martinsen, A.; Skjak-Braek, G.; Smidsrød, O.; Zanetti, F.; Paoletti, S., *Carbohydr. Polym.* **1991**, *15*, 171.
(15) Grasdalen, H., *Carbohydr. Res.* **1983**, *118*, 255.
(16) Grasdalen, H.; Larsen, B.; Smidsrød, O., *Carbohydr. Res.* **1981**, *89*, 179.
(17) De Ramos, C. M., Ph.D. Dissertation, University of Cincinnati, **1995**.
(18) Horrocks, W. W.; Sudnick, D.R., *J. Am. Chem. Soc.* **1979**, *101*, 334.
(19) (a) Stout, B. E.; De Ramos, C. M., *International Symposium on the Scientific Basis for Nuclear Waste Management*, MRS Symposium Proceeding, Vol. 353; Murakami, T.; Ewing, R. C., Eds.: Materials Research Society; Pennsylvania, **1995**.
(b) Peterman, D. R., Ph.D. Dissertation, University of Cincinnati, **1995**.
(20) (a) Marcus, Y., *Ion Solvation*; Wiley (Interscience): New York, **1985**.
(b) Marcus reports the compressed molar volume, V^∞_{ih}, of the hydrated magnesium ion as being 126.0 cm³ mol⁻¹. Using the equation $V^\infty_{ih} = Nk$ $(4\pi/3)(r^\infty_{ih})^3$ from the same text, and a packing coefficient, k, of 0.888, leads to a radius for the hydrated magnesium ion of 3.83 Å.

(21) Tanaka, T., *Polyelectrolytes Gels: Properties, Preparation and Applications*; Harland, R. S.; Prud'homme, R. K., Eds.; ACS Symposium Series 480; American Chemical Society: Washington, D.C., **1992**.

(22) (a) Habenschuss, A.; Spedding, F. H., *J. Chem. Phys.* **1979**, *70*, 2792.
   (b) *Ibid.* **1979**, *70*, 3758.
   (c) *Ibid.* **1980**, *73*, 442.
   (d) Kanno, H.; Hiraishi, J., *J. Phys. Chem.* **1984**, *88*, 2787.
   (e) Yamauchi, S.; Kanno, H.; Akama, Y., *Chem. Phys. Lett.* **1988**, *151*, 315.
   (f) Cossy, C.; Barnes, A. C.; Enderby, J. E.; Merbach, A. E., *J. Chem. Phys.*, **1989**, *90*, 3254.
   (g) Helm, L.; Merbach, A. E., *Eur. J. Solid State Inorg. Chem.* **1991**, *28*, 245.

Chapter 16

Role of Humic Substances and Colloids in the Behavior of Radiotoxic Elements in Relation to Nuclear Waste Disposal

Confinement or Enhancement of Migration

Valérie M. Moulin[1], Christophe M. Moulin[2], and Jean-Claude Dran[3]

[1]Commissariat à l'Energie Atomique, Fuel Cycle Division, Department of Waste Storage and Disposal, Service of Nuclear Waste Storage and Disposal Studies, Section of Geochemistry, BP6 92265 Fontenay-aux-Roses Cedex, France
[2]Commissariat à l'Energie Atomique, Fuel Cycle Division, DPE/SPEA/SPS/Analytical Laser Spectroscopy Group, 91121 Saclay, France
[3]CSNSM/Centre National de la Recherche Scientifique, 91405 Orsay, France

The potential role of humic substances and colloids on the fate of radiotoxic pollutants should be evaluated in particular around a nuclear waste repository. The different processes involving these entities such as complexation or sorption can strongly affect the behavior of radionuclides. In particular, results on the complexation of actinides with humic substances, investigated by laser induced fluorescence and spectrophotometry will be presented as well as data on the retention of colloids in the presence or not of heavy elements on mineral surfaces measured by Rutherford Backscattering spectrometry. From these studies, the impact of colloids and humic substances on radiotoxic element behavior will be discussed in terms of confinement or enhancement of migration in the geological media.

Due to the ubiquitous occurrence of humic substances and colloids (defined as entities of 1nm-1μm size) in natural waters, their specific properties, in particular their scavenging capacities towards metallic cations and also their well-established mobility (1-6), these organic and inorganic species could have important effects on the fate and mobility of these cations in natural systems. On one hand, the formation of organic complexes or pseudocolloidal species will modify the speciation (distribution of chemical species) of the cation of interest and its solubility. On the other hand, they can retard cation migration

0097–6156/96/0651–0259$15.00/0
© 1996 American Chemical Society

when sorbed on mineral surfaces or by filtration in the porosity of the medium. These features are particularly important in the case of radionuclides which could be released from a radwaste repository in deep geological formations (7). In this framework, two aspects have been investigated which will be reported here in order to contribute to the answer of the question: whether the interaction of radioelements with colloids or humic substances reduce or enhance their mobility. Firstly, the complexation behavior of actinide ions [An(III), An(V) and An(VI)] with humic substances has been studied by means of spectroscopic methods (laser induced fluorescence or spectrophotometry) in order to determine the speciation of radionuclides under natural water conditions representative of those encountered in granitic or sedimentary formations. Secondly, the retention of colloids and heavy elements on mineral surfaces has been investigated by ion-beam techniques (Rutherford Backscattering spectrometry) in order to determine the colloid surface coverage, the colloid detachment rate and the behavior of heavy elements in their presence. Hence, the impact of humic substances and colloids on radionuclide behavior will be evaluated and their consequences on the fate of these pollutants in the framework of radwaste disposal.

Complexation of Actinides by Humic Substances

Investigations have been devoted to the study of the complexation behavior of actinides [An(III), An(V) and An(VI)] with humic substances through the use of spectroscopic methods:
- for fluorescent cations such as lanthanides and actinides [Dy(III), Cm(III), U(VI)]: Time-Resolved Laser-Induced Fluorescence (TRLIF).
- for non-fluorescent cations such as actinides [Np(V)]: Spectrophotometry (SP).

The effect of different physico-chemical parameters, namely pH, ionic strength, presence of a competing cation and cation concentration, has been studied in order to understand complexation mechanisms.

Description of the techniques. These two methods (TRLIF and SP) applied to complexation measurements are based on the same principle: titration of the cation (in the nanomolar to micromolar range for TRLIF (Figure 1) and in the micromolar range for spectrophotometry) by the organic ligand (humic substances) at a constant pH and ionic strength, and measurement of the signal characteristic of each technique, namely the fluorescence intensity (as well as the lifetime and eventually the fluorescence spectrum modification) for TRLIF, and the absorbance for SP. In the case of TRLIF (8-9), applied to a lanthanide [Dy(III)] and an actinide [Cm(III)] cation, an increase of the fluorescence (at the wavelength of cation emission) is observed when adding the ligand, whereas for an hexavalent cation such as U(VI), a decrease of the fluorescence signal is obtained. In the case of SP (10-11), applied to two actinide cations (Am(III), Np(V)), a decrease or an increase of the absorbance corresponding respectively to the free or bound cation is observed. The analysis of such titration curves (Figure 2) permits us to obtain information on the conditional interaction constants (β) and the complexing capacities (W) of the humic materials towards the studied cation under different experimental conditions (varying pH, ionic strength and metal concentration).

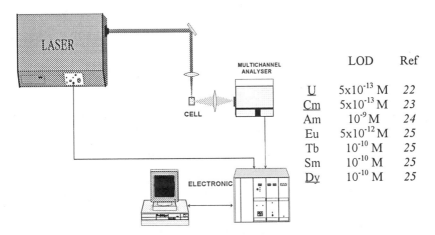

	LOD	Ref
U	5×10^{-13} M	22
Cm	5×10^{-13} M	23
Am	10^{-9} M	24
Eu	5×10^{-12} M	25
Tb	10^{-10} M	25
Sm	10^{-10} M	25
Dy	10^{-10} M	25

Figure 1. TRLIF set-up with the limits of detection (LOD) of the technique for actinide and lanthanide trace analysis in appropriate complexing media *(22-25)*.

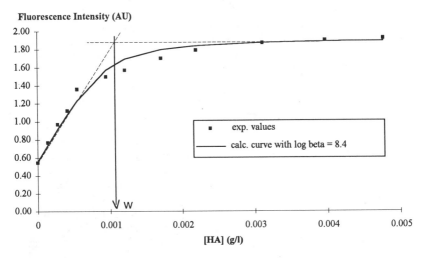

Figure 2. Typical titration curve obtained for the system Cm(III)-Aldrich humic acids by TRLIF; [Cm]=10^{-7} M, pH 4, NaClO$_4$ 0.001M.

The *single site model* is used to describe the interactions between the cation and the humic acids *(11)*. Since humic substances are heterogeneous and complex molecules, the ligand is defined as a monodentate site (A) with no particular assumption on its chemical nature. Complexes of 1:1 (metal:cation) stoichiometry are assumed to be formed according to the equilibrium:

$$M^{(z+)} \; + \; A^{(-)} \;\; \Leftrightarrow \;\; MA^{(z-1)+} \; \text{with } \beta = \left[\frac{\left[MA^{(z-1)+} \right]}{\left[M^{(z-)} \right]\left[A^{(-)} \right]} \right]$$

where

$$[A^{(-)}] = W \times C_0 - [MA^{(z-1)+}]$$

W = complexing capacity expressed in mmol/g (number of millimoles of cations bound per gram of organic ligand)
C_0 = initial concentration of the organic ligand expressed in g/l.

The complexing capacity is obtained graphically from the titration curve and β from a non-linear regression (using the Marquadt-Newton algorithm) giving the fluorescence intensity (or the absorbance) as a function of β, W and C_0.

Results and Discussion. From spectroscopic data, important features concerning the interaction of actinides with humic substances can be deduced. In the case of trivalent elements (Am^{3+}, Cm^{3+}, Dy^{3+}), the presence of one isobestic point (obtained with Am by spectrophotometry) indicates the presence of one complex *(10)*; from fluorescence data, it has been established that energy transfer via the triplet state of the organic molecules occurs explaining the increase of fluorescence intensity during the titration *(8-9)*. In the case of pentavalent elements (NpO_2^+), the presence of one isobestic point also indicates the existence of a single type of complex with the humate ligand and under our experimental conditions, no reduction of Np(V) to Np(IV) has been observed. In the case of hexavalent cations (UO_2^{2+}), the fluorescence study *(11)* shows that no reduction of U(VI) to U(IV) by humic acids under our experimental conditions occurs (no modification of the lifetime) and that the interaction of the uranyl ion with the humic molecules induces a static quenching (explaining a decrease of the signal with no fluorescence lifetime modification during the titration).

From a chemical viewpoint, *trivalent cations* form relatively strong complexes with humic acids. The interaction constants determined (Table I) are independent of pH (4-7) and ionic strength (0.1-0.001 M) but strongly dependent on the metal concentration (Figure 3) which may be related to the presence of different kinds of sites: strong and weak sites. This effect has already been observed in *(12)*. The complexing capacity (Table I) reflecting the number of sites interacting with the cation increases with pH and metal concentration but decreases with ionic strength.

Table I. Values of conditional interaction constants β (in l/mol) of tri-, penta- and hexavalent actinides with Aldrich humic acids and maximum complexing capacities (W in mmol/g)).

	[M] (M)	pH	I (M)	W_{max}	log β
Am(III)	3×10^{-5}	4.65	0.1	1.5 ± 0.3	7.0 ± 0.3
Cm(III)	3×10^{-8}	4.2	0.1	0.9 ± 0.3	9.1 ± 0.2
	1×10^{-7}	4.2-6.9	0.001-0.1	1.3 ± 0.2	8.3 ± 0.4
	5×10^{-7}	4.2-6.9	0.001-0.1	1.2 ± 0.3	7.4 ± 0.3
	1×10^{-5}	4.2-6.1	0.1	1.3 ± 0.3	6.6 ± 0.4
Dy(III)	2×10^{-6}	4-6	0.1	1.0	7.5 ± 0.2
Np(V)	6×10^{-5}	7	0.1	0.14	4.6 ± 0.2
U(VI)	4×10^{-8}	4-5	0.1	0.25	7.8 ± 0.2
	4×10^{-7}	5	"	0.7	7.2 ± 0.3

In the case of *pentavalent cations*, the conditional interaction constant (Table I) obtained for NpO_2^+ (log β = 4.6) shows a relatively low affinity of the neptunyl cation with the humic acids as it could be predicted from the charge of the ion. The interaction constant obtained for the *hexavalent cation* (U) with humic acid (Table I) is independent of pH (4-5) in the non-hydrolysis pH-range but some variation with uranium concentration is observed as for trivalent cations. Moreover, the complexation of uranium to humic substances is of the same order of magnitude than the complexation of trivalent actinides which corroborates chemical analogy between both cations.

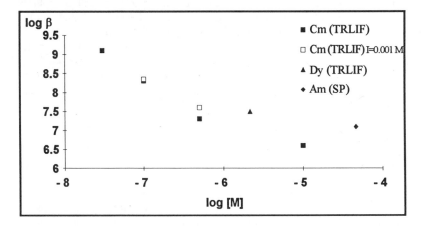

Figure 3. Effect of metal concentration on the interaction constant of trivalent actinides with Aldrich humic acids (at a ionic strength of 0.1M). TRLIF: Time-Resolved Laser-Induced Fluorescence, SP: Spectrophotometry.

Summary. Spectroscopic techniques, in particular laser induced fluorescence, appear as a good analytical tool for complexation measurements since they are non-intrusive methods, they allow to work at low level of cation concentration (in the case of TRLIF), in particular below the solubility limits permitting to cover a large range of pH and they also allows the use of low humic acid concentrations to avoid aggregation phenomena. From the conditional interaction constants measured by these techniques, the following trend of actinides for humic acids is deduced:

NpO_2^+ (log β_{Np} = 4.6) $\ll$ UO_2^{2+} (log β_U = 7.2-7.8) $\sim$ Am^{3+} (log β_{Am} = 7.1-9.1)

Consequences on actinide speciation. Speciation calculations of the actinides upon interest for radwaste disposals have been performed by considering hydrolysis, carbonate and organic complexation but *under conditions relevant to geological formations (either granitic or sedimentary)* which could be retained for nuclear waste disposals (Table II) *(11)*.

It appears that for trivalent cations (Am), organic complexes dominate their speciation in waters from granitic or sedimentary formations, except at the lowest humic acid concentration and highest partial pressure of CO_2. In the case of penta- and hexavalent actinides (Np, U), only inorganic complexes are present (hydroxide or carbonate species) in these waters. It should be emphasised that, if these speciation calculations are performed up to pH 7, organic complexation will predominate over inorganic complexation (whatever the partial pressure of CO_2 and for humic concentration equal or higher than 1 mg/l) for trivalent and hexavalent actinides. For pentavalent cations, organic complexes will control their speciation only at very high humic acid concentration (100 ppm). Moreover, it should be pointed out that if mixed complexes (organic ligand-cation-inorganic ligand (OH, CO_3)) are supposed to be formed under these conditions (Table II), the speciation will be entirely changed in favor of these latter. Hence, it shows the necessity of studying the formation of organic complexes under conditions closer to real systems (in particular high pH) which implies a need for very sensitive techniques in order to work with actinides in these alkaline media. Moreover, these kinds of investigations should also permit to confirm the existence of mixed complexes which should then modify the actinide speciation. Their existence has been reported in *(13)* in the case of Eu^{3+}.

Retention of Colloids and Heavy Elements on Mineral Surfaces

Laboratory experiments with model systems under static conditions have been aimed at the determination of the retention mechanisms of colloids and pseudocolloids (association of a heavy element with a colloid) onto mineral surfaces. This will give a better understanding of the fate of radioelements associated with colloids upon interaction with mineral surfaces as it will occur in the water flow across fissures and fractures around a radwaste repository. In these studies, polished cm-sized monoliths are used to simulate macroscopic surfaces of fine particles or as mineral surfaces. Rutherford Backscattering Spectrometry (RBS) is the technique chosen to determine accurately the amount of elements fixed on the monolith.

Table II. Main characteristics of waters representative of granitic and sedimentary formations (11)

	pH	$log\, pCO_2$ (atm)	DOC* (mg/l)
CRYSTALLINE FORMATIONS			
mine waters	7 - 8	$-3 \rightarrow -2$	1 - 10
evolved waters	8 - 9	$-5 \rightarrow -4$	1 - 10
SEDIMENTARY FORMATIONS			
Boom clay	~8.5	~ -3	~ 150

*DOC = Dissolved Organic Carbon; the humic acid concentration has been calculated from these data assuming that 50% of the organic carbon is constituted by humic substances and that these latter contain 50% of carbon.

Description of the technique. The choice of the RBS analytical technique derives from its highly quantitative character together with its good sensitivity for heavy elements (15). Moreover, it is a non-destructive method permitting an elemental depth profiling with a good depth resolution (<10nm). Nevertheless, this technique requires dried and cm-sized flat samples, has no good lateral resolution and does not yield information on the valence state of the heavy element under study. A typical RBS spectrum is presented in Figure 4.

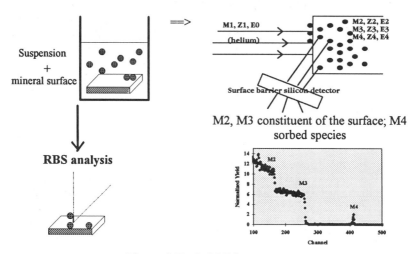

Figure 4. Typical RBS spectrum.

Due to the specificities of the RBS technique, model systems representative of colloids, heavy elements and surfaces encountered in natural systems relevant to radwaste disposal have been selected. Hence, investigations have been devoted to the study of i) colloids representative of those met in granitic or sedimentary formations *(11, 16)* such as silica, iron oxide and humic acids which may be considered as carrier colloids, ii) heavy elements as chemical analogues of radionuclides of interest such as Nd(III) or U(VI) as ionic species *(17)* or Th(IV) or Ce(IV) as true colloids representative of tetravalent actinides *(18)* and iii) monoliths representative of the minerals found in a granitic formation such as silicates (silica, biotite, muscovite), oxides (haematite) and carbonates (calcite). Binary systems [consisting of a monolith and colloids] and ternary systems [consisting of a monolith and two colloidal suspensions: colloid 1 as heavy element species and colloid 2 as a carrier colloid present in a large excess] have been considered in order to understand how colloids will modify the distribution of radionuclides (or heavy elements) at the solid/water interface. Thus emphasis has been put on the determination of the mineral surface coverage, the colloid detachment rate and the role of the carrier colloids in the behavior of heavy elements as functions of solution parameters (pH, ionic strength, colloid concentration) and surface characteristics.

Results and discussion. The binary systems, a carrier colloid [haematite: Fe_2O_3 (50-100 mg/l)] or a true colloid of heavy elements [CeO_2 (1-10 mg/l), ThO_2 (1-10 mg/l)] interacting with a mineral surface (silica, haematite), have been considered either as pseudocolloidal entities, where the monolith stands as the macroscopic surface of the carrier colloids, or as a colloid sorbing onto a mineral surface. The main features arising from these studies under the conditions pH ~5 and 0.001 M $NaClO_4$ are: i) the colloid sorption is a surficial and heterogeneous process; ii) the surface coverage, expressed in units of monolayers is inferred by assuming a spherical shape for the colloids, a specific gravity identical to that of the corresponding bulk oxide and a close packing distribution of colloids. This value is less than one monolayer of colloid (about 0.1-0.2) (Figure 5).

Moreover, the extent of colloid sorption is dependent on the colloid/surface characteristics (electric charge) and the solution parameters (pH varying from 3 to 8, ionic strength varying from 0.001 to 0.1 M): whenever the charges of the suspensions (colloids or pseudocolloids) and the monolith are opposite, sorption is favoured; this effect is particularly marked with mica as a monolith (negatively charged in the pH range studied) and with ceria colloids (positively charged) and pseudocolloids (SiO_2+CeO_2, negatively charged) (Figure 6). In the case of haematite (as colloid or as monolith), an increase of pH (3-8) produces a charge inversion (related to the point of zero charge around 7-8) leading to a decrease of colloid retention. No clear trend is observed with an increase of ionic strength since this parameter strongly affects the stability of colloidal suspensions. The surface characteristics (surface state or surface composition) influence the colloid uptake as follows: i) surface state: prehydration of the silica surface (to mimic the gel layer of silicates occurring in natural systems) does not have a marked influence on colloid sorption, ii)

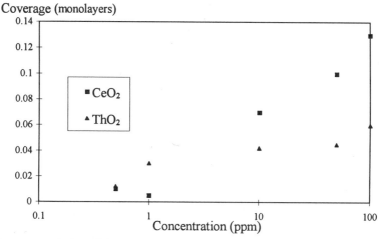

Figure 5. Variation of the surface coverage of ceria and thoria colloids on silica surface with the colloid concentration; pH 5, 0.001 M NaClO$_4$.

surface composition: colloid uptake depends on the electric charges developed at the monolith-water interface as explained above; all the silicates studied show similar retention properties for ceria and thoria colloids at a ionic strength representative of granitic environments (~0.001 M): for ceria ~ $4x10^{15}$ atoms/cm^2 and for thoria ~$5x10^{14}$ atoms/cm^2.

A particular emphasis has been placed on the detachment rate of colloids from the mineral surface. Colloid sorption is irreversible (or at least shows very slow desorption kinetics regardless of solution composition: either in electrolyte solution or in the presence of a carrier colloid such as silica or humic acids (Table III); moreover, desorption tests up to three months have not shown any colloid detachment. No marked influence of temperature on the release of retained ceria colloids has been observed between 20 and 90°C.

Table III. Results of desorption tests expressed in number of atom/cm^{2}x10^{15} concerning colloid sorption on silicate minerals (accuracy of about 30 %).

	Sorption (48h)	Desorption in NaClO$_4$ (48h)	Desorption in HA 100 ppm (48h)	Desorption in SiO$_2$ 100 ppm (48h)
Ceria colloids: 10 ppm, pH 5 , NaClO$_4$ 0.001 M				
silica	5	3	-	-
biotite	4.4/1.3	1.3/1.2	-	-
muscovite	1.6	1.2	-	-
albite	4.8/3.9	2.9/2.5	2.9	-
Thoria colloids: 10 ppm, pH 5 , NaClO$_4$ 0.001 M				
silica	0.8/0.6	0.3	0.4	0.4
biotite	0.5	0.5	0.4	0.4
muscovite	0.4/0.5/0.9	0.4	0.5	0.7

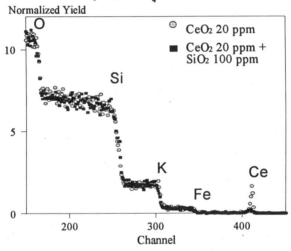

Figure 6. RBS spectra obtained after interacting ceria colloids (dots) or pseudocolloids (ceria sorbed on silica colloids, squares) with a mica surface; pH 5, NaClO₄ 0.001 M.

As for ternary systems, a study of the effect of the order of adding constituents has shown that when the carrier colloid is added after the colloid 1, there is no effect, whatever the systems. This is related to the irreversible character of the sorption of colloid 1. When a true colloid is added after interacting the surface with the carrier colloid, if the conditions are in favour of a presorbed layer (opposite charges), an increase of sorption is observed. When the pseudocolloid interacts with the mineral surface, the charge effect is important and can lead to a strong decrease in heavy element sorption (Figures 6-7).

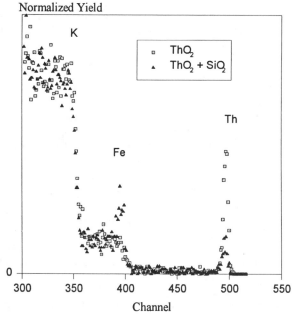

Figure 7. RBS spectra obtained after interacting thoria colloids (squares) or pseudocolloids (thoria colloids sorbed on silica colloids, triangles) with a mica surface; pH 5, NaClO₄ 0.001 M.

Summary. Ion beam techniques such as RBS are a good analytical tool for studying interface phenomena due to their multielement capability and the possibility of working on whole rock sections. Interesting information on the sorption mechanisms can be obtained from these studies: the colloid surface coverage is low (less than one monolayer); the retention mechanisms are partly controlled by the electric charges developed at the surfaces (colloid, mineral); the colloid detachment rate is very low indicating an irreversible character with respect to solution composition, or at least very slow desorption kinetics on the time scale used for the safety assessment of a nuclear waste repository; the carrier colloids can maintain the true colloids in solution (as analogues of tetravalent actinides such as Pu, Np, U). The data presented are of great interest either in the validation of theoretical assumptions (monolayer, reversibility) or in modelling colloid behavior in natural systems *(19-20)*.

Consequences for Radionuclide Behavior

It is important to outline the state of the art concerning colloids and humic substances in relation with radionuclide behavior in natural systems: we know that colloids are present in granitic groundwaters (mainly inorganic colloids and, to a lesser extent, humic substances *(21)*) and in sedimentary formations (mainly humic substances *(11)*); we know that humic substances will control speciation of trivalent cations (trivalent actinides such as Am or Pu if the redox conditions are in favor of this valence state) under conditions representative of natural waters (granitic, sedimentary); if mixed complexes are assumed to exist, other radionuclides may be present as organic pseudocolloidal entities; we know that radionuclides (in particular as true colloids which could be formed by tetravalent actinides) present strong affinities for colloids (irreversible sorption) which also have strong retention properties on mineral surfaces (irreversible sorption with a partial coverage). Thus it appears that colloids and humic substances can increase the mobility of radionuclides in natural waters but the geological matrix could also retard their mobility (due to irreversible sorption). In the case of granitic formations where fractures and fissures will constitute the migration path, the main transport mechanisms will be convection. Retardation will occur through filtration, clogging or sorption processes, whereas migration will happen under conditions which hinder the above processes. In the case of clayey formations, which constitute a porous medium, diffusion will be the major transport mechanism. Depending on the clay pore spectrum and the colloid size range, colloids may migrate provided there are enough connected channels for each particle. In this case, colloids will be retarded compared to dissolved species *(21)*. Thus, the basic question of whether colloids and humic substances will induce a confinement or an enhancement of radionuclide migration in the vicinity of a radwaste disposal is not easily answered because of the complexity and site-specific character of the processes which operate.

Acknowledgements.

This work has been performed under the shared cost programmes FI2W-CT91-0083 and FI2W-CT91-0097 of the European Union carried out within the fourth CEC R&D programme on "Management and Storage of Radioactive Waste" (1990-1994) - Part A, task 4 "Disposal of Radioactive Waste".

Literature cited

(1) Yariv S.; Cross H. *Geochemistry of colloidal systems;* Springer-Verlag: Berlin, 1979.

(2) Stumm W. *Chemistry of The Solid-Water Interface;* J. Wiley & Sons: New-York, 1992.

(3) Thurman E.M. *Organic Geochemistry of Natural Waters;* Nijhoff and Junk Publishers: Dordrecht, 1985.

(4) Buffle J. *Complexation reactions in aquatic systems;* J. Wiley & Sons: New-York, 1988.

(5) Choppin G.R. *Radiochim. Acta* **1992**, 58/59, 113.

(6) Choppin G.R.; Allard B. In *Handbook on the Physics and Chemistry of the Actinides Volume 3;* Freeman A.J.; Keller C. Eds; North Holland: 1985, pp 407-429.

(7) Moulin V.; Ouzounian G. *Appl. Geochem.* **1992**, 1, 179.

(8) Moulin C.; Decambox P.; Mauchien P.; Moulin V.; Theyssier M. *Radiochim. Acta* **1991**, 52/53, 119.

(9) Moulin V.; Tits J.; Moulin C.; Decambox P.; Mauchien P.; De Ruty O. *Radiochim. Acta* **1992**, 58/59, 121.

(10) Moulin V.; Robouch P.; Vitorge P.; Allard B. *Inorg. Chim. Acta* **1987**, 140, 303.

(11) Moulin V.; Moulin C. *Appl. Geochem.* **1995**, 10, 573.

(12) Bidoglio G.; Grenthe I.; Qi P.; Robouch P.; Omenetto N. *Talanta* **1991**, 9, 999.

(13) Moulin V.; Tits J.; Ouzounian G. *Radiochim. Acta* **1992**, 58/59, 179.

(14) Diercks A.; Maes A.; Vancluysen J. *Radiochim Acta* **1994**, 66/67, 149.

(15) Della Mea, G.; Rossi-Alvarez C.; Mazzi G.G.; Bezzon J.; Chaumont A.; Dran J.C.; Mendenhall M; Petit J.C. *Nucl. Instr. Meth. Phys. Res.* **1987**, B19/20, 943.

(16) Turrero M.J.; Gomez P.; Perez del Villar L.; Moulin V.; Magonthier M.C.; Ménager M.T. *Applied Geochem.* **1995**, 10, 119.

(17) Della Mea, G.; Dran J.C.; Moulin V.; Petit J.C.; Ramsay J.D.F.; Theyssier M. *Radiochim. Acta* **1992**, 58/59, 219.

(18) Dran J.C.; Della Mea G.; Moulin V.; Petit J.C.; Rigato V. *Radiochim. Acta* **1994**, 66/67, 221.

(19) Grindrod P. *J. Contaminant Hydrology* **1993**, 13, 167-181.

(20) Smith P.A.; Degueldre C. *J. Contaminant Hydrology* **1993**, 13, 143-166.

(21) Fourest B.; Guillaumont R.; Moulin V.; Maillard S.; Cromières L.; Giffaut E.; Merceron T. Presented at Migration 95 Conference, Saint-Malo, September 1995.

(22) Moulin C.; Beaucaire C.; Decambox P.; Mauchien P. *Anal. Chim. Acta* **1990**, 238, 291.

(23) Moulin C.; Decambox P.; Mauchien P. *Anal. Chim. Acta* **1991**, 254, 145.

(24) Thouvenot P.; Hubert S.; Moulin C.; Decambox P.; Mauchien P. *Radiochim. Acta* **1993**, 61, 15.

(25) Berthoud T.; Decambox P.; Kirsch B.; Mauchien P.; Moulin C. *Anal. Chim. Acta* **1989**, 220, 235.

Chapter 17

Immobilization of Actinides in Geomedia by Phosphate Precipitation

Mark P. Jensen, Kenneth L. Nash, Lester R. Morss,
Evan H. Appelman, and Mark A. Schmidt

Chemistry Division, Argonne National Laboratory,
Building 200, 9700 Cass Avenue, Argonne, IL 60439–4831

A method is being developed to transform actinide ions in the near surface environment to less soluble, less reactive, thermodynamically stable phosphate minerals phases through application of organophosphorus complexants. These complexants decompose slowly, releasing phosphate to promote the formation of stable phosphate mineral phases, particularly with the more soluble trivalent, pentavalent, and hexavalent actinide ions. The complexant of choice, myo-inositol(hexakisphosphoric acid) or phytic acid, is a natural product widely used as a nutritional supplement. We have determined that phytic acid decomposes slowly in the absence of microbiological effects, that crystalline phosphate minerals are formed as a consequence of its decomposition, and that the formation of actinide (lanthanide) phosphates reduces the solubility of trivalent and hexavalent metal ions under environmental conditions.

Both planned and accidental releases of radionuclides have contaminated soils and waters at many U. S. Department of Energy sites. Moreover, the purposeful disposal of materials contaminated with radionuclides in appropriately designed and maintained mixed-waste landfills has added to the contamination of the subsurface environment. For those materials possessing appreciable water solubility, serious contamination of the surrounding environs may occur via surface water runoff and percolation through underlying geologic strata. This potential pathway represents a direct route for invasion of the food chain by radioactive metal ions. Among the long-lived radioactive materials, the transuranic actinide elements represent the greatest long-term hazard.

Most radionuclides buried in waste disposal trenches are sorbed on surfaces. These surfaces are usually metallic, organic (paper and plastic), or mineral, with the surrounding geologic strata providing additional sorption opportunities. The chemical forms of the sorbed radionuclides vary widely, ranging from rather intransigent oxide films to potentially soluble residues of metal nitrates or chelated metal complexes.

0097–6156/96/0651–0272$15.00/0
© 1996 American Chemical Society

These latter species can be particularly susceptible to mobilization when in contact with natural waters. In addition, natural chelating agents like humic and fulvic acids, present at low concentrations in most groundwaters, can facilitate the migration of heavy metals, even when the metals are present in moderately insoluble forms.

Since exclusion of groundwater from landfill sites cannot be guaranteed over the lifetime of long-lived radioactive nuclides, a desirable alternative is to reduce or eliminate the potential for contaminant migration into the surrounding environment by converting the actinide ions to much more insoluble, thermodynamically stable forms. Immobilization of radionuclides in the subsurface environment can be accomplished by this method without excavating soils or pumping groundwater for treatment, and without the construction of massive barriers to restrict groundwater flow. Conversion of these species to thermodynamically stable mineral forms has the dual benefit of reducing their inherent mobility and permitting the application of thermodynamic models to predict their long-term behavior.

The chemistry of actinide ions is generally determined by their oxidation states. The trivalent, tetravalent and hexavalent oxidation states are strongly complexed by numerous naturally occurring ligands (carbonates, humates, hydroxide) and man-made complexants (like EDTA), moderately complexed by sulfate and fluoride, and weakly complexed by chloride (*1*). Under environmental conditions, most uncomplexed metal ions are sorbed on surfaces (*2*), but the formation of soluble complexes can impede this process. With the exception of thorium, which exists exclusively in the tetravalent oxidation state under relevant conditions, the dominant solution phase species for the early actinides are the pentavalent and hexavalent oxidation states. The transplutonium actinides exist only in the trivalent state under environmentally relevant conditions.

In a reducing environment, where tri- and tetravalent actinides predominate, hydroxides, oxides, hydroxycarbonates, or carbonates are the most probable stable mineral phases in near-surface waters. Under more oxidizing conditions, the higher oxidation states of uranium, neptunium, and plutonium dominate. These species typically exhibit higher solubility limits. For example, even at very low total carbonate concentrations, uranium(VI) forms strong carbonate complexes above pH 5. These carbonate complexes extend the solubility of uranium(VI), and by analogy neptunium(VI) or plutonium(VI), to very high pH. The triscarbonato uranium(VI) complex is responsible for the relatively high concentration of uranium in seawater.

Actinide(V) cations exhibit the highest aqueous solubilities of any actinide oxidation state, being quite soluble at pH's greater than 7 even in the absence of complexing anions. Over the range of environmental pH and redox potential, there is substantial evidence that the pentavalent oxidation state is thermodynamically favored for aqueous plutonium ions (*3*). Likewise, neptunium favors the pentavalent oxidation state over a wide range of conditions (*4*). Increasing the thermodynamic driving force for immobilizing neptunium(V) and plutonium(V) is critical to minimize their potential for migration into the environment.

Given the variety of important actinide oxidation states, what naturally occurring anion would best effect actinide mineralization? The lanthanides, chemically analogous to the trivalent actinides, occur naturally in three commercially important forms: monazite and xenotime, which are orthophosphate minerals, and bastnasite, which has the approximate composition $LnFCO_3$. Uranium ores may be divided into three classes: oxides, vanadates, and phosphates. Thorium (as ThO_2) is found primarily in monazite sands. Carbonate forms insoluble compounds with trivalent f-elements; however it greatly promotes the solubility of hexavalent actinides as the triscarbonato complex. Lanthanide and actinide oxides and hydroxides are insoluble, but in this case the concentration of the precipitating anion is controlled by the groundwater pH and hence is manipulated with difficulty. Nature's lesson is that phosphate mineral forms are the most reasonable choice for reducing the solubility of actinides in geomedia through induced mineralization.

Simple thermodynamic calculations based on literature data (5-12) support the choice of phosphates as the optimum mineral phases for actinide immobilization. The calculations considered every relevant species reported (5-12) that contained protons, hydroxide, or the ligand in question for each metal ion. Where necessary, equilibrium constants were corrected to 0.1 M ionic strength using the Davies equation. As an example, the calculated solubility of europium, thorium, and uranium in various media at p[H] 7.0 (p[H] = - log of the hydrogen ion concentration), 0.001 M total ligand concentration, 0.1 M ionic strength, and 25 °C are shown in Table I. Within the constraints of the calculation, the solubility of thorium is limited by $Th(OH)_4$, but the lowest europium and uranyl solubilities are observed for phosphates.

Table I. Calculated concentration of dissolved f-elements in a medium of I = 0.1 M, p[H] = 7.0, [Ligand]$_{total}$ = 0.001 M, T = 25 °C.

Metal	OH[a]	F[-]	SO_4^{2-}	CO_3^{2-}	PO_4^{3-}	SiO_4^{4-}
			-log Σ[Metal]$_{aqueous}$			
Eu(III)	5.0	7.4	sol.	6.0	11.1	< 6
Th(IV)	5.6	5.6[b]	sol.	-	≤ 5.6[b]	-
U(VI)	6.3	sol.	sol.	4.2[b]	7.8	7.3[c]

[a]p[H] = 11.0
[b]Solubility limited by M(OH)$_m$
[c]T = 30 °C

Thus the basis of this work is the introduction of phosphate into the subsurface environment in a controlled manner. The phosphate is delivered preferentially to the metal ions to be mineralized or in such a fashion that the delivery agent concentrates the metal ions. The ideal vehicle for phosphate introduction appears to be a water soluble organophosphorus complexant that can be delivered as a dissolved species or aqueous slurry. This compound would decompose under environmental conditions and release phosphate, generating microcrystalline actinide phosphate solids that combine to form crystalline deposits of actinide phosphates. Alternatively, the principal mechanism for immobilization may be coprecipitation with phosphate mineral phases of ubiquitous polyvalent metal ions (e.g., iron or calcium) since the actinides will be present at low concentrations. In addition, the residual organic portion of the organophosphorus compound should not be a chelating agent to minimize the potential for redissolution of the precipitated metal ions. Finally, the ideal organophosphorus compound should function as a cation exchanger during the time between its application and decomposition.

One very promising organophosphorus compound is the abundant, phosphate rich, natural product myo-inositol(hexakisphosphoric acid), or phytic acid. Literature reports on the chemical, environmental, and biological behavior of phytic acid abound (*13*). Phytic acid is isolated from beans and leafy vegetables and is used as the starting

$$
\begin{array}{c}
H_2O_3PO \qquad OPO_3H_2 \\
H_2O_3PO \qquad \qquad \cdots OPO_3H_2 \\
H_2O_3PO \qquad OPO_3H_2
\end{array}
$$

Phytic Acid

material in the manufacture of inositol, as a dietary supplement, as a nutrient source for microorganisms, and as a metal chelating or precipitating agent. It complexes polyvalent metal ions well and is readily hydrolyzed to release phosphate.

Experimental

Materials. Solutions of Eu, Th, and U were made by dissolving known amounts of the metal oxides in nitric or perchloric acid. The europium solution was standardized by titration with EDTA using xylenol orange indicator at pH 5.2.

Working solutions of $^{152,154}Eu^{3+}$, $^{233}UO_2^{2+}$, and $^{237}NpO_2^+$ were obtained from laboratory stocks. The Eu tracer was prepared by dilution of non-radioactive $Eu(ClO_4)_3$ to prepare a solution of known concentration which was subsequently spiked with $^{152,154}Eu$. The resulting solution contained 1.48×10^{14} cpm/mol Eu when counted by liquid scintillation. The daughter activities arising from the 0.2% ^{232}U α-activity in the ^{233}U solution were removed using a U/TEVA column (*14*). The ^{237}Np solution was oxidized to NpO_2^{2+} by $NaBrO_3$ at 70 °C, reduced to NpO_2^+ with $NaNO_2$, precipitated as $NpO_2(OH)$ by addition of NaOH, washed with water, dissolved in 1 M $HClO_4$, and passed through a Dowex-50 cation exchange column to remove a small amount of ^{238}Pu. The radiochemical purity of the stock solutions was verified by α- and γ-spectroscopy and liquid scintillation counting. The concentrations of ^{233}U and ^{237}Np were calculated from their specific activities.

Aqueous solutions of phytic acid were prepared from the dodecasodium salt (Aldrich) with a molecular weight of 1095 ± 28. The pH of the phytic acid solutions was adjusted with $HClO_4$ or HNO_3. The buffers 4-morpholineethanesulfonic acid and 4-(2-hydroxyethane)-1-piperizineethanesulfonic acid were recrystallized from aqueous ethanol before use. TRIZMA buffer was used without further purification. For the purposes of these experiments, the buffers were considered non-complexing.

Phytate Hydrolysis Kinetics. An approximately 0.1 M solution of phytic acid was adjusted to the desired pH (measured at room temperature), sealed in a glass ampoule and immersed in a thermostatically controlled water bath. Ampoules were periodically opened and sampled for colorimetric phosphate analysis as molybdovanadophosphoric acid (15). The remaining solution was transferred to a fresh ampoule, resealed, and returned to the water bath. Monitoring of the reaction continued until more than 90% of the phytate had hydrolyzed, after which a final room temperature pH was measured, unreacted phytate was determined by ^{31}P NMR, and organic products were investigated with ^{1}H NMR. Kinetic data were least-squares fitted to a three parameter first-order rate equation, except for a few measurements at the lowest temperatures where the initial reaction rates were used to evaluate the rate constants.

Preparation and Characterization of Lanthanide and Actinide Solids. Standard crystalline lanthanide and actinide phosphates were prepared by literature procedures (16-18) and characterized by X-ray powder diffraction, FTIR spectroscopy, and thermogravimetric analysis (TGA). Europium was used as an analogue of the trivalent actinides. Metal-phytate solids were generated by mixing Eu(III), U(VI), or Th(IV) nitrate solutions with 0.1 M phytic acid at pH 5 and metal:phytate ratios of 1:1 2:1, and 4:1. The metal phytates precipitated immediately. The resulting slurries were stirred at 85 °C for 30 days and sampled periodically for analysis of the solids by TGA, X-ray powder diffraction, and FTIR. The rate of phosphate release to the solution was monitored colorimetrically.

Immobilization of Lanthanides and Actinide Deposits by Addition of Phosphate. Tracer-scale amounts of representative tri-, penta-, and hexavalent f-elements were deposited on borosilicate glass surfaces to demonstrate the immobilization of small amounts of heavy-metal cations by phosphate anions. Solid samples containing 1.5×10^5 cpm $^{152,154}Eu^{3+}$ (1.0×10^{-9} mol), $^{233}UO_2^{2+}$ (3.1×10^{-8} mol), or $^{237}NpO_2^{+}$ (4.7×10^{-7} mol) were deposited from acidic nitrate, basic (hydroxide), and neutral citrate (metal:citrate = 1:1.1) solutions by evaporation. Films of octyl(phenyl)-N,N-diisobutylcarbamoylmethylphosphine oxide in 1.2 M tributylphosphate/Isopar-L (TRUEX process solvent) containing 1.1×10^5 cpm $^{152,154}Eu(NO_3)_3$ or 6.2×10^4 cpm $^{233}UO_2(NO_3)_2$ were spread on glass surfaces to simulate deposits of organic extractant solutions. We chose these conditions to represent typical forms of radionuclides in actual wastes.

Replicate solid samples were contacted with 11 ml aliquots of a simulated groundwater containing 0.1 M $NaClO_4$, 0.0005 M $NaHCO_3$, and 0, 0.0001, 0.001, or 0.01 M total phosphate, buffered at pH 5, 6, 7, and 8. The radiotracer solutions were introduced without carrier cations (i.e. only Na^+, ClO_4^-, $H_nCO_3^{n-2}$, and $H_mPO_4^{m-3}$ are present in macroscopic concentrations) except as described above for europium. The radionuclide concentration and pH of each system were monitored weekly. Radionuclide concentrations were determined by liquid scintillation counting after the samples had been filtered through 0.2 μm filters. The estimated uncertainty in the measured metal concentrations is 10%, though near the detection limits radioactive counting statistics caused the uncertainties to be greater. Each experiment was run for ten weeks. Within the first three weeks, the dissolved metal concentrations in all the systems had reached a steady-state.

Results and Discussion

Phytate Hydrolysis Kinetics. Phytate ions are known to hydrolyze rapidly in aqueous solutions near 100 °C (*19*). The kinetic results presented in Table II permit extrapolation of the hydrolysis rate to ambient temperature at various pH values. Reactions monitored to 90% or more of completion obeyed first-order kinetics. The NMR measurements indicate the principal products are phosphate and myo-inositol, $C_6H_6(OH)_6$, along with other unidentified organic species. No soluble intermediate phosphorus containing species were identified during this experiment. Prior studies have reported that myo-inositol itself decomposes at elevated temperature in near-neutral solutions to yield dark byproducts (*20*). Our solutions at pH 5.1 darkened markedly as the reaction proceeded. Solutions at pH 3.1 showed substantially less darkening, while solutions at pH 6 and above only darkened when heated at 95 °C. The latter produced a white precipitate, apparently calcium phosphate that formed from a 0.6% calcium impurity in the starting sodium phytate. The reaction rate for phytic acid hydrolysis is only weakly dependent on pH, consistent with the results of Bullock et al. who found a maximum in the rate at pH 4 (*19*). It should be noted, however, that only at pH values between 5 and 6 did the pH remain relatively constant throughout the reaction. As indicated in Table II, the pH of the more alkaline solutions decreased during the course of the reaction, while the pH of the more acidic solutions rose, tending in both cases to the pH 5-6 range. The overall reaction for phytic acid hydrolysis,

$$6\,H_2O + C_6H_6(OPO_3H_2)_6 = C_6H_6(OH)_6 + 6\,H_2PO_4^- + 6\,H^+ \tag{1}$$

releases H^+ to the medium. Thus the pH of these solutions is determined by the combined buffering effects of phytic acid and phosphate. Further degradation of the myo-inositol residue presumably produces CO_2 and H_2O which superimposes the buffering effect of carbonic acid on the system. In the absence of other competing environmental effects, this process would tend to buffer groundwater pH in the 5-7 range.

Table II. Hydrolysis of phytic acid at elevated temperatures.

Temperature °C	pH 3.1[a]	pH 5.1-5.3	pH 6.0-5.8	pH 7.0-6.1	pH 8.1-6.4
		Half-time (days)			
94.6		0.7	1.03	1.36	1.62
84.8	2.1	2.2	3.5	4.0	4.6
75.1		8.0	12.2	16.1	22
64.5		45[b]	75[c]		
55.1		193[c]	273[c]		

[a] Final pH not measured.
[b] Reaction followed to 80% of completion.
[c] Calculated from the initial reaction rate.

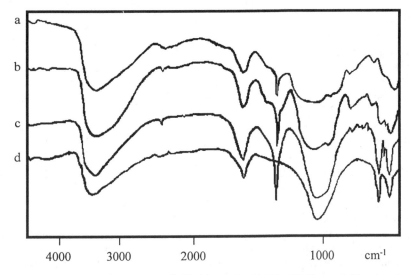

Figure 1. Infra-red spectra of Eu(phytate), $EuPO_4 \bullet H_2O$, and Eu-phytate decomposition products. (a) Eu(phytate) standard, (b) Eu-phytate mixture after 1 day, (c) Eu-phytate mixture after 29 days, (d) $EuPO_4 \bullet H_2O$ standard.

Linear plots of ln k/T vs. 1/T that permit extrapolation of the reaction rates to ambient temperature are obtained only for the reactions at pH 5.1 and 6.0. These plots each yield a value of 141 kJ/mol for the activation enthalpy, with standard deviations of 2.9 and 3.6 kJ/mol at pH 5.1 and 6.0 respectively. Extrapolation to 25 °C gives half-times of 104 ± 22 years at pH 5.1 and 156 ± 42 years at pH 6.0.

In the absence of mechanisms to accelerate phytic acid hydrolysis, the complexant will clearly persist in the environment for a long time. However, enzyme catalyzed phytate hydrolysis is known to be much faster than the thermal reaction (*21*). Thus biological activity may accelerate the hydrolysis reaction in soils. A recent report indicates that phytic acid is rapidly hydrolyzed by both aerobic and anaerobic microbiological processes in environmental samples (*22*). Complete phytate hydrolysis was observed in less than 40 days in anaerobic sediments, while half the available phosphate was released in 60 days under aerobic conditions (*22*).

Preparation and Characterization of Lanthanide and Actinide Solids. Crystalline *f*-element phosphates were prepared as standards for comparison to the solids produced in the conversion of metal phytates to phosphates. The europium standard prepared was identified by X-ray powder diffraction as hexagonal $EuPO_4 \cdot H_2O$ (JCPDS card number 20-1044), which was dehydrated at 204-234 °C and converted to monoclinic $EuPO_4$ (with the monazite structure) at 500-600 °C. The standard uranyl phosphate solid prepared was the acid phosphate, $UO_2HPO_4 \cdot 2H_2O$ (JCPDS card number 13-61). All attempts to prepare a crystalline thorium phosphate failed, though thorium solubility was low. In the latter case the solids were identified as amorphous $Th(OH)_4$ with some minor crystalline inclusions of ThO_2.

The second stage of this experiment was to demonstrate the preparation of metal phosphates during phytic acid decomposition. Infrared studies of solid samples derived from the europium-phytate experiments are shown in Figure 1. The broad features at 3500 cm^{-1} are water bands. The sharp resonance at 1300 cm^{-1} corresponds to nitrate contamination from the initial sample preparation, and this band disappears with more extensive washing of the solid. Phosphate bending modes appear in the 800-1200 cm^{-1} range. The phytate samples are characterized by a broad absorption in this region due to the multiple phosphate environments in the europium-phytate complex. After a day at 85 °C, this region of the spectrum narrows and new features appear between 400 and 500 cm^{-1}. By 29 days, the spectrum is identical to that of the hexagonal $EuPO_4$ standard (except for the nitrate band). X-ray powder diffraction results indicate a mixture of Eu(phytate) and $Eu(PO_4) \cdot H_2O$ at day 7, but only $EuPO_4 \cdot H_2O$ at 19 days. The rate of phosphate production from phytic acid at 85 °C is faster than that observed for microbial degradation of phytic acid in sediments at ambient temperature (*22*). This suggests that any increases in the phosphate release rate caused by microbial action will not hinder the formation of crystalline actinide phosphates.

Parallel experiments with uranyl-phytate mixtures produced a uranyl phosphate solid identified as $(UO_2)_3(PO_4)_2 \cdot H_2O$ by X-ray powder diffraction (*23*), not $UO_2HPO_4 \cdot nH_2O$ as expected. Neither crystalline phosphates nor phytates, were observed in the thorium-phytate mixtures, although amorphous thorium phytates were likely present initially. Hydroxide or oxide species seem to control thorium solubility.

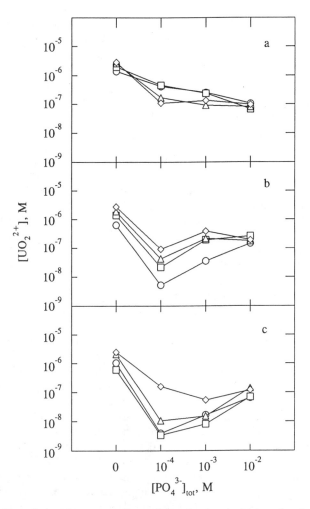

Figure 2. Experimental concentrations of dissolved uranyl ions after 5 weeks of contact between simulated groundwater at pH 5 (O), 6 (□), 7 (Δ), and 8 (◊) and (a) citrate, (b) nitrate, and (c) hydroxide deposits.

Immobilization of Lanthanides and Actinide Deposits by Addition of Phosphate.
These experiments were designed to demonstrate the reduction in the solubility of
tracer amounts of actinides that can be achieved through phosphate application. To
alleviate logistical complications in the execution of the investigation, the experiments
were conducted without phytate present. Instead, we assumed that phosphate generated
by phytate decomposition performs similarly to that introduced as $Na_xH_{3-x}PO_4$. It
should be noted that we have not measured thermodynamically rigorous metal
solubilities in phosphate media, and this data should not be used to calculate solubility
products. Yet, use of the term solubility is less cumbersome than referring to the
concentration of radionuclides in solution. We will use the term solubility to refer to
the relative concentration of dissolved metal ions. Since we have noted that thorium
solubility is controlled by oxide or hydroxide species, our investigation of radionuclide
solubility was confined to the other actinide oxidation states, represented by
$^{152,154}Eu^{3+}$, $^{233}UO_2^{2+}$, and $^{237}NpO_2^+$, which were deposited as nitrate, hydroxide, or
citrate solids, or as TRUEX films as described above.

For the europium system, the TRUEX films show little tendency to release
metal ions into the simulated groundwater with or without phosphate present. The
average solution phase concentration is 1×10^{-9} M over the range of pH and phosphate
investigated. The hydroxide and nitrate samples behave almost identically to each
other. In the absence of phosphate, europium solubility decreases from 4.4×10^{-8} M at
pH 5 to 7×10^{-10} M at pH 8. This decline results from europium hydroxide formation
in the absence of phosphate. With the addition of 0.0001 M phosphate at pH 5, the
average dissolved europium concentration drops to 1.7×10^{-9} M, strongly suggesting
that solubility is now controlled by $EuPO_4$. Increasing the pH, or adding 0.001 M
phosphate depresses europium solubility further to concentrations near the europium
detection limit of this experiment, 2×10^{-10} M. A small increase in europium solubility
is observed for both the nitrate and hydroxide deposits at the highest pH's when the
phosphate concentration is increased to 0.01 M. Greater solubility was observed for
the europium-citrate deposits than for the nitrate or hydroxide deposits with or without
phosphate. In the absence of phosphate, complexation by citrate extends the europium
solubility field to pH 8, where about half of the europium remains in solution. At pH 5
and 6, the lowest phosphate concentration (0.0001 M), is not sufficient to decrease the
europium solubility significantly. However, at or above pH 7, even the lowest
concentration of phosphate effectively removes europium from the citrate solutions.
Once again, though, the europium solubility is significantly greater for solutions of
0.01 M phosphate at pH 7 and 8 than it is for the lower phosphate concentrations.
From this set of experiments, we estimate the upper limit of europium solubility under
these conditions to be 1×10^{-9} M.

For the uranyl-TRUEX system, the average water soluble uranyl concentration
was low, 7×10^{-8} M. It appears that TRUEX bound uranyl ions are not back extracted
into the neutral pH simulated groundwater. For uranyl samples deposited as nitrate
salts or hydroxides, the solubility in the absence of phosphate is about 1×10^{-6} M at
pH 5 and pH 6, rising to 2×10^{-6} M at pH 7 and 2.6×10^{-6} M at pH 8. The increase in
solubility at higher pH is a direct result of increased uranyl-carbonate complexation.
With the introduction of phosphate, uranyl solubility is reduced at all pH's, as shown
in Figure 2. Uranyl solubility is generally lowest for 0.0001 M phosphate, rising

slightly with increasing pH and reaching a maximum of 1×10^{-7} M at pH 8, where carbonate complexation is important. The soluble uranyl concentration rises slightly with increasing phosphate concentration to a maximum of 2.5×10^{-7} M in 0.01 M phosphate. The chelating agent citric acid raises uranyl solubility at all pH's in the absence of phosphate. The data (Figure 2a) also show the expected increase in uranyl solubility with increasing pH. In the presence of both phosphate and citrate, uranyl solubility at pH 5 and 6 is higher than in the absence of the citrate, though it does drop to 1×10^{-7} M at the highest phosphate concentration. At pH 7 and 8, phosphate overcomes the complexing strength of citrate, reducing uranyl solubility to 1×10^{-7} M at all phosphate concentrations.

Neptunium(V) was expected to exhibit the greatest solubility of the metals investigated. Because NpO_2^+ is poorly extracted by the TRUEX process solvent, we did not investigate neptunium redissolution from this medium. As is true for the europium and uranyl systems, the neptunyl nitrate and hydroxide systems behave nearly identically, suggesting true equilibrium behavior governs the system. In the 0.0001 M phosphate systems, all of the neptunium introduced into the system is soluble between pH 5 and 8. At higher phosphate concentrations, the neptunium solubility decreases for pH 7 and 8, with a minimum solubility of 2×10^{-6} M at 0.01 M phosphate and pH 8. Citrate has little effect on the solubility of neptunium(V) under any conditions. The behavior of neptunium is not surprising since the pentavalent oxidation state forms the weakest actinide complexes. In addition, the mismatch of charges between NpO_2^+ and HPO_4^{2-} or PO_4^{3-} increases the likelihood that soluble anionic complexes, like $NpO_2(PO_4)^{2-}$, form. Thus, the key to immobilizing neptunium by phosphate addition may be the participation of naturally occurring countercations, such as Ca^{2+}, that can neutralize the charge of the neptunium phosphato complexes.

Thermodynamic Speciation Models.   The results observed in the radiotracer solubility experiments above are in qualitative agreement with expectations based on thermodynamic (equilibrium) control of cation solubility. The agreement between the results obtained for all systems in which the metal ion was introduced as the hydroxide or nitrate salt demonstrates thermodynamic solubility control in these experiments. If the initial hydroxide deposits were kinetically inert solids, these results would not have agreed. The solubility of uranyl citrate (in the absence of phosphate) increases with pH in accord with the increase in the free citrate concentration. Likewise, the uranyl hydroxide and nitrate samples (in the absence of phosphate) show increasing solubilities with increasing pH as a result of higher free carbonate concentrations. Europium solubilities decrease with increasing pH (again in the absence of phosphate) because the equimolar concentration of citrate, relative to europium, is unable to overcome the formation of $Eu(OH)_3$. On the basis of these observations, it is reasonable to assume that the concentration of these cations in the presence of phosphate is governed by equilibrium thermodynamics.

Calculations of metal ion speciation from literature data support the conclusions reached in the europium and uranyl systems. Literature data for the carbonate, hydroxide, and phosphate complexes of europium and uranium were used (5-12), with correction from infinite dilution to 0.1 M ionic strength via the Davies equation where necessary. The calculated fraction of metal ion precipitated as $EuPO_4$

or $(UO_2)_3(PO_4)_2$ under the experimental conditions employed above is described in Figures 3a and 3b. For solutions containing 5×10^{-4} M carbonate and 1×10^{-4} M phosphate, the europium species calculation (Figure 3a) indicates that $EuPO_4$ dominates the stability field between pH 4.5 and 10. However, the calculations for uranyl are less consistent with our observations. In the same media, the calculations (Figure 3b) indicate that $(UO_2)_3(PO_4)_2$ controls uranyl solubility from pH 4 to 7, but is overwhelmed by carbonate complexes at higher pH. Our observation is that uranyl solubility is more completely reduced by phosphate over the range of phosphate concentrations and pH's studied. Moreover, neither model shows the small increase in cation solubility at higher phosphate concentrations. This result implies that the thermodynamic data is either incomplete or erroneous.

These experiments demonstrate a significant reduction in the solubility of trivalent and hexavalent f-elements on phosphate addition, even in the presence of 1:1 citric acid complexes. Our planned studies of the effect of stronger, environmentally relevant ligands like humic and fulvic acids on phosphate immobilization of the actinides can be directed by the general thermodynamic validity of our current results. Figure 3c presents the speciation calculated for a pH 7.0, 0.1 M ionic strength solution containing 1×10^{-7} M europium, 5×10^{-4} M carbonate, and 0.1 mg/L Lake Bradford humic acid (*24*). All the europium is bound by humic acid when the phosphate concentration is less than 2 mM, where $EuPO_4$ begins to precipitate. The fraction of europium present as $EuPO_4$ increases with increasing phosphate concentration, and reaches about 75% at 10 mM phosphate. Consequently, it appears that phosphate immobilization can significantly reduce the potentially mobile concentrations of tracer amounts of the trivalent actinides even in the presence of humic acid.

Conclusions

We have shown that phytic acid readily hydrolyzes to produce phosphate with a projected lifetime of 100-150 years in the absence of microbiological effects, that actinide-phytate compounds are insoluble, and that europium and uranyl phytates are converted to phosphates within a month at 85 °C. Thorium solubility, on the other hand, is controlled by hydroxide or oxide species. Furthermore, the solubilities of radiotracer europium and uranyl are reduced by phosphate dosing of a simulated groundwater solution, even in the presence of citric acid. In the same systems, neptunium(V) solubility is only affected by 0.01 M phosphate at pH greater than 7. The results of these tracer-scale immobilization experiments indicate that phosphate mineral formation from representative deposits is under thermodynamic control.

Acknowledgments

Work performed under the auspices of the U.S. Department of Energy, Office of Environmental Waste Management, Efficient Separations and Processes Program under contract number W-31-109-ENG-38.

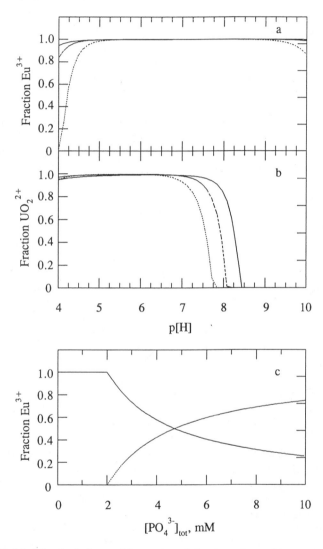

Figure 3. Model calculations of the expected fraction of metal ions precipitated as (a) $EuPO_4$ or (b) $(UO_2)_3(PO_4)_2$ with 10^{-4} M (----), 10^{-3} M (– –), and 10^{-2} M (—) total phosphate under the experimental conditions of the study, and (c) for europium in the presence of 0.1 mg/L Lake Bradford humic acid at p[H] = 7.0 and 5×10^{-4} M total carbonate; Eu(humate) (—), $EuPO_4$ (----).

Literature Cited

1. Kim, J. I. In *Handbook on the Physics and Chemistry of the Actinides*; Freeman, A. J. and Keller, C., Eds.; Elsevier: New York, NY, 1986, Vol. 4; pp. 413-455.
2. Turekian, K. K.; *Geochimica et Cosmochimica Acta* **1977**, *41*, 1139-1144.
3. Choppin, G. R.; Kobashi, A. *Marine Chem.* **1990**, *30*, 241-247.
4. Allard, B.; Kipatsi, H.; Liljenzin, J. O.; *J. Inorg. Nucl. Chem.* **1980**, *42*, 1015-1027.
5. *NIST Critical Stability Constants of Metal Complexes Database*; Martell, A. E.; Smith, R. M., Eds.; NIST Standard Reference Database 46; Department of Commerce: Gaithersburg, MD, 1993.
6. *Chemical Thermodynamics of Uranium*; Wanner, H.; Forest, I., Eds.; Chemical Thermodynamics 1; North-Holland: New York, NY, 1992.
7. Sandino, A. Ph.D. Thesis, Royal Institute of Technology, Stockholm, 1991.
8. Bernkopf, M. F.; Kim, J. I. "Hydrolysereaktionen und Karbonatkomplexierung von Dreiwertigen Americium im Natürlichen Aquatischen System"; Report RCM-02884; Institut für Radiochemie der Technische Universität München, 1984.
9. Nguyen, S. N.; Silva, R. J.; Weed, H. C.; Andrews, J. E. Jr. *J. Chem. Therm.* **1992**, *24*, 359-376.
10. Östhols, E. *Radiochim. Acta* **1995**, *68*, 185-190.
11. Fourest, B.; Baglan, N.; Guillaumont, R.; Blain, G.; Legoux, Y. *J. Alloys and Compounds* **1994**, *213/214*, 219-225.
12. Byrne, R. H.; Kim, K. H. *Geochimica et Cosmochimica Acta* **1993**, *57*, 519-526.
13. *Phytic Acid*; Graf, E., Ed.; Pilatus Press: Minneapolis, MN, 1986.
14. Horwitz, E. P.; Dietz, M. L.; Chiarizia, R.; Diamond, H.; Essling, A. M.; Graczyk, D. *Anal. Chim. Acta* **1992**, *266*, 25-37.
15. Quinlan, K. P.; DeSesa, M. A. *Anal. Chem.* **1955**, *27*, 1626-1629.
16. Hikichi, Y.; Hukuo, K., Shiokawa, *Bull. Chem. Soc. Jpn.* **1978**, *51*, 3645-3646.
17. Pekárek, V.; Benešova, M. *J. Inorg. Nucl. Chem.* **1964**, *26*, 1743-1751.
18. Tananaev, I. V.; Rozanov, I. A.; Beresnev, E. N. *Izv. Akad. Nauk SSSR, Neorg. Mat.* **1976**, *12*, 882-885.
19. Bullock, J. I.; Duffin, P. A.; Nolan, K. B. *J. Sci. Food Agric.* **1993**, *63*, 261-263.
20. Potman, W.; Lijklema, L. *Water Res.* **1983**, *17*, 411-414.
21. Phillippy, B. Q.; White, K. D.; Johnston, M. R.; Tao, S. H. *Anal. Biochem.* **1987**, *162*, 115-121.
22. Suzumura, M.; Kamatani, A. *Geochimica et Cosmochimica Acta* **1995**, *59*, 1021-1026.
23. Barten, H.; Cordfunke, E. H. P. *Thermochim. Acta* **1980**, *40*, 357-365.
24. Torres, R.; Choppin, G. R. *Radiochim. Acta* **1984**, *35*, 143-148.

ORGANIC POLLUTANT INTERACTIONS

Chapter 18

Enhancement of the Water Solubility of Organic Pollutants Such as Pyrene by Dissolved Organic Matter

Howard H. Patterson[1,2], Bruce MacDonald[1], Feng Fang[2], and Christopher Cronan[2]

[1]Department of Chemistry and [2]Department of Environmental Science and Ecology, University of Maine, Orono, ME 04469

Many factors determine the fate and transport of an organic pollutant in the environment but water solubility is certainly one of the most important. Among the environmental factors that alter the solubility of a molecule is naturally occurring dissolved organic carbon (DOC). We have hypothesized that the DOC from different sources within a watershed have different binding affinities for pollutants such as pyrene. This could lead to different rates of transport or bioavailability within the watershed. DOC samples have been isolated from a stream, adjacent wetland and nearby wooded upland sites. A fluorescence quenching method was developed to quantify the binding coefficient of the pollutants with the dissolved organic carbon. From these results a model has been constructed to determine the sites with the greatest potential to modify pollutant contamination in the environment.

Many factors determine the fate and transport of an organic pollutant in the environment but water solubility is certainly one of the most important physical properties of a molecule. While the solubility of a compound in pure water is an important guide to its overall fate, predicting the pollutant's solubility in the environment - a stream, an agricultural field, or an aquifer - is far more involved. Among the environmental factors that alter the solubility of a molecule are naturally occurring humus or organic matter. Typically, organic matter ranges in size from partially decayed fragments of plant material to particulate matter, colloids, and simple water soluble molecules. The water soluble organic carbon found in the soil is analogous to the dissolved organic matter found in an aquatic or marine environment. These two fractions will be the focus here and the more general term dissolved organic matter, DOC, will be used to refer to both sources of organic matter.

Researchers have applied a variety of techniques and probe molecules to DOC to assess the binding affinity of DOC to the probe. Carter and Suffet (1) sealed humic acid solutions in dialysis bags and then allowed the bags to equilibrate in solutions of radiolabled DDT. A DOC concentration of approximately 10 mg/L bound about 40%

0097–6156/96/0651–0288$15.00/0
© 1996 American Chemical Society

of the total DDT. Chiou et. al. (2) employed a hexane solvent extraction method to determine the quantity of free pesticide in the presence of DOC. They found the higher molecular weight DOC with lower oxygen content (therefore less polar) increased the apparent solubility of the probe molecule. Wang et. al. (3) observed the binding of the polar herbicide atrazine with a fulvic acid. The bound and unbound atrazine were separated by ultrafiltration, followed by liquid chromatography to quantify the free atrazine concentration. Here also it was found that most of the binding capacity was due to the high molecular weight fraction.

In another approach Gauthier et. al. (3) developed a technique based on fluorescence quenching to determine the equilibrium binding constants of polycyclic aromatic hydrocarbons (PAH) with DOC. PAH are efficient fluorophores, nonionic, and are usually considered to be insoluble in water. These last two properties thermodynamically drive the PAH from the aqueous phase to the less polar DOC. Also, PAHs are important examples of hydrophobic organic contaminants in the environment because these properties cause them to accumulate in the lipid deposits of higher organisms.

Herbert, Bertsch et al. (4) conducted quenching studies utilizing DOC from soils with pyrene for a probe. The DOC was divided into molecular weight fractions by ultrafiltration. Their results indicated that the more hydrophobic, high molecular weight fraction was responsible for the major portion of the probe binding. Backus and Geschwend (5) studied the quenching of perylene by organic matter and calculated that the organic matter they use could potentially double the total amount of contaminant transported.

The previous literature studies referenced above document the potential for DOC to affect the fate of an organic contaminant in the environment. In our research we have hypothesized that the DOC from different sources within a watershed have different binding affinities for pyrene. This could lead to different rates of pyrene transport or bioavailability within the watershed. DOC was isolated from a stream, an adjacent wetland, and nearly wooded upland sites. A fluorescence quenching method was utilized to determine the binding coefficient of pyrene with the different DOC samples.

Experimental

Sample Collection and Workup. Samples were collected from a watershed in the Penobscot Experimental Forest, Bradley, Maine. Organic horizon material was collected from a deciduous and coniferous site. Approximately 12 L of water was taken from the Blackman stream. Finally, a wetland sample was taken from a sedge marsh bordering the stream. Holes 20 cm wide were dug which promptly filled with sediment laden water which was collected for analysis.

The samples were worked up as in Figure 1. The Blackman Stream sample was clean enough for immediate filtration. Gelman A/E glass fiber filters, 47 mm diameter, were used with vacuum filtration followed by Rainin 0.45 μm, 47 mm diameter Nylon-66 filters. The sedge marsh sample required centrifugation at 8,000 rpm for 20 minutes to remove the coarse organic particulates. The supernatant was then passed through two stacked glass fiber filters followed by 0.45 μm filtration.

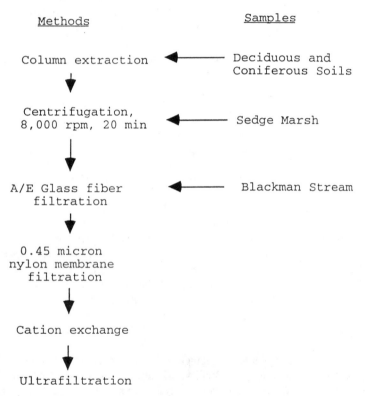

Figure 1: Summary of the steps used in the isolation of DOC from the two upland samples (Deciduous and Coniferous), a wetland (sedge marsh) and an aquatic site (Blackman Stream).

The coniferous and deciduous samples were first packed in columns and then leached with water. Two columns were set up for each sample. Between 550 to 800 grams of organic horizon material were packed to a net height of 100 cm in each column and then the columns were filled with enough distilled deionized water (1.7 to 2.0 L) to cover the samples. Columns were equilibrated for 5 h, after which they were drained at a rate of less than 5 ml/min. The slow drainage rate was to limit the colloidal material in the leachate. Leachates were dark brown in color with a combined final volume for each sample of 2 to 2.5 L. After all samples had passed through the 0.45 μm filtration, subsamples were analyzed for metals by ICP spectroscopy. Samples were passed through cation exchange columns to remove metals.

It should be pointed out that in the experiments reported herein, we have attempted to limit the colloidal material. We recognize that the potential impact of colloidal material on these studies is an important area of research to pursue.

Pyrene Solution Preparation.   Pyrene was purchased from Aldrich Chemical Company and used directly. A phosphate buffer, pH 6 upon dilution was used (6). For the fluorescence quenching experiments an aqueous pyrene stock solution was prepared from a 2.05×10^{-4} solution of pyrene in methanol. 1.5 ml of the methanolic pyrene solution was mixed with 498.5 ml of pH 6 buffer water to give a concentration of 6.15×10^{-7} M aqueous pyrene. The pyrene fluorescence intensity tended to be more stable if the solution were allowed to mix 1 hour before use.

Fluorescence Quenching Experiments.   The binding affinity of pyrene with DOC samples was measured by fluorescence quenching using a Perkin Elmer MPF-4 spectrafluorimeter. The excitation wavelength was 334.5 nm with the emission maximum at 374 nm. For these experiments a series of DOC stock solutions at 7 different concentrations was prepared from each DOC source. An aliquot was placed in a cuvette and the absorbance and DOC fluorescence measured. Next, an aliquot of aqueous pyrene was added to the cuvette for the quenching experiment and the fluorescence intensity recorded at three and four minutes after mixing. If the two measurements differed a third measurement was made and the two closer measurements retained. We have found that the fluorescence intensity did not change to an appreciable amount (<1%) after three minutes. Thus, we assumed that the system is at equilibrium after three minutes. After collection of the experimental data corrections were made for dilution effects, subtraction of background fluorescence and correction for inner filter effects (8).

Results and Discussion

Fluorescence Quenching Model.   Fluorescence quenching is sometimes broadly defined as a process which reduces the fluorescence intensity. There are at least three ways this may occur. First, by complexation the concentration of the fluorophore may be reduced. This is called static quenching. Second, the lifetime of the fluorophore may be reduced. This is collisional or dynamic quenching. Third, as the total absorbance of the solution is increased the intensity of the excitation is significantly reduced over the pathlength. This is termed the inner filtering effect.

The process of static quenching provides a method for the quantitative determination of the binding of a PAH to DOC. In aqueous solution the PAH may bind to the DOC present with the equilibrium.

$$PAH + DOC \Leftrightarrow PAH\text{-}DOC \tag{1}$$

and

$$K_b = \frac{[PAH - DOC]}{[PAH][DOC]} \tag{2}$$

Also,

$$K_b = \frac{P_b}{P_f[DOC]} \tag{3}$$

where P_b and P_f refer to the bound and free forms of the PAH. It should be pointed out that the equilibrium constant defined in equation (2) is not a thermodynamically rigorous equilibrium constant but rather a "device" for comparison of the binding ability of "DOC" for pyrene. As stated earlier, DOC represents a wide variety of organic molecules and not a single molecular species. The formulation of equation (3) presumes that $[DOC] \gg [PAH]$.

If we write the mass balance equation with P_t equal to the total concentration of the PAH we have (ignoring loss due to absorption on the vessel walls)

$$P_t = P_f + P_b \tag{4}$$

solving equation 3 for P_b and substituting into equation 4 gives:

$$P_t = P_f + P_f K_b[DOC] \tag{5}$$

We can divide this through by P_t. With no DOC present to quench fluorescence, the fluorescence intensity, F_0, is proportional to P_t. F is defined as the fluorescence intensity in the presence of DOC with a concentration denoted by [DOC]. F is proportional to P_f, so

$$\frac{P_t}{P_f} = \frac{F_o}{F} = 1 + K_b[DOC] \tag{6}$$

Thus, a plot of F_0/F versus [DOC] should be linear with a slope equal to the binding constant, K_b, and an intercept equal to 1. Equation 6 is the Stern-Volmer equation in which the formation of a nonfluorescent complex has reduced the effective concentration of the fluorophore.

Fluorescence Data Analysis and Inner Filter Corrections. Figure 2 shows the fluorescence intensity for F_{obs} versus the concentration of the DOC in ppm for the four samples. The F_{obs} plots show that increasing DOC concentrations leads to a decrease

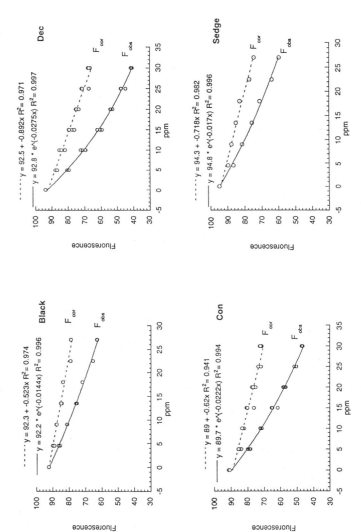

Figure 2: Observed fluorescence, Fobs, and corrected fluorescence, Fcor, versus ppm DOC for the aquatic site sample (Blackman stream -- **Black**), the two upland samples (Coniferous -- **Con**, and Deciduous -- **Dec**), and the wetland sample (Sedge marsh -- **Sedge**).

in F_{obs}. This is to be expected on the basis of the quenching action of the DOC but also may be related to inner filter effects.

A troublesome problem in dealing with fluorescence quenching by DOC is the inner filter effect. DOC solutions can absorb light strongly leading to two problems. First, F_0 can decrease significantly as it crosses the cuvette. There is a decreasing density of photons for the fluorophore to absorb and the fluorescence intensity is no longer proportional to the fluorophore concentration. This attenuation of the excitation intensity is primary inner filtering. Secondly, the photons emitted via fluorescence can also be absorbed by the DOC chromophores, reducing the number of photons the emission photomultiplier receives. Also, in this case fluorescence intensity is no longer proportional to fluorophore concentration. This attenuation of the emission intensity is secondary inner filtering. To properly interpret the decrease in fluorescence intensity of a DOC quenching experiment, the effects of inner filtering must be properly corrected.

Equation 7 below gives a correction factor which has been used to correct for inner filter effects. This equation has been discussed elsewhere (7)

$$ cf = \frac{2.3A_{ex}\Delta x 10^{A_{ex}x_1}}{1-10^{-A_{ex}\Delta x}} \cdot \frac{2.3A_{em}\Delta y 10^{A_{em}y_1}}{1-10^{-A_{em}\Delta y}} \tag{7} $$

The first and second factors correspond to the primary inner filter and secondary inner filter correction factors, respectively. For Δx and Δy it was decided to use a value of 0.12 cm and have x_1 and y_1 equal to 0.3 cm.

Figure 3 shows plots of absorbance measured at 334 and 374 nm versus the concentration of the DOC in ppm for the four samples. This data is used in the correction factor in equation 6. The coniferous and deciduous samples had the largest slopes (absorptivities). Also inspection of the slopes shows that the absorptivities at 334 nm are about twice as large as those at 374 nm for all the samples.

Figure 4 shows Stern-Volmer plots for the four samples showing F_0/F plotted against the DOC concentrations. Also, a Stern-Volmer plot of deciduous sample before correction for inner filter effects is shown. This indicates quite clearly the nonlinearity of the Stern-Volmer plots before correction for inner filter effects. Table I is a summary of the K_b values obtained for the four samples.

Table I. Average binding constants, K_b, and standard deviations for the two upland samples (Deciduous and Coniferous), a wetland (sedge marsh) and an aquatic site (Blackman Stream).

	Mean K (L/Kg)	Standard Deviation	[DOC] (ppm)	% P_b
Blackman	6.76×10^3	8.5×10^2	8	5.13
Coniferous	1.51×10^4	7.1×10^3	178	72.9
Deciduous	1.55×10^4	2.7×10^3	209	76.4
Sedge marsh	9.30×10^3	5.3×10^2	31	22.4

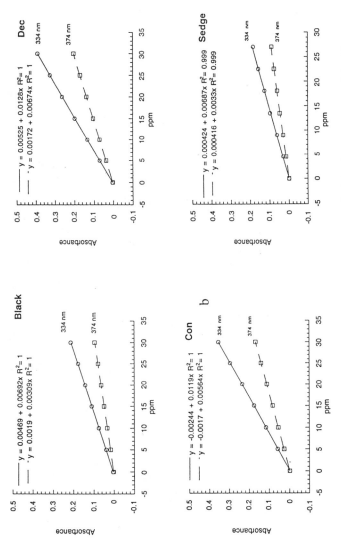

Figure 3: Absorbance versus ppm for the aquatic site sample (Blackman stream -- **Black**), the two upland samples (Coniferous -- **Con**, and Deciduous -- **Dec**), and the wetland sample (Sedge marsh -- **Sedge**). Slope of regression line is the absorptivity, ε, per ppm of DOC.

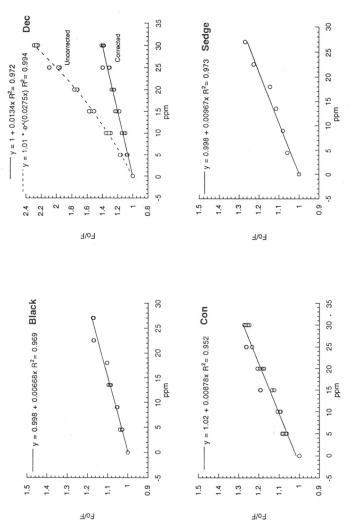

Figure 4: Stern-Volmer plots with correction for inner filter effects for the aquatic site sample (Blackman stream -- **Black**), the two upland samples (Coniferous -- **Con**, and Deciduous -- **Dec**), and the wetland sample (Sedge marsh -- **Sedge**). The plot for deciduous sample also shows the curve fitting without correction for inner filter effects.

Environmental Implications of the Equilibrium Binding Constants. Let us consider the fate of pyrene only from the standpoint of the role of DOC in the aquatic and upland environments. We can rearrange equation 3 to give

$$P_b = P_f K_b [DOC] \tag{8}$$

Note that the amount of pyrene bound is directly proportional to the concentration of free pyrene, K_b, and the concentration of DOC in the environment. Using the relation $P_t = P_f + P_b$ and substituting this into equation 8 gives

$$\frac{P_b}{P_t} = \frac{K_b[DOC]}{1 + K_b[DOC]} \tag{9}$$

We see that the fraction of bound pyrene to total pyrene depends on the magnitude of K_b for pyrene and the DOC concentration. As this product approaches values much greater than 1, the percentage of bound pyrene approaches 100%.

There is no rigorous way to compare the DOC concentrations between an aquatic or marsh environment and an upland environment. However one rough method of comparison is to use the concentration of DOC leached from the soil columns. With this as a basis, the Deciduous sample contains 26 times as much DOC as the Blackman Stream sample. Furthermore, the upland samples have the highest K_b values. These facts suggest that the upland environments will bind greater percentages of pyrene. Table I contains calculations based on equation 9 for the percentage of pyrene bound in each environment. For these samples, we see that the amount of DOC in the environment plays a greater role than the value of K_b.

It should be pointed out that another very important soil component capable of pyrene transport is colloidal organic matter (9). In our studies discussed herein the colloidal fraction was removed from the collected samples by filtration because colloidal materials may interfere with fluorescence and absorption analyses. For the sedge marsh wetlands sample, which had a binding capacity of 9.3×10^3 L/Kg, and a DOC concentration of 24.5 ppm, the colloidal fraction might be particularly important. A marsh soil wetlands sample is probably entirely organic matter. The material was very difficult to process through the initial filtration in part due to colloidal matter. Considering the binding capacities, DOC concentrations and the amount of colloidal material, the marsh environment may have the capacity to move quantities of pyrene as water moves through it. This is an area of investigation we plan to pursue in the future.

Summary

The method of fluorescence quenching has been applied to DOC isolated from a stream, wetlands marsh, as well as coniferous and deciduous sites in a local watershed. Uncertainties in the inner filter correction parameters and instability of an aqueous pyrene stock solution introduced some variability into the final binding affinity values. The coniferous and deciduous sites had the largest binding affinities and the greatest amount of DOC per unit volume. The stream and marsh samples had the lower binding affinities and lower DOC concentrations. This implies the wooded sites have the greatest potential to modify the fate of pyrene contamination in the environment. It should be pointed out, though, that colloidal organic matter was filtered away and this component may be of major importance for wetland sites; thus, further studies should be made to deduce the importance of colloidal matter in wetland sites.

Acknowledgments

We wish to acknowledge the financial support of the University of Maine Water Resource Program of the United States Geological Survey at the Department of the Interior (Grant No. 14-08-001-G2023)

References

(1) Carter, C. W.; Suffet, I. H. *Binding of DDT to Dissolved Humics* Environmental Science and Technology 1982, 16: p. 735-740.
(2) Chiou, C. T.; Malcolm, R. L.; Brinton, T. I.; and Kile, D. E. *Water Solubility Enhancement of Some Organic Pollutants and Pesticides by Dissolved Humic and Fulvic Acids* Environmental Science and Technology, 1986, 20: p. 502-508.
(3) Gauthier, T.D.; Shane, E. C.; Guerin, W. F.; Seitz, W. R.; and Grant, C. L. *Fluorescence Quenching Method for Determining Equilibrium Constants for Polycyclic Aromatic Hydrocarbons Binding to Dissolved Humic Materials* Environmental Science and Technology, 1986, 20: p. 1162-1186.
(4) Herbert, B. E.; Bertsch, P. M.; and Novak, J. M. *Pyrene Sorption by Water-Soluble Organic Carbon* Environmental Science and Technology, 1993, 27: p. 398-403.
(5) Backus, D. A.; Geshwend, P. M. *Fluorescence Polycyclic Aromatic Hydrocarbons as Probes for Studying the Impact of Colloids on Pollutant Transport in Groundwater* Environmental Science and Technology 1990, 24: p. 1214-1223.
(6) *Handbook of Chemistry and Physics (*66th Edition) pg. D-145.
(7) MacDonald, B. C.; Lvin, S. J., and Patterson, H. H. *Correction of Fluorescence Inner Filter Effects and the Partitioning of Pyrene to Dissolved Organic Carbon* Submitted to Analy. Chimica Acta, 1996.

Chapter 19

Covalent Binding of Aniline to Humic Substances

Comparison of Nucleophilic Addition, Enzyme-, and Metal-Catalyzed Reactions by ^{15}N NMR

Kevin A. Thorn[1], W. S. Goldenberg[1,2,4], S. J. Younger[1,2,5], and E. J. Weber[3]

[1]U.S. Geological Survey, Mail Stop 408, 5293 Ward Road, Arvada, CO 80002
[2]Department of Chemistry, University of Colorado, Boulder, CO 80309
[3]U.S. Environmental Protection Agency, 960 College Station Road, Athens, GA 30605

The covalent binding of ^{15}N-labelled aniline, in the presence and absence of catalysis by horseradish peroxidase and birnessite, to the fulvic and humic acids isolated from the IHSS Elliot silt loam soil, has been examined by a combination of liquid and solid state ^{15}N NMR. In the absence of catalysts, aniline undergoes nucleophilic addition reactions with the carbonyl functionality of the fulvic and humic acids and becomes incorporated in the form of anilinohydroquinone, anilinoquinone, anilide, heterocyclic, and imine nitrogens. In the presence of peroxidase and birnessite, aniline undergoes free radical coupling reactions together with nucleophilic addition reactions with the fulvic and humic acids. Among the condensation products unique to the catalyzed reactions are azobenzene nitrogens, iminodiphenoquinone nitrogens, and nitrogens tentatively assigned as imidazole, oxazole, pyrazole, or nitrile. The incorporation of aniline into the organic matter of the whole Elliot silt loam soil and the IHSS Pahokee peat most closely resembles the noncatalyzed nucleophilic addition reactions, as determined by solid state ^{15}N NMR.

Aniline is the parent compound of the aromatic amines, which are used in the synthesis of agrochemicals, dyes, and pharmaceuticals. There is a concern that aromatic amines may be released into the environment during production processes or incomplete treatment of industrial waste streams. Additionally, aromatic amines can enter the environment from the reduction of azo dyes, polynitroaromatic munitions (e.g. TNT), and dinitro herbicides, and from hydrolytic degradation of several classes of pesticides, including the phenylurea, phenylcarbamate, and acylanilide herbicides.

[4]Current address: Department of Chemistry, University of Wisconsin, Madison, WI 53706
[5] Current address: Medical College of Wisconsin, 8701 Watertown Plank Road, Milwaukee, WI 53226–0509

One possible fate of aromatic amines in the environment is covalent binding to naturally occurring organic matter, primarily humic substances, in soils, sediments, and natural waters. Covalent binding is thought to occur through nucleophilic addition of the amine group with the carbonyl functionality of the humic substances, through phenoloxidase enzyme or metal catalyzed reactions between aromatic amines and humic substances, or a combination of all three (1-9).

In our previous reports, we measured the kinetics of the covalent binding of aniline to soil and aquatic fulvic and humic acids from the IHSS, and used ^{15}N NMR to follow the incorporation of ^{15}N-labelled aniline into these samples in the absence of catalysis by enzymes or metals (8,9). In this study, we use ^{15}N NMR to examine the effects of catalysis by peroxidase and birnessite on the binding of aniline to the soil fulvic and humic acids, and then follow the incorporation of aniline into the organic matter of the whole soil and the IHSS Pahokee peat using solid state NMR. The overall objective of this work is to gain a better understanding of the reaction pathways of aromatic amines in soils, waters, and at the sediment-water interface, and to determine what consequences the binding mechanisms have for the long term immobilization of the amines by organic matter.

Background

Reaction of Aniline with Organic Functional Groups. In the absence of catalysis by enzymes or metals, aniline undergoes nucleophilic addition reactions to quinone and other carbonyl groups in humic substances to form both heterocyclic and nonheterocyclic condensation products (9). In aqueous solution, aniline undergoes 1,4-addition (Michael addition) to both 1,2- and 1,4-quinones (10-14). The reaction of aniline with 1,4-benzoquinone, from the oxidation of hydroquinone, and with 4-methyl-1,2-quinone, from the oxidation of 4-methylcatechol, are illustrated here.

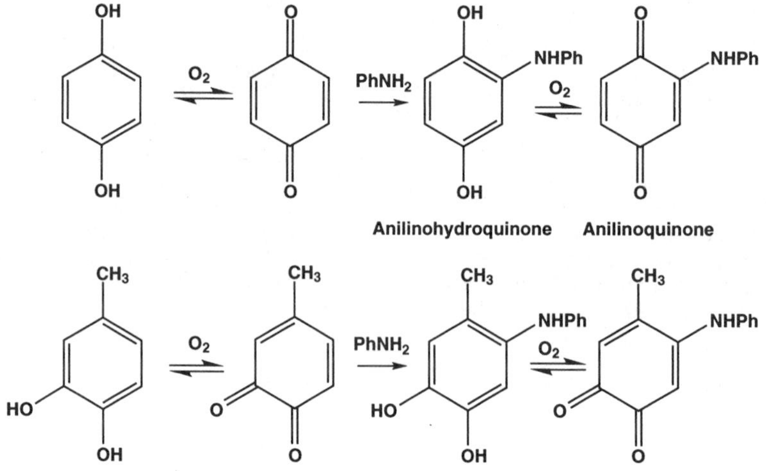

Anilinohydroquinone   Anilinoquinone

The 1,2-addition of aniline with quinones or ketones to form imines (Schiff bases) is less favorable in aqueous solution than in organic solvent, because the overall equilibrium favors hydrolysis in aqueous or partially aqueous solvents. In the case of sterically hindered quinones in aqueous solution, however, 1,2-addition by aniline is more favorable, and in some instances becomes the dominant mode of attack. As an illustration, 4-methylaniline was reported to undergo both 1,4- and 1,2-addition to 2,6-dimethyl-p-benzoquinone, resulting in a 3:1 product ratio of anilinoquinone to imine (15).

Aniline undergoes aminolysis reactions with esters and amides to form anilides.

Some examples of the reaction of aniline with other combinations of carbonyl groups to form N-heterocycles, such as N-phenyl indoles, have been discussed previously (9). ^{15}N NMR chemical shifts of these and other classes of nitrogen compounds relevant to this study are presented in Figure 1.

The phenoloxidase enzymes (peroxidases, tyrosinases, and laccases) are known to occur extracellularly in soils (16,17), and have been considered to play a role in catalyzing the coupling of aromatic amines to humic substances (3). Peroxidases have been studied most extensively. They use hydrogen peroxide to promote the one electron oxidation of aromatic amines and of phenolic moieties within humic substances to form free radicals. (12, 18).

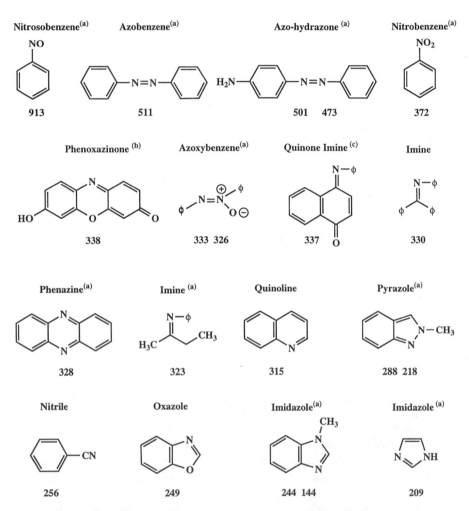

Figure 1. ^{15}N NMR Chemical Shifts of Nitrogen Compounds Representing Condensation Products of Aniline with Carbonyl and Other Functional Groups. Chemical Shifts determined in DMSO-d$_6$ in this lab unless otherwise noted. (a) From references 37,46,47 and 48. (b) Determined in solid state. (c) Tentative assignment.

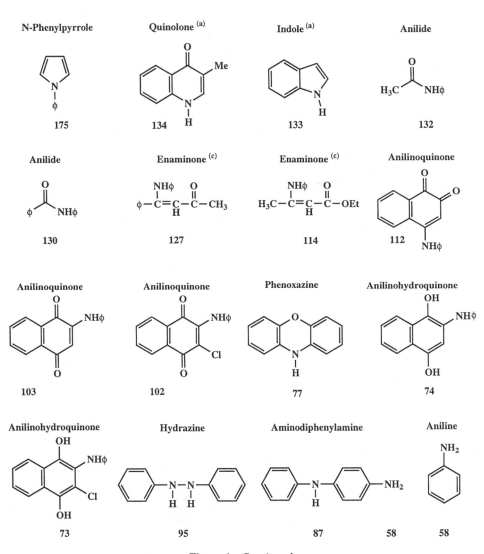

Figure 1. *Continued*

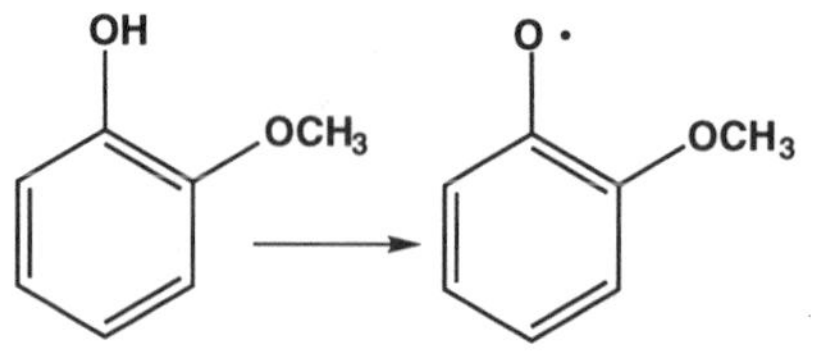

Thus, peroxidases may potentially effect direct free radical coupling reactions between aniline and humic substances or create additional substrate sites within the fulvic or humic acid molecules for nucleophilic addition by aniline. A model for the latter pathway can be found in the work with guaiacol and 4-chloroaniline by Simmons et al. (19). Peroxidase catalyzed the coupling of guaiacol, itself not a substrate for nucleophilic addition, into the extended quinonoid dimer which subsequently underwent nucleophilic attack by the chloroaniline:

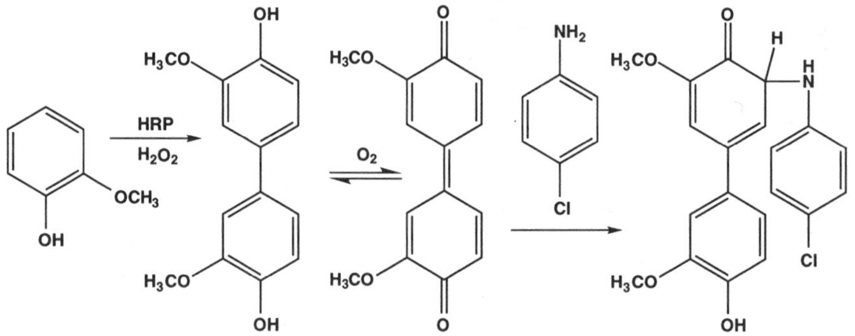

One can speculate that peroxidase may effect inter- or intramolecular free radical coupling reactions between phenolic moieties in humic molecules to form quinone groups. Also included in this latter pathway of creating additional substrate sites would be the action of tyrosinases in converting monophenolic groups into catechol groups (12). The ability of peroxidase to catalyze the coupling of 2,4-dichlorophenol, the oxidation product of 2,4-dichloroaniline, to a peat humic acid was recently confirmed by Hatcher et al. (20) using ^{13}C NMR.

Manganese dioxides operate in a manner similar to the phenoloxidase enzymes by promoting the one electron oxidation of anilines and phenols with the concomitant reductive dissolution of the manganese dioxide (Mn IV) to soluble manganese (Mn II) (21). In their work with guaiacol and 4-chloroaniline, Simmons et al. (19) reported that peroxidase, laccase, tyrosinase, and manganese dioxide all catalyzed formation of the same five co-oligomers as initial reaction products.

Experimental

Samples and Reagents. The Elliot Silt Loam Soil (Joliet, Ill.; mollic horizon; 2.92 % organic C) and Pahokee peat (Ocachobee, Fla.; 44.48% organic C) were purchased from the International Humic Substances Society (IHSS). The reference soil fulvic and humic acids, isolated from the Elliot silt loam soil, were also purchased from the IHSS. Aniline, 99 atom % ^{15}N, was purchased from ISOTEC. (Use of trade names in this report is for identification purposes only and does not constitute endorsement by the U.S. Geological Survey or U.S. Environmental Protection Agency.) Horseradish peroxidase (EC1.11.1.7; 53 purpurogallin units/mg solid) was purchased from Sigma. Birnessite was prepared by the method of McKenzie (22).

Reaction of Fulvic and Humic Acids with $\emptyset^{15}NH_2$. The H^+-saturated fulvic or humic acids (400-500 mg) were added to 150-200 ml of H_2O, dissolved by raising the pH to 6 with 1 N NaOH, and charged with 200 μl of $\emptyset^{15}NH_2$. The reaction solutions were stirred at room temperature for 4-5 days. The samples were then resaturated with H^+ by passing the solutions through a Dowex MSC-1 cation exchange resin, and freeze dried. The freeze dried powders were redissolved in 2-3 ml DMSO-d_6 for NMR analysis.

Reaction of Fulvic and Humic Acids with $\emptyset^{15}NH_2$ in the Presence of Peroxidase. The IHSS soil fulvic acid (400 mg) was dissolved in 500 ml of H_2O and titrated to pH 6 with 1 N NaOH. Fifty mg of horseradish peroxidase, 200 μl of $\emptyset^{15}NH_2$, and 5 ml of H_2O_2 were then added to the solution. The reaction was allowed to proceed for 48 hours. The reaction solution was then passed through the H^+-saturated MSC-1 exchange resin and freeze dried. The enzyme exchanged on the resin and was thereby removed from the fulvic acid solution. The humic acid was reacted similarly: 800 mg humic acid; 400 μl $\emptyset^{15}NH_2$; 100 mg peroxidase; 11 ml H_2O_2; 3 days. Approximately 400 mg of the final humic acid preparation was dissolved in 2 ml DMSO-d_6 for NMR analysis.

Reaction Blank of $\emptyset^{15}NH_2$ and Peroxidase. Twenty five mg peroxidase, 100 μl of $\emptyset^{15}NH_2$, and 3 ml of H_2O_2 were added sequentially to 100 ml of pH 6 phosphate buffer, and stirred for 6.5 hours. The reaction was then quenched by acidifying to pH 2 with acetic acid. The colored reaction products were sorbed onto a C_{18} solid phase extraction cartridge, rinsed with distilled deionized water, and then eluted with methanol. The methanol was removed using rotary evaporation and the reaction products redissolved in 2 ml of DMSO-d_6 for NMR analysis.

Reaction of Humic Acid with $\emptyset^{15}NH_2$ in Presence of Birnessite. Four hundred mg of the soil humic acid was added to 200 ml of H_2O and adjusted to pH 6 with 1N NaOH. Two hundred μl of $\emptyset^{15}NH_2$ and 50 mg of birnessite were added and the reaction mixture stirred for 6 days. The reaction solution was then dialyzed, filtered through a sintered glass funnel, H^+- saturated on the cation exchange resin,

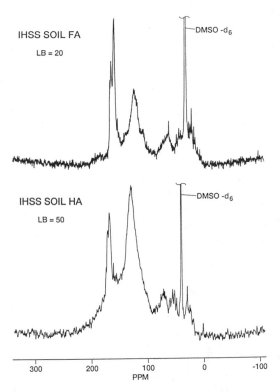

Figure 2. Quantitative Liquid Phase ^{13}C NMR Spectra of Unreacted IHSS Soil Fulvic and Humic Acids. LB=line broadening in Hz.

and freeze dried. The sample was then redissolved in DMSO-d_6 for NMR analysis. Enough dissolved manganese (Mn II) remained in the final humic acid preparation that no addition of paramagnetic relaxation reagent was necessary for the recording of the quantitative liquid phase ^{15}N NMR.

NMR Spectrometry. Liquid phase ^{13}C and ^{15}N NMR spectra were recorded on a Varian XL300 NMR spectrometer at carbon and nitrogen resonant frequencies of 75.4 and 30.4 MHz, respectively, using a 10 mm broadband probe. Quantitative ^{13}C NMR spectra of the unreacted fulvic and humic acid samples were recorded in DMSO-d_6, 99.9 atom % [12]C, as previously described (23). INEPT (24) and ACOUSTIC (25) ^{15}N NMR spectra were recorded on the aniline-reacted fulvic and humic acids. Refocussed INEPT (proton decoupled) spectra were recorded as previously described (9). ACOUSTIC spectra, with the exception of the birnessite catalyzed sample, were recorded with the use of paramagnetic relaxation reagent (100-200 mg chromium (III) acetylacetonate). Acquisition parameters included an 18,656.7 Hz spectral window (613.7 ppm), 0.5-s acquisition time, 45° pulse angle, 2.0-s pulse delay, and τ delay of 0.1 ms. Neat formamide in a 5 mm NMR tube, assumed to be 112.4 ppm, was used as an external reference standard for all spectra. ^{15}N NMR chemical shifts are reported in ppm downfield of ammonia, taken as 0.0 ppm.

Solid state CP/MAS (cross polarization/magic angle spinning) spectra were recorded on a Chemagnetics CMX-200 NMR spectrometer at carbon and nitrogen resonant frequencies of 50.3 and 20.3 MHz, respectively, using a 7.5 mm ceramic probe (zirconium pencil rotors). Acquisition parameters for the ^{13}C NMR spectra of the unreacted IHSS soil and peat included a 30,000 Hz spectral window, 17.051-ms acquisition time, 2.0-ms contact time, 1.0-s pulse delay, and spinning rate of 5 KHz. Spectra were referenced to hexamethylbenzene. Acquisition parameters for the ^{15}N NMR spectra of the unreacted soil fulvic and humic acids, aniline-reacted soil humic acid (noncatalyzed), and aniline reacted IHSS soil and peat included a 30,000 Hz spectral window, 17.051-ms acquisition time, 2.0-ms contact time, 1.0-s pulse delay, and spinning rate of 3.5 KHz. Chemical shifts were referenced to glycine, taken as 32.6 ppm.

Results and Discussion

^{13}C NMR Spectra of Unreacted Samples. Quantitative liquid phase ^{13}C NMR spectra of the unreacted samples are shown in Figure 2. Peak areas of the spectra are listed in Table I together with elemental analyses. Characteristically, the humic acid has a greater aromatic carbon and lesser carboxylic acid carbon content than the fulvic acid. The naturally occurring nitrogen contents are 2.68% and 4.18% for the fulvic and humic acids, respectively. Overlap of functional groups which may serve as substrate sites for nucleophilic addition by aniline occurs within the major peak areas of the spectra. Quinone carbons (190 to 178 ppm) overlap with ketone carbons from 220 to 189 ppm, amides and esters (174 to 164 ppm) overlap with

Table I. Elemental composition and peak areas for quantitative ^{13}C NMR spectra of fulvic and humic acids.

Sample[a]	C	H	O	N	S	P	Ash
IHSS Soil FA	50.5	4.01	42.6	2.68	0.62	0.05	0.79
IHSS Soil HA	58.0	3.78	33.7	4.18	0.41	0.32	0.90

Sample[b]	220-180 ppm	180-160 ppm	160-90 ppm	90-60 ppm	60-0 ppm
IHSS Soil FA	3	25	42	12	19
IHSS Soil HA	7	16	56	10	13

[a] Elemental analyses reported on ash- and moisture-free basis

[b] Peak areas, as percent of total carbon, measured by electronic integration

carboxylic acid carbons from 175 to 159 ppm, and phenolic carbons, including hydroquinone and catechol moieties, overlap with other substituted aromatic carbons in the range from approximately 165 to 135 ppm. The application of ^{13}C NMR subspectral editing techniques to these samples indicates that aldehyde carbons, which would occur from approximately 202 to 192 ppm, are not present (23,26). This is significant in view of the fact that some researchers have considered aldehydes to be part of the carbonyl functionality in humic substances that reacts with aromatic amines (27). It is possible that latent aldehydes, e.g. aldose sugars which are free to convert from the cyclic (hemiacetal) to open chain form under the conditions of the reaction, are present in the samples and may condense with the aniline.

The solid state CP/MAS ^{13}C NMR spectra of the Elliot Silt Loam soil and Pahokee peat (Figure 3) are similar to other spectra of whole soils and peats published in the literature (28). Although it cannot be directly compared to the liquid phase spectra of the fulvic and humic acids because of uncertainties in the quantitative accuracy of the CP/MAS experiment, the solid state spectrum of the whole soil clearly shows that the large amount of carbohydrate material (~74 ppm) is the major compositional difference between it and the isolated fulvic and humic acids. One can speculate, therefore, that condensation with the reducing groups of the nonhumic carbohydrates is an additional pathway potentially available to aniline in the whole soil.

Regarding quantitation in the CP/MAS experiment, for peak areas to accurately represent the number of nuclei resonating, one of the conditions that must be met is that the time constant for cross polarization must be significantly less than the time constant for proton spin lattice relaxation in the rotating frame, T_{CH} or $T_{NH} \ll T_{1\rho}H$. Other factors affecting quantitation in CP/MAS have been discussed in several reviews (28-33). Since no analyses of the spin dynamics were performed in this study, the solid state spectra presented in this manuscript will be interpreted only semiquantitatively.

^{15}N NMR Spectra of Unreacted Fulvic and Humic Acids. The naturally occurring nitrogens in the fulvic and humic acids were examined both to identify functional groups which might be involved in the reactions with aniline, and to determine the potential for overlap between the naturally occurring nitrogens and the labelled aniline nitrogens incorporated into the samples. Naturally abundant ^{15}N nuclei were not observed in liquid phase ACOUSTIC or INEPT ^{15}N NMR spectra recorded on the unreacted samples at concentrations and numbers of transients comparable to the reacted samples. Therefore, the nitrogens observed in the liquid phase spectra of the reacted fulvic and humic acids represent only the labelled aniline nitrogens incorporated into the samples. Naturally abundant nitrogens are observed in the solid state CP/MAS ^{15}N NMR spectra of the unreacted fulvic and humic acids shown in Figure 4. Spectra of naturally abundant nitrogens in soil humic substances have only recently begun to appear in the literature (34-36). The spectra shown in Figure 4 are similar to those reported by

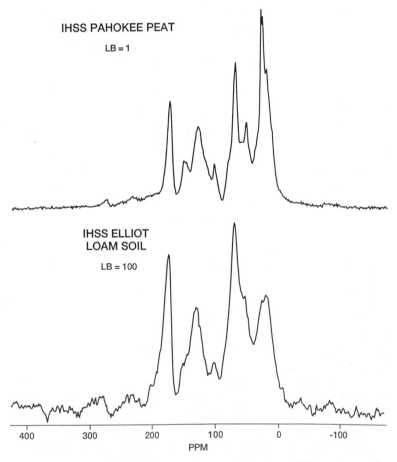

Figure 3. Solid State CP/MAS ^{13}C NMR Spectra of Unreacted IHSS Elliot Silt Loam Soil and IHSS Pahokee Peat. LB=line broadening in Hz.

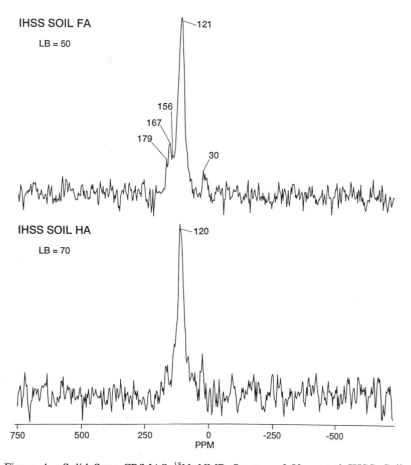

Figure 4. Solid State CP/MAS ^{15}N NMR Spectra of Unreacted IHSS Soil Fulvic and Humic Acids Showing Naturally Abundant Nitrogens. LB=line broadening in Hz.

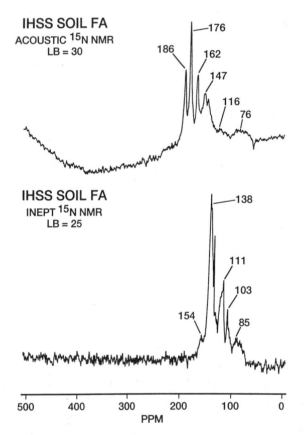

Figure 5. Liquid Phase ACOUSTIC and INEPT ^{15}N NMR Spectra of IHSS Soil Fulvic Acid Reacted with ^{15}N-labelled Aniline. (No catalyst).

Knicker et al. (34,35) in several respects. First, amide nitrogens (2° amides of amino acids involved in peptide linkages) comprise the major peaks centered at ~ 120 ppm. Secondly, the fulvic acid contains more free amino nitrogen (peak centered at 30 ppm; amino sugars and free amino groups of amino acids) than the humic acid. Thirdly, no resonances are observed downfield of approximately 185 ppm. In other words, imine, pyridine, and other sp^2 hybridized nitrogens involved in heterocyclic linkages are not present. With regard to the latter observation, however, it is important to note that some classes of nitrogens (e.g. porphyrins, phthalocyanins, and related enamino-imino systems) may not be observed in the CP/MAS experiment because of problems of molecular motion or chemical exchange at room temperature, and that low temperature experiments may be necessary for detection (37,38). One feature which distinguishes the IHSS fulvic acid spectrum from previously published spectra (34,35) is the increased resolution in the downfield shoulder of the amide peak. These nitrogens, from approximately 135 to 185 ppm, with peak maxima at 156, 167, and 179 ppm, may include for example heterocyclic sp^3 hybridized nitrogens such as indoles, pyrroles, and the imide or lactam nitrogens of nucleosides. One possible explanation for the enhanced detail in the spectrum may lie within the more extensive extraction procedure employed by the IHSS for the isolation of the fulvic acid, including processing on XAD-8 resin. Removal of free proteinaceous material from the fulvic acid extract through processing on the resin for example could serve to "unmask" the heterocyclic nitrogens comprising the downfield shoulder of the amide peak.

Although the signal to noise ratio of the naturally abundant nitrogen peaks is quite good, it is unlikely these nitrogens distort the intensities of the labelled nitrogen peaks in the solid state spectra of the aniline reacted samples, particularly the whole soil and peat, because the concentration of the ^{15}N label is so much greater than the naturally abundant ^{15}N. Detection of the naturally abundant amide groups in the soil fulvic and humic acids substantiates the argument that aminolysis reactions with aniline are a real possibility in these samples.

^{15}N NMR Spectra of Soil Fulvic Acid Reacted with Aniline. Liquid phase ^{15}N NMR spectra of the fulvic acid reacted with labeled aniline are shown in Figure 5. The ACOUSTIC spectrum represents the quantitative distribution of all nitrogens incorporated into the fulvic acid, whereas the INEPT spectrum shows only nitrogens directly bonded to protons. Comparison of the two spectra indicates that nitrogens downfield of approximately 140 ppm in the ACOUSTIC spectrum are nonprotonated and therefore represent heterocyclic nitrogens. N-phenylindole and N-phenylpyrrole structures are likely to comprise some of the heterocyclic condensation products in this region. Possible mechanisms leading to the formation of these heterocycles were discussed in our previous report (9). The major peaks in the INEPT spectrum, with maxima at 85 ppm, 103 & 111 ppm, and 138 ppm, correspond primarily to anilinohydroquinone, anilinoquinone, and anilide nitrogens, respectively. These and other possible assignments are listed in Table II.

TABLE II. SUMMARY OF ASSIGNMENTS FOR ^{15}N NMR SPECTRA OF
SAMPLES REACTED WITH ^{15}N-LABELLED ANILINE.

NONCATALYZED REACTIONS

chem shift range, ppm	assignment [a]
60 - 100	*anilinohydroquinone*, phenoxazine
100 - 122	*anilinoquinone*, enamine
122 - 148	*anilide*, enaminone, quinolone, indole
148 - 200	N-phenylindole, N-phenylpyrrole, *heterocyclic N*
300 - 350	*imine*, phenoxazinone, quinoline

PEROXIDASE AND BIRNESSITE CATALYZED REACTIONS [b]

60 - 100	diphenylamine, hydrazine
230 - 280	oxazole, imidazole, pyrazole, nitrile
310 - 360	*iminodiphenoquinone*, *imine*, azoxybenzene
470 - 525	*azobenzene*

[a] Most probable assignments indicated by italics.

[b] For samples reacted with aniline in the presence of peroxidase and birnessite, assignments from both sections apply.

The ACOUSTIC spectrum of the fulvic acid reacted with aniline in the presence of horseradish peroxidase is shown in Figure 6. The distribution of nitrogens in the region from approximately 0 to 200 ppm is significantly different than in the spectrum of the nonenzyme reaction (Figure 5). The peak at 89 ppm corresponding to the anilinohydroquinone nitrogens is the one of maximum intensity, followed by the peak at 169 ppm. In general, interpretation of this region of the spectrum is complicated by the fact that products from both enzyme catalyzed and nonenzyme reactions potentially overlap with one another. Three sets of peaks in the spectrum downfield of ~200 ppm represent reaction products unique to catalysis by peroxidase. The first, from ~ 233 to 280 ppm with maximum at 259 ppm, has not been identified but was present in the spectra of all other fulvic and humic acid samples reacted with peroxidase (39), the Suwannee River fulvic acid reacted with mushroom tyrosinase (39), and the soil humic acid reacted with birnessite (Figure 8). The types of nitrogens that occur in this chemical shift range include imidazole, pyrazole, oxazole, and nitrile. Peaks in this region were not observed in ^{15}N NMR spectra of the product mixtures from the peroxidase catalyzed reactions of aniline with guaiacol, 4-methylcatechol, ferulic acid, conniferyl alcohol, or syringic acid (39), nor in the reaction blank of aniline and peroxidase (Fig 7). They were, however, present in the peroxidase catalyzed reaction mixture of DL-3,4-dihydroxyphenylalanine (DL-DOPA) and 3-hydroxytyramine (dopamine) with aniline (39). Thus, there is an indication that amino acid residues, DOPA-, or dopamine-like structures in the fulvic acid may be involved in the reactions giving rise to the resonances centered at 259 ppm. The second peak, from 310 to 360 ppm, with maxima at 318, 334, and 345 ppm, was also present in all other humic samples reacted with peroxidase (39), and the soil humic acid reacted with birnessite (Figure 8). A plausible assignment for these peaks are the iminodiphenoquinone nitrogens observed by Simmons et al. (19) in the product mixtures from the laccase catalyzed reaction of guaiacol and 4-chloroaniline:

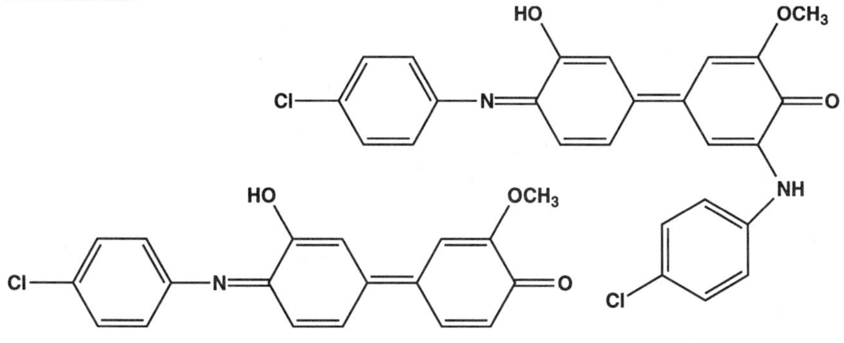

IMINODIPHENOQUINONE

These peaks may also be comprised of imine nitrogens arising from 1,2-addition of aniline to quinones formed from the peroxidase catalyzed oxidative decarboxylation

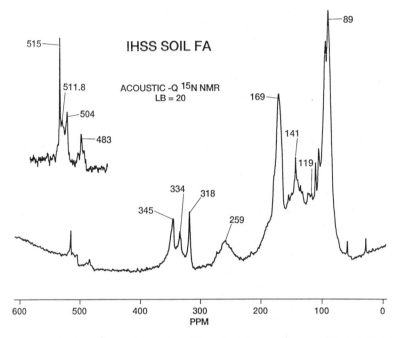

Figure 6. Liquid Phase ACOUSTIC ^{15}N NMR Spectrum of IHSS Soil Fulvic Acid Reacted with ^{15}N-labelled Aniline in the Presence of Peroxidase.

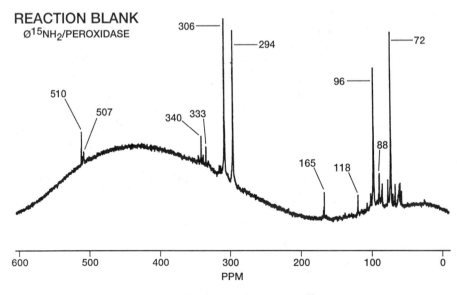

Figure 7. Liquid Phase Inverse Gated Decoupled ^{15}N NMR Spectrum of Product Mixture from Reaction of ^{15}N-labelled Aniline with Peroxidase and Hydrogen Peroxide. (Reaction blank).

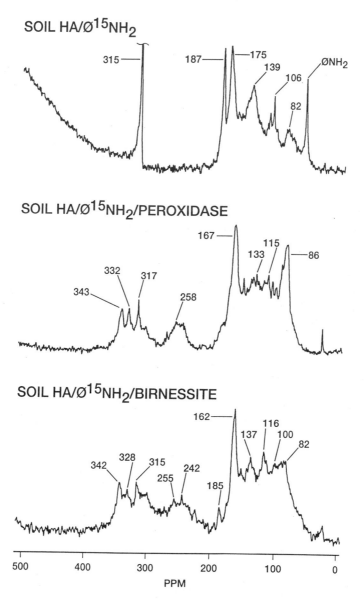

Figure 8. Liquid Phase ACOUSTIC ^{15}N NMR Spectra of IHSS Soil Humic Acid Reacted with ^{15}N-labelled Aniline Alone, with Peroxidase, and with Birnessite.

of aromatic carboxylic acids. The laccase catalyzed reaction of 4-chloroaniline with syringic acid provided an example of this pathway (40).

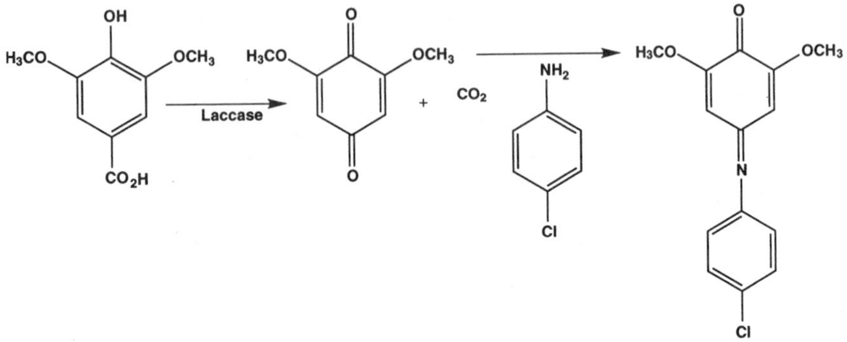

Oxidation of the syringic acid is considered to result in the formation of an unstable semiquinone, which undergoes decarboxylation with concurrent uptake of O_2 to form the 2,6-dimethoxy-1,4-benzoquinone (41). The third peak from 475 to 521 ppm, with maxima at 483, 504, and 515 ppm, corresponds to azobenzene nitrogens. These most likely arise from nitrogen to nitrogen coupling reactions between the anilino radicals and free aromatic amino groups in the fulvic acid molecules:

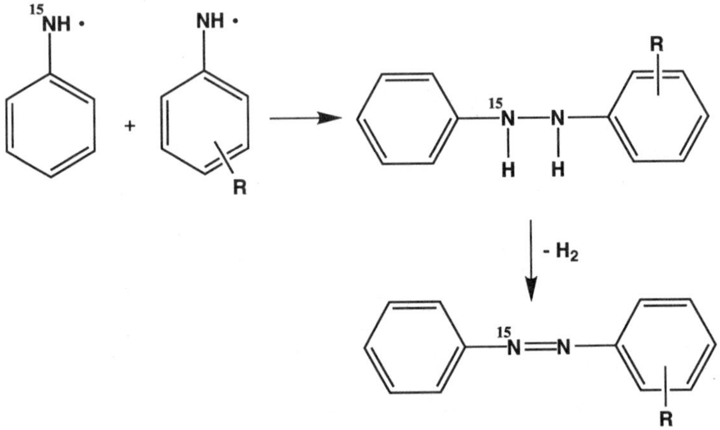

The resolution of the azobenzene nitrogens into two separate peaks centered at 483 and 512 ppm is indicative of an azo-hydrazone tautomerism:

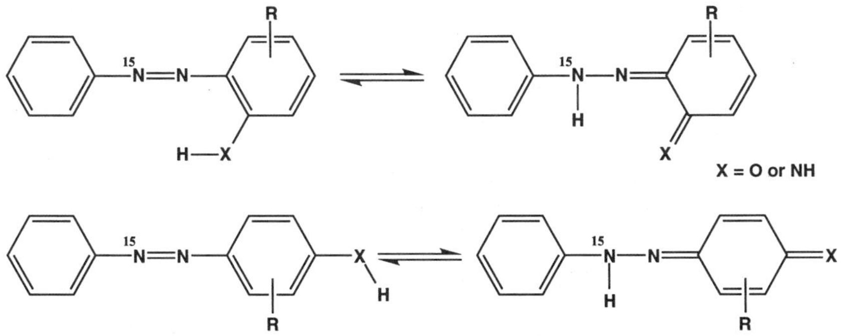

X = O or NH

The tautomerism occurs in the case of ortho or para substitution of one of the aromatic rings with an amino or hydroxyl group (42). This substitution pattern can therefore be inferred for some of the aromatic amino groups in the fulvic acid molecules undergoing nitrogen to nitrogen coupling with the aniline. A discreet peak at 510 ppm, corresponding to pure azobenzene itself, is present in the spectrum of the blank reaction of aniline with peroxidase; however, there is no indication of any products exhibiting the azo-hydrazone tautomerism in the spectrum of the reaction blank (Figure 7).

The occurrence of azobenzene nitrogens in the fulvic acid suggests that their immediate precursors prior to dehydrogenation, the hydrazine adducts, may also be present. These would occur around 95 ppm (Figure 1). The observation of nitrogen to nitrogen coupling between the anilino radicals and aromatic amino groups in the fulvic acid also suggests that nitrogen to carbon couplings between the anilino radicals and both phenolic or aromatic amino groups in the fulvic acid may be occurring as well.

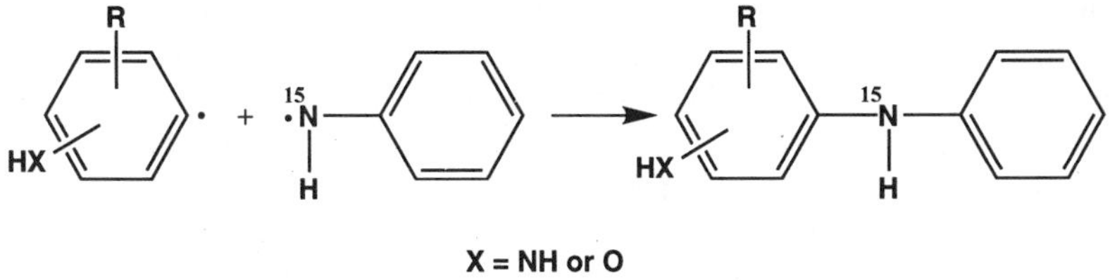

X = NH or O

The major peak at 89 ppm in the fulvic acid spectrum may be comprised in part of diphenylamine structures. (The ^{15}N NMR chemical shift of the singly protonated nitrogen in 4-aminodiphenylamine is 87 ppm (Figure 1)).

Primary aromatic amino groups, which would occur in the approximate chemical shift range from 45 to 80 ppm, were not clearly evident in the natural abundance ^{15}N NMR spectrum of the soil fulvic acid (Figure 4). Assuming aromatic amino groups are in fact the site for nitrogen to nitrogen coupling with aniline, then either the concentration of aromatic amino groups in the fulvic acid is minor, or there is a problem in detecting these groups in the CP/MAS experiment.

The absolute amounts of aniline covalently bonded to the soil fulvic acid in the presence and absence of the peroxidase were not measured. However, the relative signal to noise ratios obtained in the ^{15}N NMR spectra indicate that significantly more aniline was taken up by the fulvic acid in the enzyme catalyzed reaction. It should also be pointed out that, in the execution of the peroxidase experiment, the solution containing the fulvic acid, aniline, and peroxidase darkened instantaneously upon addition of the hydrogen peroxide, indicating significantly faster kinetics than in the nonenzyme reaction.

Reaction Blank. The liquid phase inverse gated decoupled ^{15}N NMR spectrum of the product mixture from the reaction blank (aniline + peroxidase + H_2O_2; Figure 7) shows several peaks. The distortion in the baseline is due to acoustic ringing. Some of these can be assigned to the dimers reported to result from the anodic oxidation of aniline (43) and from the oxidation of aniline by manganese dioxide (21). The peak at 88.3 ppm may represent the singly protonated nitrogen in 4-aminodiphenylamine (Figure 1), whereas the free amino group in this compound would occur within the cluster of resonances at ~ 58 ppm. The peak at 96 ppm corresponds to 1,2-diphenylhydrazine (Figure 1), and as just discussed, the peak at 510 ppm to azobenzene. Other peaks in the cluster at ~ 58 ppm may correspond to benzidine. Peaks at 333 to 340 ppm may correspond to azoxybenzene, whereas the major peaks at 294 and 306 ppm may represent the following diimine structures:

Phenazine (44) and other diimine structures as reported by Simmons et al. (45) may also occur in the imine region. Further work will be necessary to more accurately assign these peaks. In general, the relatively few number of resonances in this blank spectrum compared to the spectrum of the fulvic acid reacted with aniline in the presence of peroxidase, together with the fact that kinetically reaction of aniline with the fulvic acid is favored over coupling with itself in the presence of the peroxidase, suggest that the contribution of self condensation products of aniline to the spectrum of the fulvic acid (Figure 6) is negligible.

^{15}N NMR Spectra of Soil Humic Acid Reacted with Aniline. ACOUSTIC ^{15}N NMR spectra of the soil humic acid reacted with the labelled aniline in the presence and absence of peroxidase and birnessite are shown in Figure 8. As discussed previously, the sharp peak at 315 ppm in the spectrum of the noncatalyzed reaction appears to represent the reaction product of aniline with a contaminant or pure component in the humic acid sample (9). Vertical expansion of the spectrum revealed a broad, low intensity imine peak underlying the sharp contaminant peak. This underlying imine peak is more clearly visible in the solid state spectrum of the sample (Figure 9B), where the sharp contaminant peak is broadened out presumably as a result of chemical shift anisotropy. Imine nitrogens were also

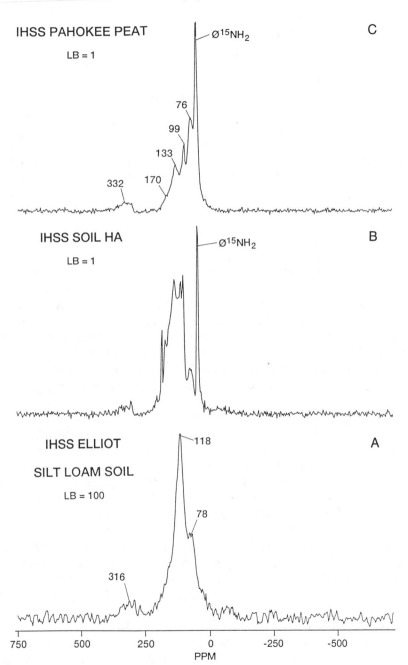

Figure 9. Solid State CP/MAS ^{15}N NMR Spectra of IHSS Elliot Silt Loam Soil (A), Soil Humic Acid (B), and Pahokee Peat (C) Reacted with ^{15}N-labelled Aniline.

clearly detected in an Iowa soil humic acid similarly reacted with aniline (9). In the absence of catalysts, the imines are most likely formed from the 1,2-addition of aniline to sterically hindered quinones. The effects of peroxidase on the incorporation of aniline into the fulvic acid are replicated in the humic acid. The distribution of nitrogens in the region from approximately 0 to 200 ppm changes in going from the noncatalyzed to the enzyme catalyzed reaction, with a significant increase in the peaks corresponding to the anilinohydroquinone or diphenylamine nitrogens (86 ppm). Major peaks from 230 to 280 ppm, corresponding to the unidentified nitrogens, and from 290 to 365 ppm, corresponding to the iminodiphenoquinone nitrogens, are present. The azobenzene nitrogens are not observed in the humic acid. The overall signal to noise of the spectrum again indicates that peroxidase increases the amount of incorporation of aniline into the humic acid.

Birnessite acts similarly to peroxidase in effecting the formation of the unidentified nitrogens and the iminodiphenoquinone nitrogens. The distribution of nitrogens from 0 to 200 ppm is again altered compared to the noncatalyzed reaction. There is an increase in the proportion of anilinohydroquinone and anilinoquinone nitrogens, although the increase of the anilinohydroquinone nitrogens is not as great as in the peroxidase catalyzed reaction. The birnessite catalyzed reaction was allowed to proceed for 6 days, whereas the peroxidase reaction ran for 3 days. The superior signal to noise achieved in the spectrum of the peroxidase reaction (comparable concentrations of sample and acquisition times) suggests that, at least under the laboratory conditions employed, peroxidase is a stronger catalyst than birnessite.

Solid State ^{15}N NMR Spectra of IHSS Peat and Soil Reacted with Aniline. Solid state CP/MAS ^{15}N NMR spectra of the peat, soil humic acid, and whole soil reacted with aniline are shown in Figure 9. The solid state spectrum of the humic acid was recorded to provide the closest comparison between it and the whole soil, as the liquid and solid state spectra are not directly comparable in terms of quantitation. (The heterocyclic nitrogens in the soil humic acid are clearly underestimated in the CP/MAS compared to the ACOUSTIC experiment.) The signal to noise ratio achieved in the spectrum of the IHSS soil is quite good (Figure 9C), demonstrating the feasibility of the whole soil experiment. The peat spectrum (Figure 9A) is comparable to the solid state spectrum of the humic acid in terms of signal to noise. The major peaks observed in the spectra of the fulvic and humic acids reacted with aniline in the absence of the catalysts are also present in the spectrum of the whole peat: anilinohydroquinones at 76 ppm, anilinoquinones at 99 ppm, anilides at 133 ppm, heterocyclic nitrogens at 170 ppm, and a significant imine peak at 332 ppm. The main difference is that less of the aniline is tied up as heterocyclic nitrogen compared to the fulvic and humic acids. Whether the proportion of heterocyclic nitrogen increases with a longer reaction time will need to be determined. The amount of imine formation is surprising. In addition to steric hindrance of the quinone functionality, it is possible that the equilibrium in the 1,2-addition of aniline to carbonyls is shifted toward imine formation within hydrophobic domains that may exist in the peat. The spectrum of the soil exhibits two major peaks corresponding to the anilinohydroquinone (78 ppm) and anilinoquinone nitrogens (118 ppm). There is also a weak but definite imine peak centered at about 316 ppm. As in the case of the peat (Figure 9A), and compared to the humic acid (Figure 9B), there appear to be fewer nitrogens incorporated into heterocyclic structures. Again, it is possible that formation of heterocycles could increase with a longer reaction time. Another possible explanation is that in both the whole soil and peat environment the potential for inter- and intramolecular condensation reactions is diminished compared to the situation in which the isolated fulvic or humic acids are dissolved in solution. This may be a result of the fact that some of the reactive functional groups in the fulvic or humic acid molecules are tied up with the inorganic constituents of the soil and thus are unavailable for reaction with aniline. As an example, catechol moieties in the humic substances could be involved in ligand exchange complexation reactions with hydrous metal oxides and, therefore, be unable to oxidize to quinones and condense with the aniline:

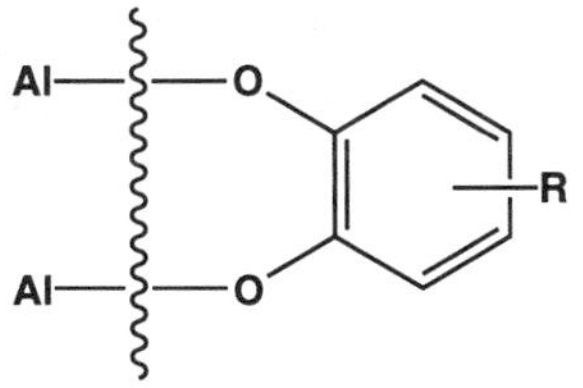

In a similar vein, the restricted mobility of the individual fulvic and humic acid molecules, also as a result of complexation with the inorganic matrix in the whole soil or peat, may limit the inter- or intramolecular condensation reactions. A final consideration is the fact that the solid state ^{15}N NMR spin dynamics may differ in going from the isolated humic acid to the whole soil, the relaxation effects of paramagnetic metal ions included in the latter.

In the ^{15}N NMR spectra of the fulvic and humic acids reacted with aniline in the presence of peroxidase and birnessite, the iminodiphenoquinone nitrogens are always associated with the presence of the unassigned nitrogens. The absence of the unassigned nitrogens (230 to 280 ppm) in the solid state spectra of the whole soil and peat tends to confirm that the imine nitrogens in these spectra are in fact the reaction products from simple 1,2-addition of aniline to carbonyl groups. The absence in the soil and peat spectra of the peaks most indicative of the peroxidase or birnessite catalyzed reactions, the azobenzene, iminodiphenoquinone, and unassigned nitrogens, therefore suggests that phenoloxidase enzyme or metal catalyzed reactions are not operable in these experimental systems. We emphasize that this conclusion pertains to these two systems. Further work will be necessary to fully assess the role of phenoloxidase enzyme or metal catalysts in whole soil systems under varying states of microbial activity or varying concentrations of hydrous metal oxides and clay minerals.

Conclusion

The liquid phase ^{15}N NMR spectra comprise the first direct spectroscopic evidence differentiating phenoloxidase- and metal-catalyzed reactions from noncatalyzed nucleophilic addition reactions of aniline with humic substances. The solid state ^{15}N NMR spectra provide the first direct evidence for nucleophilic addition of aniline to quinone and other carbonyl groups in the organic matter of whole soil and peat. The ^{15}N NMR approach has potential for further investigation of the effects of reaction conditions on the incorporation of aromatic amines into naturally occurring organic matter, and for studies on how aromatic amines covalently bound to organic matter may ultimately be re-released or remineralized, either chemically or microbially.

Acknowledgment

We thank Angus McGrath, Lawrence Berkeley Laboratory, for supplying the birnessite and information on the execution of the birnessite reactions. This research was supported from U.S. EPA Grants DW14935164, DW14935652, and USDA Grant 92-34214-7352.

Literature Cited

1.   Adrian, P.; Andreux, F.; Viswanathan, R.; Freitag, D.; Scheunert, I. *Toxicol. Environ. Chem.* **1989**, *20-21*, 109-120.
2.   Marco, G.J.; Novak, R.A. *J. Agr. Food Chem.* **1991**, *39*, 2101-2111.
3.   Bollag, J.-M. *Environ. Sci. Technol.* **1992**, *26*, 1876-1881
4.   Bartha, R.; You, I-S.; Saxena, A. *Pestic. Chem.: Hum. Welfare Environ., Proc. Int. Congr. Pestic. Chem., 5th* **1983**, *3*, 345-350.
5.   Saxena, A.; Bartha, R. *Soil Biol. Biochem.* **1983**, *15*, 59-62.
6.   Saxena, A.; Bartha, R. *Soil Sci.* **1983**, *136*, 111-116.
7.   Saxena, A.; Bartha, R. *Bull. Environ. Contam. Toxicol.* **1983**, *30*, 485-491.
8.   Weber, E.J.; Spidle, D.L.; Thorn, K.A. *Environ. Sci. Technol.* In press.
9.   Thorn, K.A.; Pettigrew, P.J.; Goldenberg, W.S.; Weber, E.J. *Environ. Sci. Technol.* In press.
10.   Kutyrev, A.A. *Tetrahedron* **1991**, *47(38)*, 8043-8065.
11.   Monks, T.J.; Hanzlik, R.P.; Cohen, G.M.; Ross, D.; Graham, D.G. *Toxicol. Appl. Pharmacol.* **1992**, *112*, 2-16.
12.   Peter, M.G. *Angew. Chem. Int. Ed. Engl.* **1989**, *28*, 555-570.
13.   Finley, K.T. In *Chemistry of the Quinonoid Compounds, Part 2*; Patai, S., Ed.; John Wiley and Sons: London, **1974**; Chapter 17.
14.   Pattai, S; Rappaport, Z. In *The Chemistry of the Quinonoid Compounds*; Wiley: New York, **1988**; Vol. 2, Parts 1&2.
15.   Ononye, A.I.; Graveel, J.G. *Environ. Toxicol. Chem.* **1994**, *13*, 537-541.
16.   Burns, R.G. *Soil Enzymes*; Academic Press: New York, **1978**
17.   Burns, R.G. In *Interactions of Soil Minerals with Natural Organics and Microbes*; Huang, P.M., Schnitzer, M., Eds., SSSA Special Publication No. 17, Madison, WI: **1986**, pp 429-451.

18. Barr, D.P.; Aust, S.D. *Environ. Sci. Technol.* **1994**, *28*, 78A-87A.
19. Simmons, K.E.; Minard, R.D.; Bollag, J.-M. *Environ. Sci. Technol.* **1989**, *23*, 115-121.
20. Hatcher, P.G.; Bortiatynski, J.M.; Minard, R.D.; Dec, J.; Bollag, J.M. *Environ. Sci. Technol.* **1993**, 27, 2098-2103.
21. Laha, S.; Luthy, R.G. *Environ. Sci. Technol.* **1990**, *24*, 363-373.
22. McKenzie, R.M. *Miner. Mag.* **1971**, *38*, 493-502.
23. Thorn, K.A.; Arterburn, J.B.; Mikita, M.A. *Environ. Sci. Technol.* **1992**, *26*, 107-116.
24. Morris, G.A.; Freeman, R. *J. Am. Chem. Soc.* **1979**, *101*, 760-762.
25. Patt, S.L. *J. Magn. Reson.* **1982**, *49*, 161-163.
26. Thorn, K.A.; Folan, D.W.; MacCarthy, P. Characterization of the IHSS Standard and Reference Fulvic and Humic Acids by Solution State ^{13}C and ^{1}H NMR. *Water-Resour. Invest.* (U.S. Geol. Surv.) **1991**, No. 89-4196.
27. Parris, G.E. *Environ. Sci. Technol.* **1980**, *14*, 1099-1106.
28. Wilson, M.A. *NMR Techniques and Applications in Geochemistry and Soil Chemistry*; Pergamon Press: Oxford, **1987**.
29. *NMR of Humic Substances and Coal - Techniques, Problems, and Solutions*; Wershaw, R.L.; Mikita, M.A., Eds.; Lewis Publishers: Chelsea, MI, **1987**.
30. Wind, R.A.; Maciel, G.E.; Botto, R.E. *In Magnetic Resonance of Carbonaceous Solids*; Botto, R.E., Sanada, Y., Eds.; Advances in Chemistry Series 229; American Chemical Society: Washington, D.C.; **1993**; pp.3-26.
31. Axelson, D.E. *Solid State Nuclear Magnetic Resonance of Fossil Fuels: An Experimental Approach*; Multiscience: Montreal, Canada, **1985**.
32. Snape, C.E.; Axelson, D.E.; Botto, R.E.; Delpeuch, J.J.; Tekely, P.; Gerstein, B.C.; Pruski, M.; Maciel, G.E.; Wilson, M.A. *Fuel* **1989**, *68*, 547.
33. Kinchesh, P.; Powlson, D.S; Randall, E.W. *European J. Soil Sci.* **1995**, *46*, 125-138.
34. Knicker, H.; Frund, R.; Ludemann, H.-D. *Naturwissenschaften* **1993**, *80*, 219-221.
35. Knicker, H.; Frund, R.; Ludemann, H.-D. In *Humic Substances in the Global Environment and Implications on Human Health*; Senesi, N., Miano, T.M., Eds.;Elsevier: Amsterdam, **1994**, pp 501-506.
36. Zhuo, S.; Wen, Q. *Pedosphere* **1993**, *3*, 193-200.
37. Witanowski, M; Stefaniak, L; Webb, G.A. In *Annual Reports on NMR Spectrometry*; Webb, G.A., Ed.; Academic Press, Harcourt Brace Jovanovich: London, **1993**; Vol. 25.
38. Earl, W.L. In *NMR of Humic Substances and Coal - Techniques, Problems, and Solutions*; Wershaw, R.L.; Mikita, M.A, Eds.; Lewis Publishers: Chelsea, MI, **1987**, pp. 167-187.
39. Thorn, K.A.; Unpublished Results.
40. Bollag, J.M.; Minard, R.D.; Liu, S.Y. *Environ. Sci. Technol.* **1983**, *17*, 72-80.
41. Flaig, W.; Beutelspacher, H.; Rietz, E. In *Soil Components, Volume 1, Organic Components*; Gieseking, J.E., Ed.; Springer-Verlag: New York, 1975.
42. Lycka, A. Multinuclear NMR of Azo Dyestuffs, In *Annual Reports on NMR Spectrometry*; Webb, G.A., Ed.; Academic Press, Harcourt Brace Jovanovich: London, **1993**; Vol. 26.
43. Sharma, L.R.; Manchanda, A.K.; Singh, G.; Verma, R.S. *Electrochim. Acta* **1982**, *27(2)*, 223-233.
44. Pillai, P.; Helling, C.S.; Dragun, J. Chemosphere **1982**, *11*, 299-317.

45.   Simmons, K.E.; Minard, R.D.; Bollag, J.M. *Environ. Sci. Technol.* **1987**, *21*, 999-1003.

46.   Levy, G.C.; Lichter, R.L. *Nitrogen-15 Nuclear Magnetic Resonance Spectrometry*; John Wiley and Sons: New York, **1979**.

47.   Martin, G.J.; Martin, M.L.; Gouesnard, J.P. *15N-NMR Spectroscopy*; Springer-Verlag: New York, **1981**.

48.   Witanowski, M; Stefaniak, L; Webb, G.A. *In Annual Reports on NMR Spectrometry*; Webb, G.A., Ed.; Academic Press, Harcourt Brace Jovanovich: London, **1986**; Vol. 18.

INDEXES

Author Index

Affiliation Index

Subject Index

Highlights from ACS Books

Good Laboratory Practice Standards: Applications for Field and Laboratory Studies
Edited by Willa Y. Garner, Maureen S. Barge, and James P. Ussary
ACS Professional Reference Book; 572 pp; clothbound ISBN 0–8412–2192–8

Silent Spring Revisited
Edited by Gino J. Marco, Robert M. Hollingworth, and William Durham
214 pp; clothbound ISBN 0–8412–0980–4; paperback ISBN 0–8412–0981–2

The Microkinetics of Heterogeneous Catalysis
By James A. Dumesic, Dale F. Rudd, Luis M. Aparicio, James E. Rekoske,
and Andrés A. Treviño
ACS Professional Reference Book; 316 pp; clothbound ISBN 0–8412–2214–2

Helping Your Child Learn Science
By Nancy Paulu with Margery Martin; Illustrated by Margaret Scott
58 pp; paperback ISBN 0–8412–2626–1

Handbook of Chemical Property Estimation Methods
By Warren J. Lyman, William F. Reehl, and David H. Rosenblatt
960 pp; clothbound ISBN 0–8412–1761–0

Understanding Chemical Patents: A Guide for the Inventor
By John T. Maynard and Howard M. Peters
184 pp; clothbound ISBN 0–8412–1997–4; paperback ISBN 0–8412–1998–2

Spectroscopy of Polymers
By Jack L. Koenig
ACS Professional Reference Book; 328 pp;
clothbound ISBN 0–8412–1904–4; paperback ISBN 0–8412–1924–9

Harnessing Biotechnology for the 21st Century
Edited by Michael R. Ladisch and Arindam Bose
Conference Proceedings Series; 612 pp;
clothbound ISBN 0–8412–2477–3

From Caveman to Chemist: Circumstances and Achievements
By Hugh W. Salzberg
300 pp; clothbound ISBN 0–8412–1786–6; paperback ISBN 0–8412–1787–4

The Green Flame: Surviving Government Secrecy
By Andrew Dequasie
300 pp; clothbound ISBN 0–8412–1857–9

For further information and a free catalog of ACS books, contact:
American Chemical Society
Customer Service & Sales
1155 16th Street, NW
Washington, DC 20036
Telephone 800–227–5558

Bestsellers from ACS Books

The ACS Style Guide: A Manual for Authors and Editors
Edited by Janet S. Dodd
264 pp; clothbound ISBN 0–8412–0917–0; paperback ISBN 0–8412–0943–X

Understanding Chemical Patents: A Guide for the Inventor
By John T. Maynard and Howard M. Peters
184 pp; clothbound ISBN 0–8412–1997–4; paperback ISBN 0–8412–1998–2

Chemical Activities (student and teacher editions)
By Christie L. Borgford and Lee R. Summerlin
330 pp; spiralbound ISBN 0–8412–1417–4; teacher ed. ISBN 0–8412–1416–6

Chemical Demonstrations: A Sourcebook for Teachers,
Volumes 1 and 2, Second Edition
Volume 1 by Lee R. Summerlin and James L. Ealy, Jr.;
Vol. 1, 198 pp; spiralbound ISBN 0–8412–1481–6;
Volume 2 by Lee R. Summerlin, Christie L. Borgford, and Julie B. Ealy
Vol. 2, 234 pp; spiralbound ISBN 0–8412–1535–9

Chemistry and Crime: From Sherlock Holmes to Today's Courtroom
Edited by Samuel M. Gerber
135 pp; clothbound ISBN 0–8412–0784–4; paperback ISBN 0–8412–0785–2

Writing the Laboratory Notebook
By Howard M. Kanare
145 pp; clothbound ISBN 0–8412–0906–5; paperback ISBN 0–8412–0933–2

Developing a Chemical Hygiene Plan
By Jay A. Young, Warren K. Kingsley, and George H. Wahl, Jr.
paperback ISBN 0–8412–1876–5

Introduction to Microwave Sample Preparation: Theory and Practice
Edited by H. M. Kingston and Lois B. Jassie
263 pp; clothbound ISBN 0–8412–1450–6

Principles of Environmental Sampling
Edited by Lawrence H. Keith
ACS Professional Reference Book; 458 pp;
clothbound ISBN 0–8412–1173–6; paperback ISBN 0–8412–1437–9

Biotechnology and Materials Science: Chemistry for the Future
Edited by Mary L. Good (Jacqueline K. Barton, Associate Editor)
135 pp; clothbound ISBN 0–8412–1472–7; paperback ISBN 0–8412–1473–5

For further information and a free catalog of ACS books, contact:
American Chemical Society
Customer Service & Sales
1155 16th Street, NW, Washington, DC 20036
Telephone 800–227–5558

T

1 Month